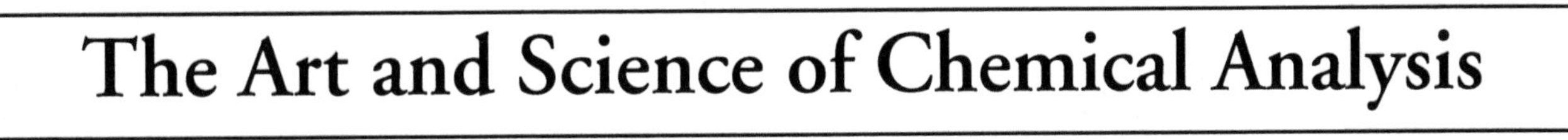
The Art and Science of Chemical Analysis

The Art and Science of Chemical Analysis

Editor: Alana Wood

New York

Published by NY Research Press
118-35 Queens Blvd., Suite 400,
Forest Hills, NY 11375, USA
www.nyresearchpress.com

The Art and Science of Chemical Analysis
Edited by Alana Wood

International Standard Book Number: 978-1-64725-452-0 (Hardback)

Cataloging-in-Publication Data

The art and science of chemical analysis / edited by Alana Wood.
p. cm.
Includes bibliographical references and index.
ISBN 978-1-64725-452-0
1. Analytical chemistry. 2. Instrumental analysis. 3. Chemistry, Physical and theoretical. I. Wood, Alana.
QD75.22 .A78 2023
543--dc23

Contents

Preface

I am honored to present to you this unique book which encompasses the most up-to-date data in the field. I was extremely pleased to get this opportunity of editing the work of experts from across the globe. I have also written papers in this field and researched the various aspects revolving around the progress of the discipline. I have tried to unify my knowledge along with that of stalwarts from every corner of the world, to produce a text which not only benefits the readers but also facilitates the growth of the field.

Chemical analysis refers to the study of the chemical composition and structure of different types of substances. The different techniques of chemical analysis are used to determine the exact chemical information of a substance. Chemical analysis is a part of analytical chemistry, which is a branch of chemistry involved in the study of instruments and methods for the separation, identification and quantification of matter. Some of the major chemical analysis procedures include isotopic analysis and nuclear magnetic resonance (NMR), field flow fractionation (FFF), mass spectrometry (MS), calorimetry, and surface plasma resonance (SPR). NMR is an analytical procedure which studies the structure of molecules and the interaction of different molecules. Advancements in the field of analytical chemistry find applications in biomedicine, quality control of industrial manufacturing, environmental monitoring, and forensic science. The book aims to shed light on chemical analysis procedures and its applications. It consists of contributions made by international experts. The book aims to serve as a resource guide for students and experts alike.

Finally, I would like to thank all the contributing authors for their valuable time and contributions. This book would not have been possible without their efforts. I would also like to thank my friends and family for their constant support.

Editor

1

The Identification of Cotton Fibers Dyed with Reactive Dyes for Forensic Purposes

Daria Śmigiel-Kamińska [1], Jolanta Wąs-Gubała [2], Piotr Stepnowski [1] and Jolanta Kumirska [1,*]

[1] Faculty of Chemistry, University of Gdańsk, ul. Wita Stwosza 63, 80-308 Gdansk, Poland; d.smigiel-kaminska@phdstud.ug.edu.pl (D.Ś.-K.); piotr.stepnowski@ug.edu.pl (P.S.)

[2] Institute of Forensic Research, Microtrace Analysis Section, Westerplatte 9, 31-033 Krakow, Poland; jwas@ies.krakow.pl

* Correspondence: jolanta.kumirska@ug.edu.pl

Academic Editors: Constantinos K. Zacharis and Paraskevas D. Tzanavaras

Abstract: Some of the most common microtraces that are currently collected at crime scenes are fragments of single fibers. The perpetrator leaves them at a crime scene or takes them away, for example, on their clothing or body. In turn, the microscopic dimensions of such traces mean that the perpetrator does not notice them and therefore usually does not take action to remove them. Cotton and polyester fibers dyed by reactive and dispersion dyes, respectively, are very popular within clothing products, and they are hidden among microtraces at the scene of a crime. In our recently published review paper, we summarized the possibilities for the identification of disperse dyes of polyester fibers for forensic purposes. In this review, we are concerned with cotton fibers dyed with reactive dyes. Cotton fibers are natural ones that cannot easily be distinguished on the basis of morphological features. Consequently, their color and consequently the dye composition are often their only characteristics. The presented methods for the identification of reactive dyes could be very interesting not only for forensic laboratories, but also for scientists working in food, cosmetics or pharmaceutical/medical sciences.

Keywords: cotton fibers; dyed fibers; reactive dyes; forensic analysis; chromatographic methods; spectroscopic methods

---

1. Introduction

The experts from forensic laboratories are able to establish relationships between people, objects and a crime scene on the basis of the study of microtraces, very often in the form of fragments of single fibers, which sometimes also help to reconstruct the circumstances of an event. This means that even the small fragment of a fiber can be used as evidence in legal proceedings regarding such events as homicides, sexual crimes, assaults, or road accidents. The main purpose of identifying and comparing single fibers is to determine their physicochemical features and then classify them and determine if they originate from a known source. In the case of fibers belonging to the same type, e.g., cotton fibers, color plays an important role in their comparative study. Colors may be specified by 15 different color formats such as RGB, CMYK, HSV, HSL, CIELab, Android, Decimal, and YUV. For example, as an RGB triplet, a color is specified according to the intensity of its red, green, and blue components, each represented by eight bits. Thus, there are 24 bits used to specify a web color within the sRGB gamut, and 16,777,216 colors that may be so specified. The RYB color format or color space builds the colors from a combination of red, yellow, and blue. Each of these color values defines the intensity of the color as an integer value from 0 to 255. The value 255 stands for 100% intensity for the color value, so an RYB color of 0, 255, 0 would describe yellow.

In the forensic laboratory, color is established firstly with the use of optical microscopy techniques, and then, on this basis, the evidential fibers are qualified for further research using more advanced analytical methods. So far, it has mainly concerned UV-Vis microspectrophotometry [1], Raman spectroscopy [2], and more traditional and low-cost chromatographic methods, such as thin-layer chromatography (TLC) [3]. Single cotton fibers comprise the most common group of microtraces, and these fibers have been part of the human environment since ancient times. This is confirmed by population studies conducted so far on extraneous fibers present on different surfaces [4] and, for example, studies showing that such fibers accumulate even in regions as remote as the Arctic [5].

Reactive dyes, including a number of their improved structures and mixtures, are used to dye these particularly widespread fibers. There are a limited number of chromatographic methods settled for the comparison of cotton fibers dyed with reactive dyes, and no review paper has been published concerning this subject more broadly.

The analysis of the textile market shows that cotton dyed with reactive dyes and polyester dyed with disperse dyes are the most popular fibers within clothing products. In our recently published review paper [6], we summarized the possibilities for the identification of polyester fibers dyed with disperse dyes for forensic purposes. In this review, we are concerned with cotton fibers dyed with reactive dyes. Therefore, apart from the specificity of reactive dyes used for dyeing cotton in the textile industry, we intend to discuss the possibilities of using specific research methods in the analysis of this type of materials, i.e., chromatographic and spectroscopic ones.

2. Literature Review Sections

2.1. *Reactive Dyes in the Textile Industry—An Overview*

Textile dyeing of fibers, yarns, or fabrics is the aqueous application of color, most commonly using synthetic organic dyes. Dye and auxiliary processing chemicals are incorporated into the textiles in this process so as to achieve a uniform depth of color and color fastness properties suitable for the end use. For the latter, various types of dyes and chemical additives as well as various technological processes are used [7–11].

When dyeing on wool, silk, cotton, and regenerated cellulosic fibers, reactive dyes offer excellent fastness properties. They are also a very effective class of modern synthetic dyes due to the offered range of shades and flexibility in use. Reactive dyes may be broadly defined as chromophores containing pendant groups that are capable of forming covalent bonds with nucleophilic sites in fibrous substrates. If these bonds remain stable under the conditions of laundering, the final user can expect excellent the wash fastness properties of the coloration. Lewis gave detailed information on the development of both the chemistry of reactive systems and the chemical technology related to the application of reactive dyes to different types of fibers [12].

Reactive dyes have many parts: a color part, a solubilizing part, and a reactive part [13–21]. They form covalent bonds with the hydroxyl groups of the fibers, as a result of which the dyes are characterized by a good resistance to wet factors [13–21]. Compounds used to develop a bond between a textile fiber and a dye are generally called mordants. The structure of reactive dyes can be presented as a diagram, as shown in Figure 1, and an example of reactive dyes is shown in Figure 2 [22].

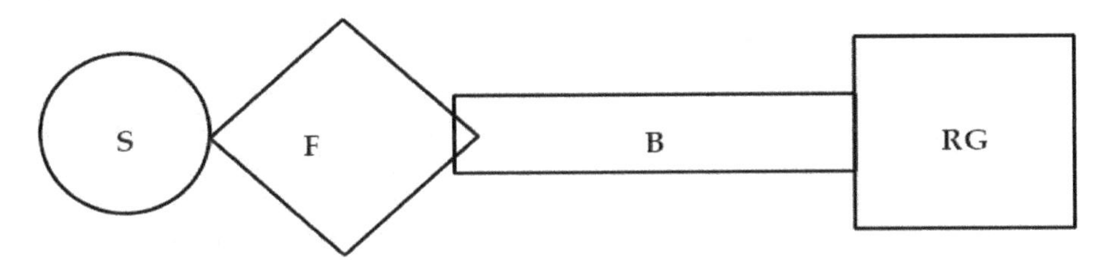

S solubilising group: sulfone (SO_3Na) or carboxy (COONa) group or combination of both

F chromophoric group: an azo, metal-complex azo, antraquinone, triphenyl dioxazines and formazan molecules or phthalocynanine residue, etc.

B bridging group: –NH, -O-, -NHCO-, $-OCH_3-$, $-SO_3-$, etc.

RG reactive group: reacts chemically with the functional group of the fibre with the formation of covalent bond between the dye and the fibre

Figure 1. The general formula of reactive dyes.

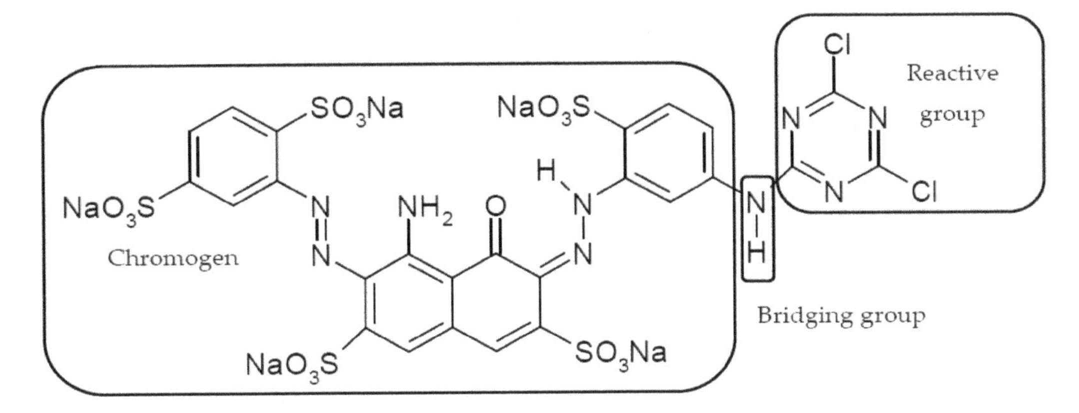

Figure 2. Molecular structure of a reactive dye—C. I. Reactive Blue 109.

This class of dyes has become increasingly popular since the 1950s, when the first commercial reactive dyes for cotton were introduced, and to date, a number of improved dye structures and mixtures have been introduced into practice. However, the use of these dyes results in disagreeable levels of dissolved solids and oxygen demand in effluent, and the remaining unfixed dye on cotton and the hydrolyzed dyes also contribute to this pollution. Khatri et al. [23] reviewed the options to progress the sustainability of the dyeing process through the improvement of dyeing machinery and processes, the chemical modification of cotton fiber prior to dyeing, the development of reactive dyes, the use of biodegradable organic compounds in dyebath formulations, and effluent treatment processes. The use of two reactive groups in a dye molecule, resulting in higher fixation efficiencies (bi-functional reactive dyes), and incorporating more than two reactive groups into the dye molecule, which should theoretically further increase the fixation efficiency (polyfunctional dyes), have been indicated as example advances for reducing effluent pollution. For example, other options are reactive dyes fixable at neutral pH, acid-fixing reactive dyes, low-salt reactive dyes, and cationic reactive dyes [23].

Reactive dyes can be characterized by the following five main types [15].

1. MCT/VS bi-functional dyes—these are reactive dyes containing a monochlorotriazinyl group and a vinyl sulfone group, and they are useful for the dyeing of cellulose fibers.
2. VS dyes—these are reactive dyes containing a vinyl sulfone group, and they are useful for the dyeing of cellulose fibers.

3. MCT/MCT bi-functional dyes—these are reactive dyes containing two monochlorotriazinyl groups and characterized by high fixation in dyeing polyester/cotton blends.
4. MCT dyes—these are reactive dyes containing a monochlorotriazinyl group.
5. DCT dyes—these are reactive dyes containing a dichloro triazinyl group.

Reactive dyes cover the full color palette with vivid shades. Therefore, the following color groups can be presented: yellows, oranges, reds, purples, navy blues and blacks, blues. Reactive yellows are azo dyes such as aromatic amines, pyrazolates, and pyridones and their derivatives. The chromophore system of the dye molecule consists of one or two isolated azo bonds. Reactive oranges are azo dyes that are mainly derivatives of 6-amino-1-hydroxynaphtalene-3-sulfonic acid. Reactive reds are dyes such as N-acylated H acids (1-amino-8-hydroxynaphtalene-3,6-disulfonic acid) and often contain one or two isolated azo bonds. The next reactive dyes are purples. These dyes are mainly copper complexes of N-acylated H acid (o,o′-hydroxyazo). Reactive dyes such as navy blues and blacks are diazo compounds. These dyes are obtained by double coupling with H acid [13].

Table 1 shows the characteristic groups of reactive dyes.

Table 1. Examples of reactive dyes with different reactive groups.

Reactive Group	Chemical Structure
Monochlorotriazine	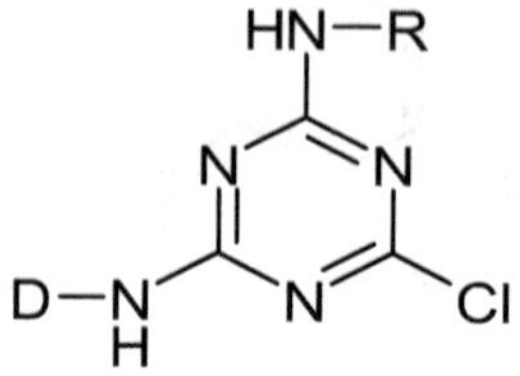 HN—R; N, N; D—N(H), N, Cl
Dichlorotriazine	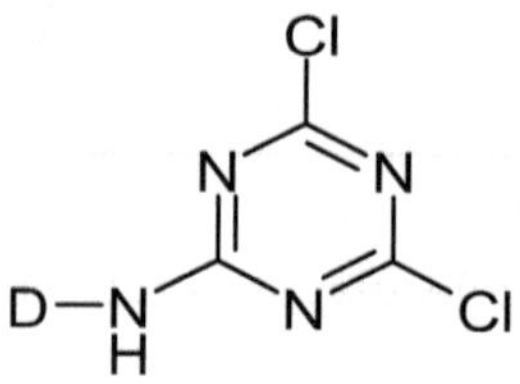 Cl; N, N; D—N(H), N, Cl
Dichloropyrimidine	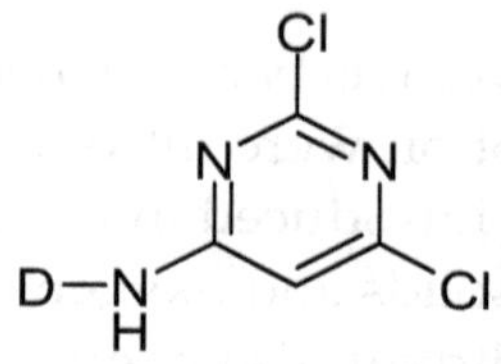 Cl; N, N; D—N(H), Cl
Dichloroquinoxaline	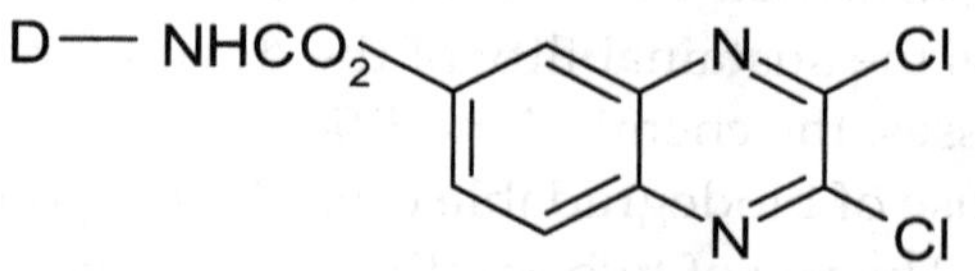 D— $NHCO_2$; N, Cl; N, Cl
Aminochlorotriazine	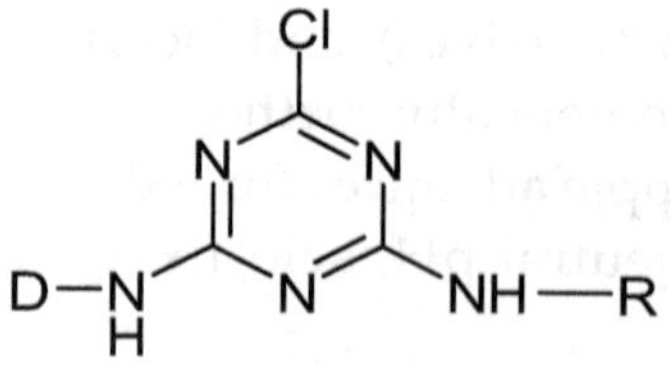 Cl; N, N; D—N(H), N, NH—R
Monofluorotriazine	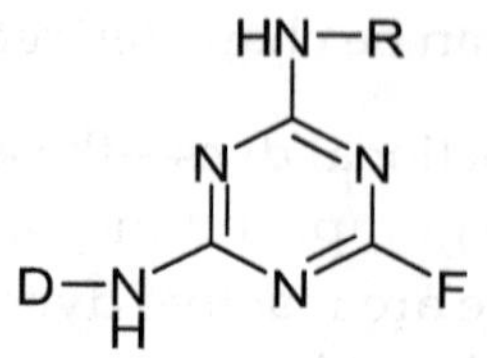 HN—R; N, N; D—N(H), N, F

Table 1. *Cont.*

Fluorochloropirymidine	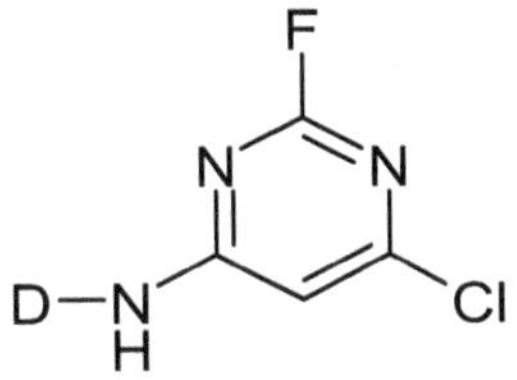
Aminofluorotriazine	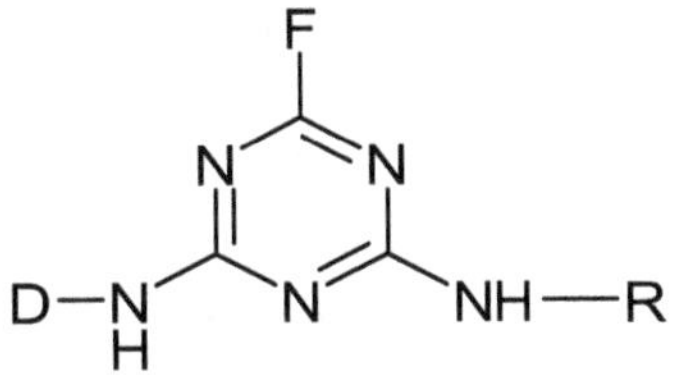
Vinyl Sulphone	$D{-}SO_3{-}CH{=}CH_2$
Sulphatoethylsulphone	$D{-}SO_2CH_2CH_2OSO_3Na$

Reactive dyes can be classified in several ways. The first is based on the types of the reactive group: (a) halogen (triazine group, pyridimine group), (b) activated vinyl compound (vinyl sulfone, vinyl acrylamide, vinyl sulfoamide) [13,24]. The next classification is based on the reactivity of the dyes: (a) lower reactive dye/medium reactive dye, (b) higher reactive dye. They can also be classified based on a parameter such as the dyeing temperature: (a) cold brand, (b) medium brand, (c) hot brand [15].

Reactive dyes are the youngest and the most important dye class for cellulosic material. They offer a wide range of dyes of many colors [15].

The Formation of a Reactive Dye–Fiber Bond

The necessary condition for a chemical reaction to occur is the physical contact of reactants. The process of the application of reactive dyes on cotton fibers consists of two steps. The first step is the sorption of dyes onto cellulose fibers from the dyeing bath. Reactive dyes contain anionic sulfone groups. These groups make it difficult for dye molecules to enter into physical contact with negatively ionized cellulose fiber. In order to overcome the electrostatic barrier of dye–fiber, the addition of electrolytes is necessary. The most commonly used electrolytes are sodium chloride or disodium sulfate (VI). The process of the sorption of the dye from the bath onto the fiber is carried out in a nearly neutral environment. In the process, dyeing is established between the dye concentration in solution and the fiber.

The second step in the dyeing process is the reaction with the cellulose fiber. The reaction proceeds in the presence of an alkali due to the dye–fiber reaction mechanism. Dye, containing a chlorine of a fluorine atom (triazine, pyrimidine, quinoxaline) in the molecule, reacts with cellulose according to the nucleophilic aromatic substitution mechanism of S_N2 [13,18,20]. The reaction mechanism of a reactive dye with cellulose is shown in Figure 3 [25] and in Figure 4 [25].

$$\mathrm{Cel{-}OH} \overset{OH^-}{\rightleftharpoons} \mathrm{CelO^-} + \mathrm{H_2O}$$

Figure 3. Formation of cellulosate anione.

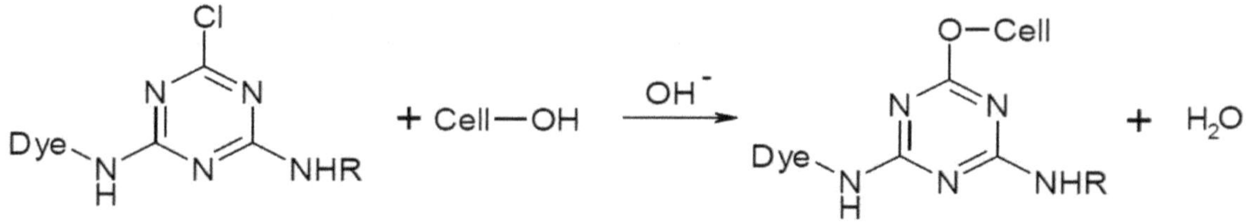

Figure 4. Reaction of a reactive dye with cellulose.

Dyes containing a vinylsulfonyl moiety as a reactive system or a 2-sulfate ethylsulfonyl moiety, which is a precursor of the vinylsulfonyl moiety, react with cellulose fiber according to the reaction mechanism of nucleophilic attachment to an unsaturated bond [13,18,20]. The example of such a reaction is shown in Figure 5.

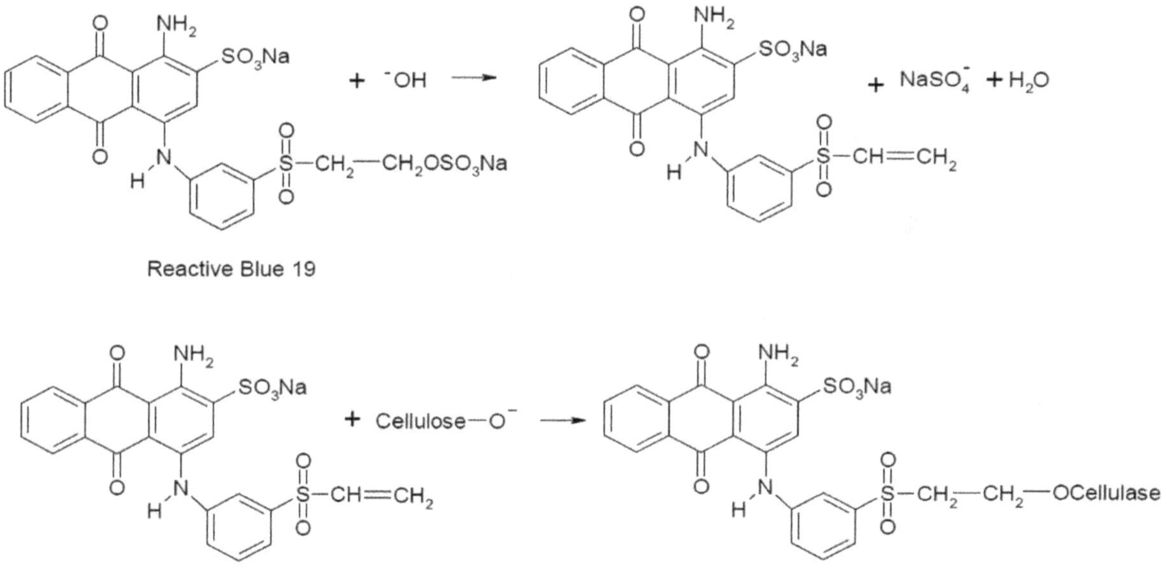

Figure 5. Reaction of Reactive Blue 19 with cellulose.

2.2. *Extraction of Reactive Dyes from Dyed Textile Fibers for Forensic Purposes*

2.2.1. Cleavage of Reactive Dyes from Dyed Cotton Textiles

The conducted studies on the population of fibers have shown that cotton is the most common textile fiber on internal and external surfaces [26–28]. Untypical to other textiles, cotton fibers can be dyed using three types of dyes: direct, vat, and reactive. Among them, reactive dyes are the most resistant to extraction, because between dye and the cotton fiber, the covalent bond is formed [29,30]. The conventional extraction methods are too weak to cleavage this bond. Reactive dyes can be released from cotton by disrupting the fiber [16]. In order to remove reactive dyes from the fiber, an aqueous solution of a strong base is typically used. The alcohol groups on the glucose unit in the cellulose backbone of cotton act as a weak acid and are ionized under alkaline conditions [29]. During hydrolysis, except for cleavage of the above-mentioned covalent bond, other reactions such as the cleavage of amide bonds and other chemical bonds in the dye molecule are possible. For this reason, the various structural changes can take place, and in many cases, multiple reaction products from a single dye molecule could be produced. The mechanism of the extraction of reactive dyes from cotton, involving dye hydrolysis due to excess NaOH, is shown in Figure 6.

Another method for the separation of a reactive dye from cotton fibers is enzymatic digestion using cellulase solution. However, in this method, a covalent bond between the reactive dye and the cotton fiber is still present, and after enzymatic digestion, a molecule of the dye connected with one or more glucose units is obtained. More details are presented in Sections 2.2.2 and 2.3.

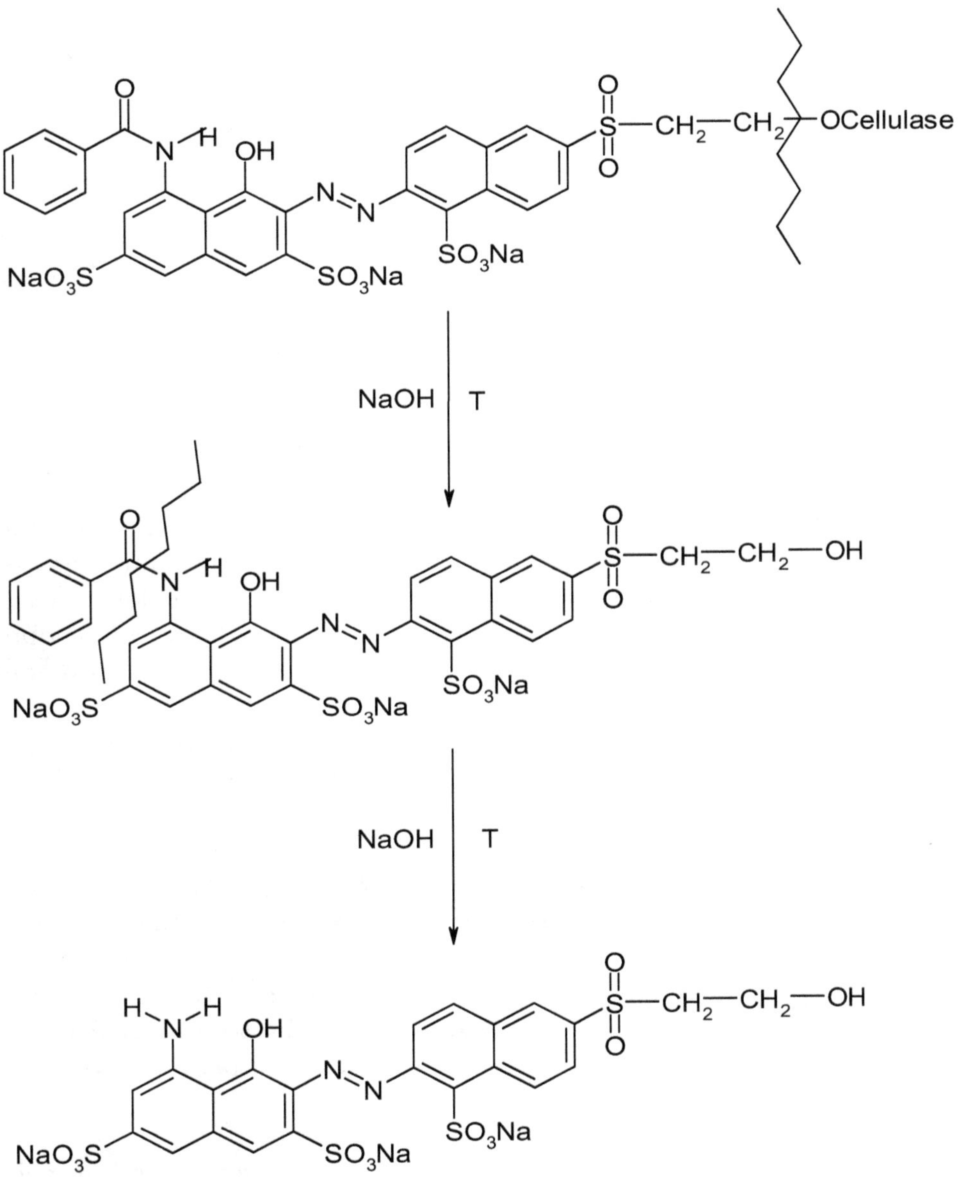

Figure 6. Mechanism of the extraction of Reactive Red 180 from cellulose, involving dye hydrolysis due to excess NaOH.

2.2.2. Extraction of Reactive Dyes from Dyed Cotton Fibers

Reactive dyes are different from other classes of dyes because they form a covalent band with cotton. This makes reactive dyes the most substantive of dyes used on cotton because it provides excellent resistance to washing [31]. However, the extraction of reactive dyes from cotton requires the alkaline hydrolysis of the covalent bond at optimal temperatures or the application of enzymatic digestion.

Table 2 shows the extraction procedure for the isolation of reactive dyes from dyed cotton, which has been described in the literature [3,14,29,32–39].

Method No. 1—As the extraction system, the authors used [32]: sodium–water–poly(vinylpyrrolidone) (PVP), hydrogen bromide, 60% aqueous sulfuric acid, and 1.5% aqueous sodium hydroxide. All solvents were used at 100 °C, with only sulfuric acid used at room temperature.

The best results of extraction were reported for sodium hydroxide. Out of 50 reactive dyes, four were successfully extracted using NaOH extraction. However, in the publication, the extraction time was not mentioned [32].

Method No. 2—The material researched was 0.25–5.5 cm^2 pieces of black cloth (natural fibers such as cotton or wool). Then, 1.5% NaOH was added to the samples, and extraction was carried out for 25 min at 100 °C. Next, 10% (*v*/*v*) methanol was added to the samples [33]. The dry residues (after evaporating the solvent) were dissolved in a 300 μL water/methanol (1:9, *v*/*v*). The samples were diluted with water (1:10). Always, each sample was filtered through 0.45-μm polytetrafluoroethylene (PTFE) filters [33].

Method No. 3—A single fiber 2–15 mm long was extracted in a heat-sealed FEP tube. Then, 5 μL of 0.27 M NaOH and 50% MeOH were added to the research material, and the extraction process was provided at 100 °C (the extraction time was modified depending on the extraction efficiency of the dye). The ca. 20 resin (H^+ form) particles with 5 μL of methanol/water (1:1) mixture were used to remove NaOH or KOH. After 3–5 min of extraction, the pH-neutralized extract was transferred to a glass microvial and evaporated to dryness [34].

Method No. 4—The authors [29] prepared 10-cm threads of cotton fiber to optimize conditions of macro-scale extraction. The samples were placed into 500-μL glass inserts and put in a 96-well plate system. Solvents were added to the samples using an automated liquid-handling workstation and programmed operations for the specific fiber–dye combinations. The extraction of reactive dyes from cotton was provided using: (1) 29.7% aqueous ammonium hydroxide, (2) 1.5% aqueous solutions of sodium hydroxide, and (3) barium hydroxide. The extraction time and temperature were not presented. Alkaline and alkaline earth cations were removed from the extract using three methods: (1) cation exchange resin (Dowex®HCR-W2, H+ form, spherical beads, 16–40 mesh); (2) solid-phase extraction (SPE) Waters Oasis HLB 6-cm^3 cartridges were used; (3) the precipitation of barium carbonate by ammonium bicarbonate [29].

Method No. 5—Fibers of 10 mm, 5 mm, and 1 mm lengths, dyed with Reactive Yellow 160, Reactive Blue 220, or Reactive Orange 72 were used as the investigated materials [14]. These fibers were put in 200 μL conical vials, and 50 μL of 0.1875 M sodium hydroxide water solution was added. The vials were sealed to prevent evaporation. The extraction was carried out at 100 °C for 60 min. After that, 25 μL of 0.375 M hydrochloric acid (equimolar to the sodium hydroxide) was added, and subsequently, 25 μL of 10 mM ammonium acetate (adjusted to pH 9.3) was introduced for the dissolution of the reactive dyes [14].

Method No. 6—A 3 mg fabric strip was cut from a 2 × 2 cm textile and immersed in a 5 mL glass vial. Next, 1 mL of 0.15% sodium hydroxide solution was added, and the vial was placed in a heating module, which was subsequently closed. After 1 h at 80 °C, the vial was removed from the module and cooled to room temperature. Next, 30 μL of 1 N hydrochloric acid solution was added for neutralization. The obtained extract was filtered through a 0.2 μm polytetrafluoroethylene filter into an HPLC vial [35].

Method No. 7—A sample 10 mm long fiber was immersed in 10 μL of NaOH solution and cooled at 4 °C for 4 h. After the 4 h cooling time, the NaOH solution was removed. The fiber was washed with acetic acid solution once and with cellulase solution twice. Next, 10 μL of the cellulase solution was added to the rinsed fiber and the sample was mixed in a thermo mixer (Eppendorf Comfort, 50 °C, 550 rpm) for 20 h. Finally, the sample was centrifuged (5000 rpm, 5 min), and 10 μL of methanol was added [36].

Method No. 8—The authors developed three procedures for the isolation of different types of dyes from different types of fibers [37]. One of them was applied for the isolation of reactive dyes from cotton or viscose fibers. In this case, a fiber of 10 mm in length was placed in a vial, and 10 mL of 3 M NaOH was added. After 4 h at 4 °C, the fiber was washed using acetic acid solution (0.5 M) and cellulase solution (1.1 U/mg, 0.01 g in 10 mL of acetic acid solution at pH 5). This procedure was repeated. Next, the rinsed fiber was submerged in 10 mL of cellulase solution and placed in a thermo mixer (Eppendorf Comfort, 50 °C, 550 rpm, 20 h). Finally, after centrifugation (5000 rpm, 5 min), 10 mL of methanol was added to the extract [37].

Method No. 9—The authors based this on the procedure described in Collective Work, European Fibers Group [38]. As the sample, a 2 × 1 cm strip of material was cut from each piece of clothing. From each strip, four thread lengths of 1 cm were prepared. The thread sample was introduced into a 1.5 mL Eppendorf tube and 50 μL of NaOH solution was added. Next, the Eppendorf tube was placed in an ice pocket and put in a refrigerator for 4 h. After 4 h, the NaOH was removed. Firstly, the thread sample was washed using 50 μL of 0.5 M acetic acid solution, and then twice with the application of 150 μL of a 1.6 g/L cellulase solution in acetate buffer (pH 5). After the rinsing procedure, the thread sample was again covered with 150 μL of a cellulose solution and was mixed in a thermomixer for 20 h at 45 °C, 500 rpm. Next, the sample was centrifuged for 5 min at 7000 rpm. Góra and Wąs-Gubała [3] modified the above-described extraction procedure and applied *Aspergillus niger* and *Trichoderma reesei* cellulose. The extraction of reactive dyes using the *Aspergillus niger* cellulase solution was carried out at 50, 55, and 60 °C with the application of a concentration of solutions three and ten times greater than those reported in the literature. A thermomixer was replaced by a water bath and an ultrasound. The same procedure was repeated with the use of *Trichoderma reesei* cellulase. Thus, the extraction conducted was also carried out for eight threads taken from each textile product, and a double volume of each reagent was used. The extract was evaporated to dryness, and the dry residue was dissolved in 150 μL of acetate buffer (pH 5) [3].

Method No. 10—The authors used two methods for the extraction of reactive dyes from fabrics: chemical and enzymatic digestion [39]. Before chemical treatment, the textiles were washed with 2 mL of water, 2 mL of methanol, and 2 mL of acetonitrile (three times with each solvent) in order to remove impurities. The 3 mg fabric samples (three replicates from different sections of the degraded fabrics and one control fabric sample) were placed in 5 mL Fisher glass vials, and 1 mL of 1.5% NaOH solution was added to each vial. Next, the vials were heated at 80 °C for 1 h with constant stirring. After cooling down, the extracts were neutralized by the addition of 300 μL of 1 M HCl solution and were filtered through polytetrafluoroethylene (PTFE) syringe filters into HPLC vials. Before enzymatic digestion, 3 mg of the control fabric and degraded samples were prewashed by the application of firstly, 1 mL of water and gently shaken by hand, secondly, 1 mL of methanol, and thirdly, 1 mL of acetonitrile. Finally, the washed fabric samples were dried at room temperature (RT) and were subjected to enzymatic treatment. After prewashing, the 3 mg fabric sample was transferred to a vial. Then, 100 μL of 3 M NaOH solution was added, and the vial was placed in a grip seal bag and in a container with ice and cooled for 4 h. Next, the NaOH solution was removed, and 500 μL of 0.5 M acetic acid was added. After incubation for 1 min, the acetic acid was removed, and 1.5 mL of buffer solution (0.1 M sodium acetate, pH 5 with acetic acid) was added to the extract for a period of 1 min. After this time, the buffer solution was removed, and 1 mL of enzyme solution (90 g of cellulase in 50 mL of buffer) was added. The vial was tightly closed and placed in a shaking bath for 24 h at 50 °C. Finally, the vial was sonicated for 30 s. Next, the extract was filtered and transferred to an HPLC vial [39].

According to the data presented above, the most commonly used method for the isolation of reactive dyes from cotton fibers was chemical treatment [32], especially alkaline hydrolysis [14,29,32–35], followed by enzymatic extraction (enzymatic digestion) [3,36,37,39] (Figure 7).

Table 2. Overview of extraction procedure for isolation of reactive dyes from dyed cotton fibers described in the literature.

No. Method	Fiber/Dyes	Cleavage of Covalent Bonds of Reactive Dyes With Functional Groups on the Cotton Fibers	Extraction Procedure					Lit.
			Solvents	Fiber Length	Volume	Temperature	Time/Additional Information	
1	Dyes extracted from manufacturers' pattern cards and casework materials (50 different reactive dyes)	Hydrolysis	1. Sodium sulfide–water–poly(vinylpyrrolidone) (PVP) 2. Hydrogen bromide 3. 60% aqueous sulfuric acid 4. 1.5% aqueous sodium hydroxide	n.d.	n.d.	Sulfuric acid at room temp.; other solvents at 100 °C	n.d.	[32]
2	Black cotton/Names of reactive dyes not presented	Alkaline hydrolysis	1.5% NaOH; 10% MeOH (*v*/*v*)	0.25–5.5 cm^2 of black cloth (cotton)	n.d.	100 °C	25 min Most of the solvent was evaporated and the precipitate was dissolved into 300 µL of water–methanol (1:1, *v*/*v*). Next, the samples were diluted with water (1:10) and filtered through 0.45-/zm PTFE filters and analyzed using capillary zone electrophoresis with UV detection	[33]
3	Cotton 35% and Polyester 65%/black/Cibacron (reactive dyes) Dianix (disperse dyes) Cotton/brown/ Brown A-96138, reactive dyes: 1.66% Procion Yellow H-EXL 0.62% Procion Crimson H-EXL 1.72% Procion Blue H-ERD Cotton/marine/ Marine A-9514, reactive dyes: 0.96% Procion Yellow H-E4R 1.6% Procion Red H-EXL 5.16% Procion Navy H-ER	Alkaline hydrolysis	0.27 M (1.5%) NaOH; 50% MeOH	A single fiber of 2–15 mm	5 µL	100 °C	The extraction time varies depending on the extraction efficiency of the dye(s). Resin H(+) form was used to remove NaOH. Next, the samples were analyzed using micellar electrokinetic capillary chromatography	[34]

Table 2. *Cont.*

No. Method	Fiber/Dyes	Cleavage of Covalent Bonds of Reactive Dyes With Functional Groups on the Cotton Fibers	Extraction Procedure					Lit.
			Solvents	Fiber Length	Volume	Temperature	Time/Additional Information	
4	Cotton dyed fabrics and standard dyes: Reactive Blue 21, Reactive Yellow 160, Reactive Orange 72, Reactive Blue 19, Reactive Yellow 176, Reactive Violet 5, Reactive Black 5, Reactive Blue 250, Reactive Red 198, Reactive Blue 220, Reactive Red 180, Reactive Red 239/241	Alkaline hydrolysis in automated extraction system	Solvent systems: 1. 29.7% aqueous ammonium hydroxide, 2. 1.5% aqueous solutions of sodium hydroxide, 3. barium hydroxide	10-cm threads of fiber (yarns consisting of a bundle of twisted fibers)	500 µL	n.p.	n.d. Three methods to remove alkaline and alkaline earth cations from the extract: (1) Cation exchange resin using Dowex® HCR-W2, H+ form, spherical beads, 16–40 mesh; (2) Solid-phase extraction (SPE) using Waters Oasis HLB 6-cm^3 cartridges. (3) Precipitation reactions using ammonium bicarbonate to precipitate barium carbonate. Next, the samples were analyzed using capillary electrophoresis	[29]
5	Reactive Yellow 160, Reactive Blue 220, Reactive Orange 72	Alkaline hydrolysis	1.5% NaOH	Fibers of lengths 10 mm, 5 mm, and 1 mm	50 µL	100 °C	1 h The resulting solution was then treated with 25 µL of 0.375 M hydrochloric acid (equimolar with the sodium hydroxide), 25 µL of 10 mM ammonium acetate adjusted to pH 9.3 to neutralize any excess sodium hydroxide remaining prior to UPLC-DAD-MS/MS	[14]
6	Cotton/Reactive Blue 19	Alkaline hydrolysis	0.15% sodium hydroxide solution	3 mg fabric strip was cut from 2 × 2 cm fabric samples	1 mL	80 °C	1 h After 1 h, the vail was cooled to room temperature, neutralized with 30 µL of 1 N hydrochloric acid solution, filtered with a 0.2 µm polytetrafluoroethylene filter and analyzed using HPLC-DAD-HRMS	[35]
7	Cotton/Reactive Yellow 145 Cotton/Reactive Red 120 Cotton/Reactive Orange 16	Enzymatic extraction	10 µL of NaOH solution; 4 °C, 4 h. Next, the NaOH solution was removed and the fiber was rinsed in acetic acid solution and twice in cellulase solution. Then, the fiber was submerged in 10 µL of cellulase solution and mixed in a thermo mixer (Eppendorf Comfort, 50 °C, 550 rpm).	10 mm piece of fiber	10 µL	50 °C	20 h Afterward, the samples were centrifuged (5000 rpm, 5 min), and 10 µL of methanol was added. Next, the samples were analyzed using HPLC-DAD-MS	[36]

Table 2. *Cont.*

No. Method	Fiber/Dyes	Cleavage of Covalent Bonds of Reactive Dyes With Functional Groups on the Cotton Fibers	Extraction Procedure					Lit.
			Solvents	Fiber Length	Volume	Temperature	Time/Additional Information	
8	Cotton/Reactive Orange 122 Cotton/Reactive Red 195	Enzymatic extraction	10 µL of NaOH solution (3 M, 4 °C, 4 h. Next, the NaOH solution was removed and the fiber was rinsed twice in acetic acid solution (0.5 M) and in cellulase solution 0.01 g in 10 mL acetic acid solution at pH 5). Then, the fiber was submerged in 10 µL of cellulase solution and mixed in a thermo mixer (Eppendorf Comfort, 50 °C, 550 rpm)	10 mm of fiber	10 µL	50 °C	20 h Afterward, the samples were centrifuged (5000 rpm, 5 min), and 10 µL of methanol was added. Next, the samples were analyzed using HPLC-DAD-MS	[37]
9	21 pieces of red clothing, of a similar shade of color, made solely of cotton or with addition of other type of fibers polyesters, modal, elastane/Names of reactive dyes not presented	Enzymatic extraction Cellulase from 1. *Aspergillus niger* 2. *Trichoderma reesei ATCC 26921*	Basic procedure: 50 µL NaOH solution for 4 h at 4 °C. Next, the fibers were washed first in 50 µL of 0.5 M acetic acid solution and then twice in 150 µL of a 1.6 g/dm3 cellulase solution in acetate buffer (pH = 5). Next, the fibers were again covered with 150 µL of a cellulose solution 1. Concentration of cellulase solutions were three and ten times greater than in basic procedure 2. Concentration of cellulase solutions were three and ten times greater than in basic procedure and double volume of reagents. The extraction was carried out in bath water and the ultrasounds	4 (1) and 8 (2) threads of a length of 1 cm	150 µL	50 °C; 55 °C, 60 °C, respectively	20 h After that centrifugation 5 min, 7000 rpm. Next, the samples were analyzed using TLC coupled with VSC	[3,38]

Table 2. *Cont.*

No. Method	Fiber/Dyes	Cleavage of Covalent Bonds of Reactive Dyes With Functional Groups on the Cotton Fibers	Extraction Procedure					Lit.
			Solvents	Fiber Length	Volume	Temperature	Time/Additional Information	
10	Degraded Cotton/Reactive Red 198 Degraded Cotton/Reactive Black 5 Degraded Cotton/Reactive Blue 49 Degraded Cotton/Reactive Orange 35	Two methods: Chemical Treatment Method Before treatment, the degraded fabrics were washed three times with different solvents (2-mL water, 2-mL methanol, and 2-mL acetonitrile) in turn to remove impurities that may interfere with treatment.	1 mL of 1.5% NaOH solution with constant stirring.	3 mg of fabric samples	1 mL	80 °C	1 h After 1 h, the sample was cool down, neutralized by adding 300 µL of 1 M HCl solution, and filtered with a polytetrafluoroethylene (PTFE) syringe filter. Next, the samples were analyzed using HPLC-HRMS	[39]
		Enzymatic Digestion Method Before treatment, the degraded fabrics were washed by 1 mL of water, 1 mL of methanol, and 1 mL of acetonitrile. Finally, the washed fabric samples were dried at RT and were ready for enzymatic treatment.	100 µL of 3 M NaOH solution was added to the vial, and the vial was placed in a container containing ice for 4 h. Then, the NaOH solution was removed, and 500 µL of 0.5 M acetic acid was added and incubated for 1 min; then, acetic acid was removed and 1.5 mL of buffer solution (0.1 M sodium acetate, pH 5 with acetic acid) was added and kept for 1 min; the buffer solution was removed and 1 mL of enzyme solution (90-g cellulase in 50-mL buffer) was added. The vials were sealed and placed in a shaking bath	3 mg of fabric samples	1 mL	50 °C	24 h After 24 h, the vials were removed from the shaker and sonicated for 30 s, the solutions were filtered with a polytetrafluoroethylene (PTFE) syringe filter. Next, the samples were analyzed using HPLC-HRMS	

n.d.—not determined/not presented.

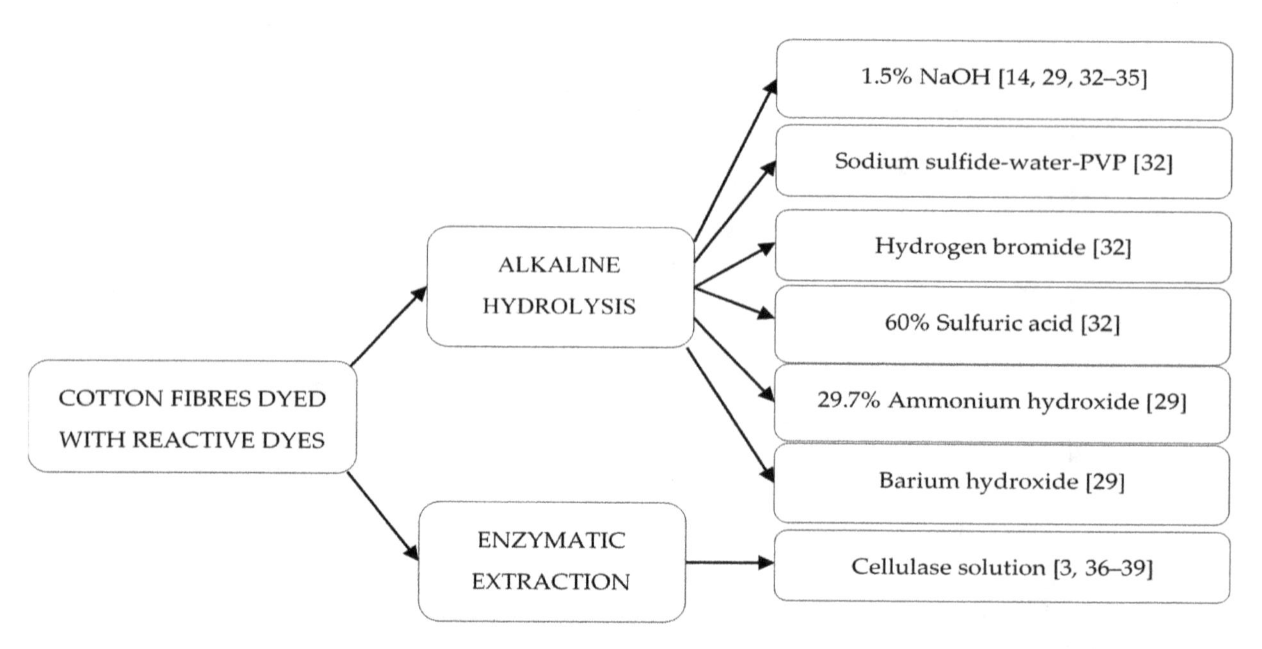

Figure 7. Reagents applied for the isolation of reactive dyes from dyed cotton fibers.

The reagent most commonly used for the cleavage of reactive dyes from dyed cotton was 1.5% NaOH [14,29,32–35], but in a few cases, sodium sulfide-water-poly(vinylpyrrolidone) [32], hydrogen bromide [32], 60% aqueous sulfuric acid [32], 29.7% aqueous ammonium hydroxide [29], and barium hydroxide [29] were also applied. Investigations were performed for such textile materials as a piece of clothing 0.25–5.5 cm^2 [33]; 3 mg [35,39], 10 cm threads of fiber [29]; and single fibers in a 1–15 mm length range [14]. Alkaline hydrolysis was conducted for 25 min [33] or 60 min [14,35,39]; the applied extraction temperature was 80 °C [35,39] or 100 °C [14,33,34]. Depending on the size of the investigated samples, the volume of NaOH solution was different: 5 μL [34], 50 μL [14], 500 μL [29], and 1 mL [35,39]; the data for other solvents were not presented (Table 2, Figure 7).

The first step of enzymatic extraction was similar to alkaline hydrolysis as presented above [3,36,37,39], with only a different volume of the added NaOH solution: 10 μL [36,37], 50 μL [3], and 100 μL [39]. After this step, the investigated material was usually rinsed in acetic acid and cellulase solution [36] or cellulase solution in acetic acid at pH 5 [37] or cellulase solution in acetate buffer solution at pH 5 [3,39]. Next, cellulase solution was added to the sample and enzymatic extraction was carried out using different procedures: 20 h with stirring (500 rpm) at 50 °C [36,37], 20 h in a water bath or ultrasound at 50, 55, and 60 °C [3], and 24 h in a shaking bath at 50 °C [39]. Investigations were performed for such textile materials as 3 mg of fabric samples [39], threads of a length of 1 cm [3], and 10 mm of fiber [36,37]. Depending on the size of the investigated sample, the required volume of cellulase solution was 10 μL [36,37], 150 μL [3], and 1 mL [39].

To summarize, the extraction of reactive dyes from cotton was mainly carried out by alkaline hydrolysis and enzymatic extraction. The best results were obtained for 1.5% NaOH (alkaline hydrolysis) and cellulase solution (enzymatic extraction). However, alkaline hydrolysis was shorter than the enzymatic extraction.

2.2.3. Extraction of Unknown Reactive Dyes from Dyed Cotton and Viscose

The extraction procedures presented above are dedicated for cotton fibers dyed with reactive dyes. However, most textiles can be and are dyed with multiple dyes. This fact may be useful to forensic scientists, but it also may lead to complications in the analysis of dyes in unknown samples. Different extraction systems and schemes for the classification of different dyes have been proposed in the literature. All of the systems and schemes have the same goal; they are intended to lead to the identification and classification of dyes. Among them, Laing and co-workers [40] for the first time proposed a scheme for the identification of reactive dyes on cotton. It was an exclusionary procedure. In this procedure, reactive dyes were not extracted by organic solvents, but when the fibers were treated with a reducing agent (sodium dithionite in sodium hydroxide), the original dyes, if azo in nature, were irreversibly decolorized. Thus, this procedure allows azo reactive dyes to be distinguished from other dye classes, but the identification of reactive dyes by TLC was impossible [40].

Experts from the former Forensic Science Service of England proposed a general, comprehensive, and widely used extraction scheme for the isolation of different types of dyes from dyed fibers [16]. It was also presented by Lewis [41], and this scheme is shown in Figure 8.

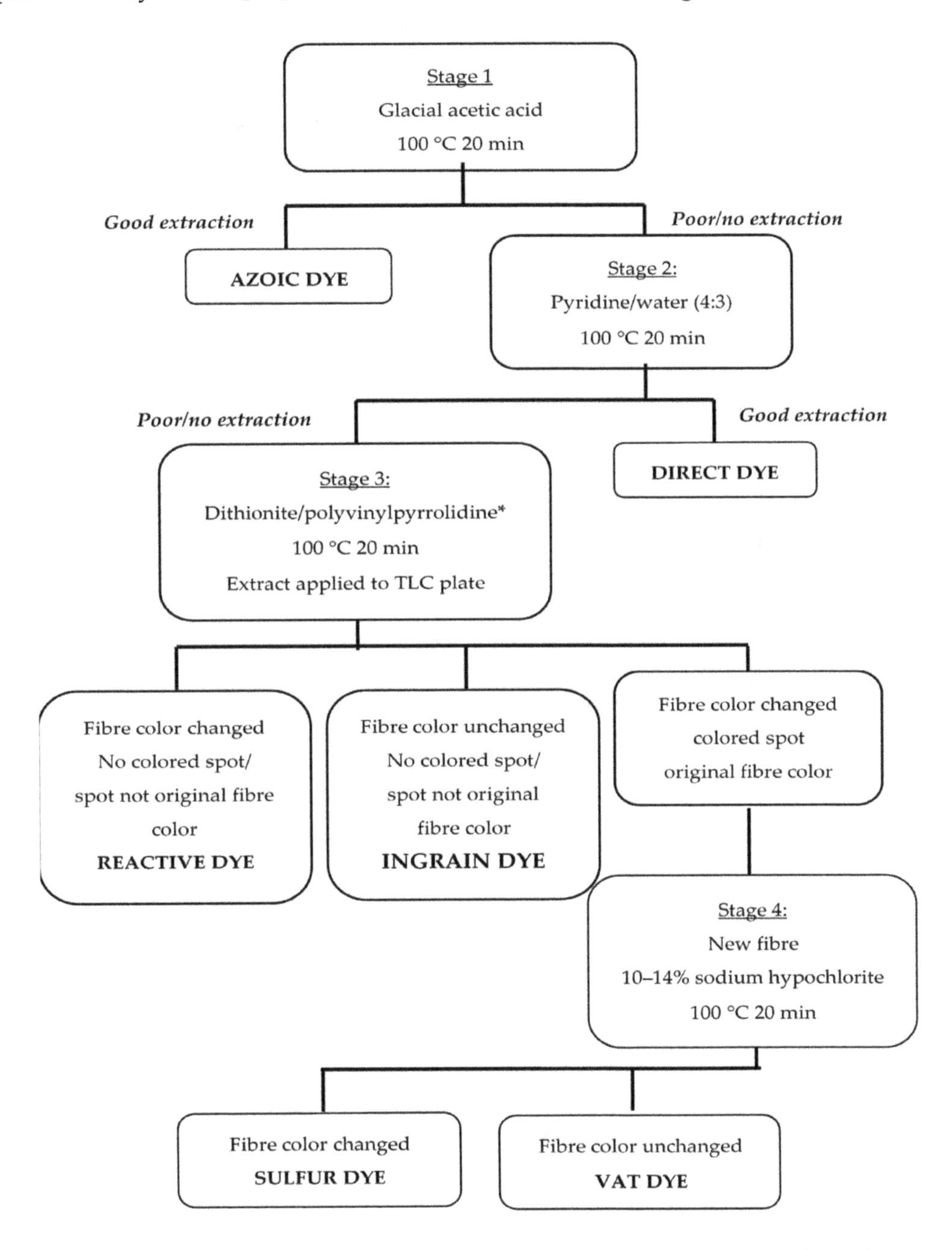

Figure 8. Classification scheme for the extraction of dyes from cotton and viscose fibers.

The extraction of dyes from fibers is a very simple process. A single fiber should be put in a glass tube, which should be sealed at one end. Around 10 μL of solvent should be added to the tube. The tube should be sealed and heated in the oven according to the time and temperature described in the extraction procedure. Details of the extraction scheme for dyes isolated from cotton and viscose fibers are shown in Figure 8 [41].

2.3. Identification of Dyed Cotton Fibers for Forensic Purposes Based on Chromatographic Techniques

Fiber traces, being the subject of forensic analyses, are usually no more than a few millimeters long and they usually contain 2 to 200 ng of dye [42]. For this reason, sensitive analytical methods must be applied for the identification of reactive dyes isolated from dyed cotton fibers. Tables 3 and 4 show the methods developed for the identification of reactive dyes based on chromatographic techniques [14,35–37,39,43–46]. A description of the chromatographic conditions is included in Table 3; the selected qualification and quantification parameters are in Table 4. A general scheme of the identification of extracted reactive dyes for forensic purposes based on chromatographic techniques is presented in Figure 9.

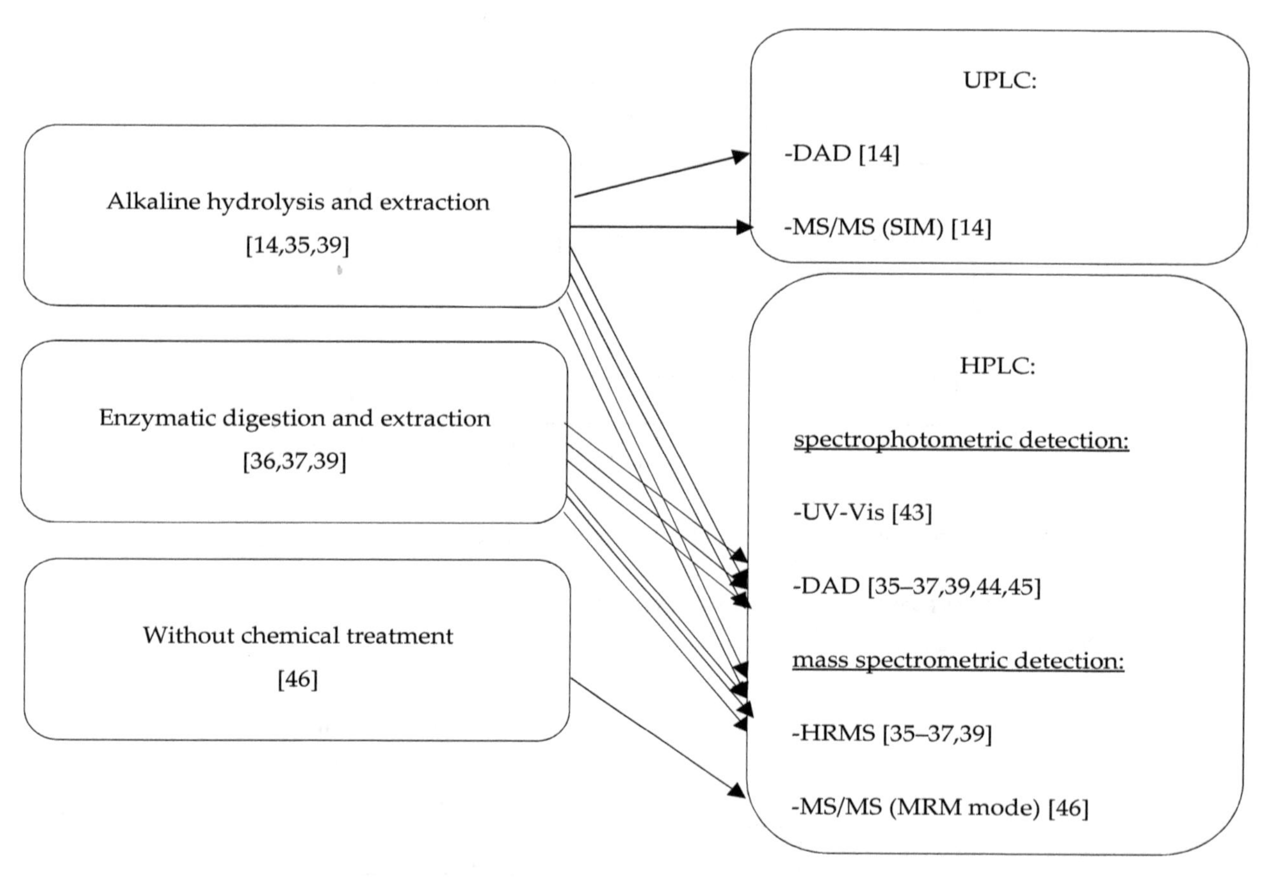

Figure 9. Chromatographic methods applied for the identification of extracted reactive dyes for forensic purposes.

Table 3. Overview of chromatographic methods described in the literature for the identification of reactive dyes extracted from cotton fibers (chromatographic conditions).

No. Method	Analytes	Technique	Column	The Mobile Phase	Gradient Program	The Mobile Phase Flow Rate	Injection	Lit.
1	Table 2 No. 5	UPLC	BEH C18 column (1.7 µm, 2.1 × 50 mm, Waters Acquity UPLC®) heated to 40 °C	A—10 mM ammonium acetate (pH 9.3) B—acetonitrile	UPLC-DAD-MS/MS 0 min 95% A; 5% B 0–2 min 50% A; 50% B 2–3 min 95% A; 5% B 3–5 min 95% A; 5% B	0.4 mL/min	10 µL	[14]
2	Table 2 No. 6	HPLC-DAD-HRMS	a Zorbax Eclipse Plus C18 (2.1 × 50 mm, 3.5 µm) column Pre-column a Zorbax Eclipse Plus C18 narrow bore guard column (2.1 × 12.5 mm, 5 µm) temp. 40 °C	A—20 mmol/L ammonium formate with formic acid in water (pH = 4) B—70/30 MeOH/ACN	3% B from 0 to 1 min, 3%–60% B from 1 to 1.5 min, 60%–90% B from 1.5 to 7 min, holding 90% B from 7 to 9 min then returning to 3% B at 9 to 9.5 min. A 4 min post-run of 3% B was used to re-equilibrate the column	0.5 mL/min	10 µL	[35]
3	Table 2 No. 7	HPLC	Grom-sil 120 ODS-5 ST (3 µm, 2 × 150 mm) Grace Davison Discovery Sciences, Deerfield, USA Precolumn AJO-4286 Temp. 22 °C	A—10 mM ammonium acetate in water: MeOH (95:5, *v*/*v*) B—25 mM ammonium acetate in ACN: MeOH (50: 50, *v*/*v*)	50% B (0–53 min) 100% B (53–67 min)	n. d.	10 µL	[36]
4	Table 2 No. 8	HPLC	Grom-sil 120 ODS-5 ST (150 × 2.0 mm i.d., 3 µm) Grace Davison Discovery Sciences, Deerfield, USA Precolumn AJO-4286 Temp. 22 °C	A—10 mM ammonium acetate in water: MeOH (95:5, *v*/*v*) B—25 mM ammonium acetate in ACN: MeOH (50:50, *v*/*v*)	Linear gradient Total time 78 min	n. d.	20 µL	[37]
5	Table 2 No. 10	HPLC	Zorbax Eclipse Plus C18 (2.1 × 50 mm, 3.5 µm) column with a Zorbax Eclipse Plus C18 narrow bore guard column (2.1 × 12.5 mm, 5 µm); Temp. 40 °C	A—water with 20-mM ammonium formate and formic acid (pH = 4) B—70:30, *v*/*v*) methanol/acetonitrile	3% B from 0 to 1 min, 3%–60% B from 1 to 1.5 min, 60%–90% B from 1.5 to 7 min, holding at 90% B from 7 to 9 min, and 3% B at 9 to 9.5 min. A 4-min post run of 3% B before the next run was performed.	0.5 mL/min	10 µL	[39]

Table 3. *Cont.*

No. Method	Analytes	Technique	Column	The Mobile Phase	Gradient Program	The Mobile Phase Flow Rate	Injection	Lit.
6	Cibacron Yellow F-4G Cibacron Blue F-R	Ion-pair HPLC	Nucleosil 100–5, C18, 150 × 4.6 mm I.D. analytical column Macherey-Nagel GmbH & Co.KG	A 47:53 *v/v* mixture of acetonitile and 0.05 M ammonium acetate buffer containing 1 mM trimethylammonium bromide (CTAB) ion-pairing agent	isocratic mode	0.8 mL/min each dye eluted separately 0.6 mL/min a mixture of dyes	n.d.	[43]
7	The C.I. Reactive Red 195, C.I. Reactive Yellow 145, C.I. Reactive Blue 221	HPLC	BDS Hypersil C18 column (150 mm × 4.6 mm, 5 µm)	A—90% deionized water and 10% acetonitrile containing 0.1% ammonium acetate (of buffer of pH 6). B—90% acetonitrile and 10% deionized water containing 0.1% ammonium acetate of buffer of pH 6	0.00 min 100% A; 0% B 10.00 min 0% A; 100% B 10.10 min10 0% A; 0% B 20.00 min 0% A; 0% B	0.75 mL/min	20 µL	[44]
8	The C.I. Reactive Red 195, C.I. Reactive Yellow 145, C.I. Reactive Blue 221	HPLC	BDS Hypersil C18 column (150 mm × 4.6 mm, 5 µm)	A—90% deionized water and 10% acetonitrile containing 0.1% ammonium acetate (of buffer of pH 6). B—90% acetonitrile and 10% deionized water containing 0.1% ammonium acetate of buffer of pH 6	0.00 min 100% A; 0% B 10.00 min 0% A; 100% B 10.10 min10 0% A; 0% B 20.00 min 0% A; 0% B	0.75 mL/min	20 µL	[45]
9	Reactive orange 16 (RO16)	HPLC-MS/MS	Symmetry C18 (50 mm × 1.0 mm I.D., 3.5 µm, Ireland)	(A) 0.1% Acetic acid in water (B) 0.1% Acetic acid in acetonitrile	From 95:5 A:B (*v/v*) to 24:76 A:B (*v/v*)A (0–15 min), 100% B (for 0.1 min and retention for 2 min), 95:5 A:B (*v/v*) (for 0.1 min and retention for 2 min)	0.3 mL/min	10 µL	[46]

n.d.—not determined/not presented.

Table 4. Overview of chromatographic methods described in the literature for the identification of reactive dyes extracted from cotton fibers (selected qualification and quantification parameters).

No.	Analytes	Technique	Detector	Scanned Wavelength Range	Retention Time	Value [*m/z*]	LOD	Lit.
1	Table 2 No. 5	UPLC	DAD MS(ESI) tandem quadrupole mass spectrometer MS/MS—Selected Reaction Monitoring (SRM) mode	300–700 nm for Reactive Yellow 160, Reactive Blue 220, and Reactive Orange 72 at 405 nm, 610 nm, and 478 nm, respectively	Several peaks between 0.75 and 2.75 min within which multiple additional components are present	*m/z* 652, 572, 554, and 614 from the analysis the Reactive Yellow 160; Reactive Blue 220 did not did not appear in UPLC-MS; *m/z* 572, 474, 492, and 417 from the analysis of Reactive Orange 72	DAD—0.33–1.42 ppb MS/MS—3 pg to 83 pg 1 mm extract of Reactive Yellow 160 was detected and arguably quantifiable	[14]
2	Table 2 No. 6	HPLC	DAD ESI(-)-(Q-TOF)MS	200–800 nm. The main wavelengths for absorbance analysis 254 nm, 620 nm, and 660 nm; the quantitative analysis at 620 nm	4.6 min RB19-OH Degraded products were also monitored	Q-TOF MS RB19-OH a theoretical *m/z* 501.0432; monitored *m/z* 501.0432	0.12 ± 0.07 μg/mL (DAD) n.d. (HRMS)	[35]
3	Table 2 No. 7	HPLC	Diode array detection (DAD) MS(ESI)-LTQ Orbitrap (HRMS)	200–800 nm λ_{max} 494 nm for Reactive Orange 16 λ_{max} 417 nm for Reactive Yellow 145 λ_{max} 540 nm for Reactive Red 120 150–2000 *m/z*	Between 14.5 and 24.8 min	Calculated mass [*m/z*] e.g., Reactive Orange 16 pure *m/z* 572.00980 *m/z* 474.04242 Reactive Orange 16 fiber *m/z* 774.14806 *m/z* 816.15863 (the dye molecule connected to two cellulose units)	e.g., Reactive Orange 16 DAD—37.3 μg/L powder/0.06 mm fiber HRMS—4.2 μg/L powder/0.011 mm fiber	[36]
4	Table 2 No. 8	HPLC	DAD MS(ESI)-LTQ Orbitrap (HRMS)	200–800 nm 150–2000 *m/z*	Reactive Orange 122 16.4 min 17.5 min 19.9 min Reactive Orange 195 16 min 14.2 min 17.5 min	Reactive Orange 122 584.5629 $[M+2glu-2H]^{2-}$ 746.616 $[M+4glu-2H]^{2-}$ -(red dye) 624.5407 $[M+2glu-2H]^{2-}$ -(yellow dye) -(blue dye)	n.d.	[37]

Table 4. *Cont.*

No.	Analytes	Technique	Detector	Scanned Wavelength Range	Retention Time	Value [*m/z*]	LOD	Lit.
5	Table 2 No. 10	HPLC	Diode array detection (DAD) MS(negative ESI)-QTOF (HRMS)	200 to 800 nm. The main wavelengths for absorbance analyses were set to 254, 515, 610, 620, and 660 nm 200–800 nm (λ_{max} 515 nm for RR198 λ_{max} 620 nm for RBlk5 λ_{max} 600 nm for RB49 λ_{max} 430 nm and 254 nm for RO35 100–1600 *m/z*	RR198 45.6 min 51.0 min	The hydrolyzed form of dye RR198 at *m/z* 397.5020 (doubly charged deprotonated ion) The hydrolyzed form of dye RBlk5 at *m/z* 370.5093 (doubly charged deprotonated ion) The hydrolyzed form of dye RB49 at *m/z* 397.5354 (doubly charged deprotonated ion) The hydrolyzed form of dye RO35 at *m/z* 363.5312 (doubly charged deprotonated ion)	n.d.	[39]
6	Cibacron Yellow F-4G Cibacron Blue F-R	Ion-pair HPLC	UV absorbance	275 nm	Blue 4.70 min Hydrolyzed Blue 3.44 min Yellow 3.32 min Hydrolyzed Yellow 2.38 min Blue in mixture (m) 7.54 min Hydrolyzed Blue (m) 5.07 min Yellow in mixture (m) 4.57 min Hydrolyzed Yellow (m) 3.16 min		n.d.	[43]
7	The C.I. Reactive Red 195, C.I. Reactive Yellow 145, C.I. Reactive Blue 221	HPLC	DAD	200–800 nm λ_{max} 540 nm Reactive Red 195 λ_{max} 420 nm Reactive Yellow 145 λ_{max} 610 nm Reactive Blue 221	Total runtime of 20 min e.g., several peaks with different RT belonging to different dye forms for Disperse Red 195 t_R = 1.77 min; t_R = 2.75 min; major peak at t_R = 5.54 min attributed to the active form of Reactive Red 195. In the opposite, the Reactive Yellow 145 and Reactive Blue 221 show only one big peak respectively at RT 6.71 min and RT 6.64 min		n.d.	[44]

Table 4. *Cont.*

No.	Analytes	Technique	Detector	Scanned Wavelength Range	Retention Time	Value [*m*/*z*]	LOD	Lit.
8	The C.I. Reactive Red 195, C.I. Reactive Yellow 145, C.I. Reactive Blue 221	HPLC	DAD	200–800 nm λ_{max} 540 nm Reactive Red 195 λ_{max} 420 nm Reactive Yellow 145 λ_{max} 610 nm Reactive Blue 221	Total runtime of 20 min e.g., several peaks with different RT belonging to different dye forms for Disperse Red 195 t_R = 1.79 min, 2.2 min, 2.34 min, 2.92 min, 3.55 min and 4.11 min—the inactive forms; two major peaks at t_R = 4.73 min and t_R = 5.96 min attributed to the partially hydrolyzed forms of Reactive Red 195. Peak at t_R = 5.33 min attributed to the completely hydrolyzed form. In the opposite, the Reactive Yellow 145 and Reactive Blue 221 showed only one big peak respectively at RT 2.74 min and RT 2.71 min of the inactivated forms and at 5.15 min (RY145), 5.24 min, 5.91 min (RB221) of the partially hydrolyzed forms and at t_R = 6.30 min attributed to the completely hydrolyzed form of RB221		n.d.	[45]
9	Reactive orange 16 (RO16)	HPLC-MS/MS	MS(ESI) quadrupled type tandem MS MS/MS–MRM mode	n.d.	5.09 min	Molecular weight 617.54 285.4 $[M\text{-}2H]^{2-}$ 285.4→236.85 285.4→264	MS/MS—2.10 ng/mL	[46]

n.d.—not determined/not presented.

2.3.1. Chromatographic Conditions

High-performance liquid chromatography (HPLC) coupled with spectrophotometric detectors such as UV-Vis [43] and diode array (DAD) [35–37,39,44,45], as well as coupled with high-resolution mass spectrometric (HRMS) [35–37,39] and tandem mass spectrometric (MS/MS) detectors [46] was the most often used technique for the identification of reactive dyes for forensic purposes (Figure 9, Tables 3 and 4). Usually, the HPLC-MS apparatus was equipped with both detection systems [35–37,39] (Figure 9; Tables 3 and 4). Chromatographic analyses were usually performed using reverse-phase C18 columns [35,39,43–46] of a length from 50 to 150 mm. On the other hand, Carey et al. and Schotman and co-workers applied a Grom-sil 120 ODS-5 ST chromatographic column with a length of 150 mm [36,37]. In most cases, the chromatographic separation of reactive dyes was carried out under gradient programs using such mobile phases as an ammonium formate aqueous phase and formic acid in water (pH 4): a ratio of 70/30 methanol/acetonitrile organic phase [35]; ammonium acetate in water/methanol (95/5): ammonium acetate in acetonitrile/methanol (50/50) [36,37]; ammonium formate and formic acid in water (pH 4): methanol/acetonitrile (70/30) [39]; ammonium acetate in water/acetonitrile (90/10) (pH 6): ammonium acetate in water/acetonitrile (10/90) [44,45]; acetic acid with water/acidified acetonitrile [46]. In one case, the isocratic conditions using a 47:53 *v/v* mixture of acetonitrile and ammonium acetate buffer containing trimethylammonium bromide (CTAB) as an ion-pairing agent was applied [43]. The mobile phase flow rate was in the range of 0.3 mL/min to 0.8 mL/min; the time of analysis was from 9.5 [39] to 78 min [37]; the injection volume was 10 or 20 μL.

The next chromatographic technique applied for the identification of reactive dyes for forensic purposes was ultra-performance liquid chromatography (UPLC) coupled with DAD and MS/MS detectors [14] (Figure 9; Tables 3 and 4, Method No. 1). The separation of analytes was done using a C18 column with a length of 50 mm, and 10 mM of ammonium acetate (pH 9.3) and acetonitrile as mobile phases in gradient conditions. The mobile phase flow rate was 0.4 mL/min, the time of analysis was 5 min, and the injection volume was 10 μL [14].

2.3.2. Qualification and Quantification Analysis of Extracted Reactive Dyes for Forensic Purposes

As mentioned in Section 2.2, the necessity to cleavage the covalent bond formed between the dye and cotton fibers using alkaline hydrolysis or by enzymatic digestion results in various structural changes of reactive dyes and in the formation of multiple reaction products from a single dye molecule. For this reason, prior to chromatographic analyses, the synthesis and the study of partially and completely hydrolyzed forms of selected reactive dyes is very useful. The knowledge of the structures of their conjugates with glucose units formed during enzymatic digestion is also very important because they could be used as standards during the qualification and quantification analysis of reactive dyes extracted from dyed cotton fibers.

Synthesis of Hydrolyzed and/or Enzymatic Digestion Dye Standards

Nayar and Freeman (2008) analyzed the commercial and hydrolyzed forms of six reactive dyes (C.I. Reactive Red 2, Reactive Red 24:1, Reactive Orange 72, Reactive Blue 19, C.I. Reactive Blue 4, C.I. Reactive Red 120) using negative ion fast atom bombardment (FAB) and negative ion electrospray (ES) mass spectrometry [47]. These dyes were selected as representatives of the structural types of currently used reactive dyes. Negative ion FAB and ES mass spectrometric analyses have been used to determine the level of hydrolysis following the dyeing of cotton with reactive dyes [47].

The synthesis of hydrolyzed dye standards of C.I. Reactive Blue 19 was also performed by Sultana et al. [35]. The obtained hydrolyzed dye standard (RB19-OH) was characterized via DAD and high-resolution MS, and it was used for the quantification and qualification of dyes on biodegraded samples.

Feng et al. [39] also decided to synthesize hydrolyzed dye standards and enzymatic digestion dye standards having cellobiose units of four commonly used reactive dyes, C.I. Reactive Black 5, C.I.

Reactive Red 198, C.I. Reactive Blue 49, and C.I. Reactive Orange 35, in order to use them as standards for comparison in the study of the biodegradation of reactive dyes on cotton jersey fabrics buried in soil. The structures of the degradation products were determined using the HPLC-HRMS technique [39].

Identification and Quantification of Extracted Reactive Dyes for Forensic Purposes

Hoy demonstrated successful alkaline hydrolysis and extraction of reactive dyes from single 10 mm, 5 mm, and 1 mm cotton fibers dyed with Reactive Yellow 160, Reactive Blue 220, or Reactive Orange 72, and they analyzed the obtained extracts using the UPLC-DAD-MS/MS technique (Tables 3 and 4, Method No. 1) [14] (Figure 9). UPLC-DAD and UPLC-(ESI)MS/MS analyses of standard solutions of reactive dyes and extracted (treated with NaOH) reactive dyes confirmed that additional reactions occur during the extraction process. For example, the UPLC-DAD chromatogram for standard Reactive Orange 72 showed a single peak corresponding to the dye with an absorbance maximum of 478 nm, whereas the UPLC-DAD chromatogram for extracted dye showed two peaks, both of which increased in retention, and the main peak had an absorbance maximum of 473 nm. Hoy suggested that both peaks corresponded to the dye and one of them corresponded to the incomplete base hydrolysis of Reactive Orange 72. An analysis of the Reactive Orange 72 dye standard using UPLC-MS/MS confirmed that Reactive Orange 72 is actually a mixture of derivatives of the dye, which are characterized by *m/z* values at 572 (retention time (RT) 13.92 min), *m/z* 492 (RT 14.14 min), *m/z* 474 (RT 15.87 min) and *m/z* 417 (RT 14.47 min). On the other hand, the UPLC-MS/MS analysis of Reactive Orange 72 treated with NaOH confirmed the presence of only one product at *m/z* 450 (RT 13.24 min); the second peak observed during the UPLC-DAD measurement did not appear. The Reactive Yellow 160 standard was also a mixture of derivatives at *m/z* 652 (RT 16.61 min), *m/z* 572 (RT 17.50 min), *m/z* 554 (20.77 min), and *m/z* 614 (RT 19.92 min); whereas Reactive Blue 220 was not observed in UPLC-MS/MS. The amount of dye on each fiber was determined based on the peak area on the chromatogram acquired for each dye at the maximum wavelength (405, 610, and 478 nm, respectively for Reactive Yellow 160, Reactive Blue 220, and Reactive Orange 72) and comparison with standard dye mixtures. The limits of detection (LOD) of the investigated reactive dyes using UPLC-MS/MS were in the range from 3 to 83 pg (based on 10 μL injections), confirming the possibility of performing quantitative analysis of 1 mm fiber extracts. The proposed UPLC-DAD method gave the detection limits of 0.33–1.42 ppb and the quantitation limits of 1.00–4.30 ppb. Trace fiber extractions confirmed the possibility of the detection and quantification of 1–10 mm extracts of Reactive Yellow 160. The obtained UPLC-MS/MS results for Reactive Orange 72 and Reactive Yellow 160 suggested that many reactive dyes could be mixtures of multiple compounds [14].

Sultana and co-workers investigated C.I. Reactive Blue 19 (RB19) extracted from cotton fabrics biodegraded in soil in laboratory conditions over intervals of 45 and 90 days using the HPLC-DAD-HRMS method [35]. High-resolution mass spectrometric detection (ESI(-)-(Q-TOF) was applied to characterize the isolated dye degradation products; DAD detection was used to quantify the intact dye and the degradation product isolated from the fabric samples. As mentioned in this section, prior to chromatographic analysis, the hydrolyzed dye standard (RB19-OH) was synthesized and characterized via DAD (λ_{max} at 620 nm, RT 4.6 min) and high-resolution MS (theoretical *m/z* value 501.0432; value monitored in the mass chromatogram *m/z* 501.0432) [35]. The DAD analysis of the obtained extracts at 620 nm confirmed the presence of two compounds on biodegraded cellulosic fabrics: RB19-OH at the RT of 4.6 min, and the desulfonated degradation product formed after losing an $-SO_3$ group from RB19-OH at the RT of 5.2 min (the last signal was observed also during HPLC-HRMS as the deprotonated singly charged molecule at *m/z* 421.0858). Thus, the isolated dye was successfully quantitatively analyzed via the HPLC-DAD method by synthesizing a hydrolyzed form of the dye and creating calibration curves (limit of quantification 0.4 ± 0.2 μg/mL). Moreover, the HPLC-HRMS analysis showed that the degradation product was formed by losing an $-SO_3$ group from the intact hydrolyzed form of the dye [35].

The identification of Reactive Orange 16, Reactive Yellow, and Reactive Red 120 extracted from cotton fibers (Tables 3 and 4, Method No. 3) was based on retention times, and the mass accuracy was recorded using a high-resolution MS detector (deviation generally < 2 ppm) [36]. The HPLC analyses with the application of both detector systems (DAD, MS) were performed for each standard dye solution and for each dye extracted from a fiber, with a continuous record of acquired spectra. In this way, ample data were recorded. The absorption spectra recorded for a standard dye solution in the range of 200–800 nm proved that λ_{max} for Reactive Orange 16 is 494 nm, for Reactive Yellow 145 λ_{max} is 417 nm, and for Reactive Red 120 λ_{max} is 540 nm. In each case, the chromatograms obtained for pure standard dyes and those acquired after the hydrolysis of dyed fibers using the cellulase method were compared. The RTs observed for the powder dye references and extracted reactive dyes were different. For example, the RTs of pure Reactive Orange 16 were 19.4 min (*m/z* 572.009800) and 24.8 min (*m/z* 474.04242); whereas for Reactive Orange 16 fiber, the RTs were 17.6 min (*m/z* 774.14806) and 19.7 min (*m/z* 816.15863), respectively. This means that the recovered dye was still linked to one or more cellulose units and thus differed from the chemical structure of the unreacted dye. This altered the RT and the observed molecular mass. For example, the signal at *m/z* 816.15863 observed for Reactive Orange 16 fiber was attributed to the structure containing the dye molecule connected to two cellulose units. The LODs were established as the minimum concentration of dissolved dye powder (μg/L) and the minimum fiber length required to identify the dye. For example, the LOD values determined using the DAD detector established for Reactive Orange 16 were 37.3 μg/L and 0.06 mm, while for the MS detector, they were 4.2 μg/L and 0.011 mm, respectively [36].

Schotman and co-workers identified dyes present on fibers or textiles submitted to forensic examination using the HPLC–DAD–MS method (Tables 3 and 4, Method No. 4) [37]. Among seven investigated samples (case samples), four were made from cotton. Reactive Orange 122 was detected in the sample of case 2, while Reactive Orange 195 was detected in the sample of case 3. Usually, one or more of the dyes were identified in the obtained extracts, and the number of identified dyes was from a few to around 50 in case 8. Such data as the retention time, DAD spectrum, and mass spectrum allowed a useful "fingerprint" of the dye to be provided, which could at least be very useful to compare fibers from a known source with recovered traces [37].

Feng at al. applied the HPLC-DAD-HRMS method for the determination and characterization of products obtained from four reactive dyes (C.I. Reactive Black 5 (RBlk5), C.I. Reactive Red 198 (RR198), C.I. Reactive Blue 49 (RB49), and C.I. Reactive Orange 35 (RO35)) present on cotton jersey fabrics subjected to biodegradation in soil (Tables 3 and 4, Method 5) [39]. The optimized chemical and enzymatic procedures allowed the cotton fabric to be digested, and they removed RBlk5, RR198, and RB49 from the dyed material. As mentioned in this section, hydrolyzed reactive dyes and reactive dyes having cellobiose units were synthesized and used as standards for comparison in this study. The determination of the structures of the biodegraded products, based on exact mass measurements supported by the DAD measurements (retention time and absorbance spectra), allowed the possible degradation pathways of the reactive dyes in degraded cotton fabrics to be proposed [39].

Zotou and co-workers investigated the hydrolysis kinetics of two reactive fluorotriazinic dyes (Cibacron Yellow F-4G and Cibacron Blue F-R) separately and in a 1:1 mixture, in the presence as well as in the absence of cotton, on a laboratory scale using reversed-phase ion-pair HPLC (Tables 3 and 4, Method No. 6) [43]. The two forms of the dyes, hydrolyzed and non-hydrolyzed, were separated satisfactorily (different RTs). However, when these two dyes were studied as a mixture, the hydrolyzed Blue co-eluted with the non-hydrolyzed Yellow. For this reason, in order to obtain complete separation, the flow rate of the mobile phase during the HPLC analysis has to be decreased from 0.8 to 0.6 mL/min [43].

Chemchame and co-workers also studied the hydrolysis of reactive dyes (Tables 3 and 4, Methods No 7 and No. 8) [44,45]. In these studies, four standard dye solutions of heterobi-functional reactive dyes containing monochlorotriazine/β-sulfatoethylsulfone (C.I. Reactive Red 195, C.I. Reactive Yellow 145, and C.I. Reactive Blue 221) were prepared and subjected to hydrolysis using NaOH concentrations fixed at pH 11 at 98 °C for 90 min [44] or using NaOH concentrations fixed at pH 10 at 60 °C for 30 min [45]. Among the three methods tested for quantifying unfixed forms of bi-functional reactive dyes used in dyeing cotton fibers (spectrophotometric approach, colorimetric approach, and HPLC technique), the last one was the most useful to identify and quantify the hydrolyzed and inactive forms of dye in a residual soaping bath.

Hu and co-workers developed a sensitive HPLC-MS/MS method using Multi Reaction Monitor (MRM) mode for the analysis of different fiber dyes, among them for the analysis of cotton fibers dyed with reactive dyes (Tables 3 and 4, Method No. 9) [46]. Unfortunately, in this study, the extraction of reactive dyes from cotton fibers was not presented (such information was presented only for disperse dyes extracted from polyester fibers). The HPLC-MS/MS analyses were performed both in positive ion mode (the molecular ions $[M + H]^+$ and $[M - Cl]^+$ were observed) as well as in negative ion mode ($[M - H]^-$, $[M - Na]^-$, and $[M - 2Na]^{2-}$ were observed). Reactive Orange 16 (RO16) (calculated molecular mass at *m/z* 617.54) produced a molecular ion in the form of $[M{-}2Na]^{2-}$ at *m/z* 285.4, which was subjected to fragmentation to an ion at *m/z* 236.85 and an ion at *m/z* 264, as observed in MRM mode. The RT of reactive dye RO16 was 5.09 min; the limit of detection during MS/MS measurements was 2.10 ng/mL [46].

2.4. Examination of Dyed Textile Fibers for Forensic Purposes Based on Routinely Used Spectroscopic Techniques

Routine methods of identifying and comparing textile fibers in forensic science are based on the principles of microscopy and spectroscopy. Information derived from microspectrophotometry and vibrational spectroscopy investigations has contributed to the benefit of forensic examinations of dyed textile fibers. This is emphasized by examples taken from real case studies and targeted scientific research in this field [16,48].

UV–visible microspectrophotometry (UV-Vis MSP) is used for objective observations of colored fibers because it is nondestructive, repeatable, and unlike other methods that require the extraction of the dye, it involves little sample preparation. Despite this, it is rather used for identifying and comparing the spectral characteristics of a sample, not identifying particular dyes or mixtures of dyes [49].

The discrimination of single cotton fibers dyed with reactive dyes of the same manufacturer, as well as the possibility of assessing the concentration of a dye in examined fibers were verified with the use of UV-Vis MSP [50]. Woven cotton fabrics dyed with different concentrations of one-compound reactive dyes (the commercial name Cibacron®) were examined, and the results obtained indicated that all of the analyzed samples were distinguishable from each other with the use of this technique. The detection limit was 0.18% of the concentration of a dye in the textile sample. However, the authors observed intra-sample and inter-sample variation, as well as the dichroism effect.

Subsequently, the same authors presented an assessment of the applicability of UV-Vis MSP together with Raman spectroscopy in the examination of textile fibers dyed with mixtures of synthetic, also reactive dyes [51]. The MSP study conducted in the 200 to 800 nm range and three types of excitation sources, 514, 633, and 785 nm, used during Raman examinations, were applied for the examination of single cotton fibers, which were dyed with binary or ternary mixtures of reactive dyes. UV–Vis MSP showed limited possibilities for the discriminatory analysis of cotton fibers dyed with a mixture of reactive dyes, where the ratio of the concentration of the main dye used in the dyeing process to the minor one was higher than four. The results show that the capability of distinguishing dye mixtures was similar for both spectroscopic methods used.

Buzzini and Massonnet evaluated the potential and limitations of Raman spectroscopy on a broad range of fiber types and colors, including those containing reactive dyes [52]. For fiber samples collected from 180 textiles, the results obtained using Raman spectroscopy were compared to those collected with the use of traditional methods of textile fibers examination, i.e., bright field, double polarization, fluorescence and comparison microscopy, UV-Vis MSP, and TLC. This study presented that Raman spectroscopy can play a complementary role in the routine analytical sequence of forensic fiber analysis to detect and identify the dye composition. Combining data obtained with several laser wavelengths allowed for the further discrimination of pairs previously indistinguishable with the use of light microscopy and UV-vis MSP. Despite the lower discriminating power calculated for Raman data when single laser wavelengths were considered, additional discriminations were observed. It was concluded that an instrument equipped with several laser lines is necessary for efficient use [53].

Was-Gubala and Machnowski [54] confirmed the difference between cotton and regenerated cellulose fibers resulting from variations in the degree of polymerization and supramolecular structure during Raman research [54]. The examined fibers were dyed with eight reactive dyes, among others, and their spectra were obtained with the use of three excitation sources: 514, 633, and 785 nm. For more than 80% of the obtained spectra, the presence of dye bands was confirmed. Bands originating from cotton and viscose were usually predominant in the spectra of fibers with a lower dye concentration, while in the other case, the majority of the bands originated from the dye. However, the dye concentration in cotton and viscose fibers did not directly influence the interrelationships between the intensity of characteristic dye bands and their concentration in the fiber samples in every case.

Techniques that have increased Raman sensitivity in the last decade are surface-enhanced Raman spectroscopy (SERS) and surface-enhanced resonance Raman scattering (SERRS). However, they are more complex and require more experience in sample preparation than Raman spectroscopy itself; therefore, there are not many reports of their use in forensic practice, for example for the examination of reactive dyes in cotton fibers yet. Nevertheless, in other scientific fields, they are used for dyes and textiles research [55–60].

While the use of infrared microscopy is almost natural to determine the composition of fibers, its use to identify fiber dyes may be appropriate in highly colored fibers and is likely to be below the pale fiber detection limits [61]. This stems from the lack of sensitivity of infrared absorbance to components that represent less than approximately 5% of a sample and the characteristically low level of dye concentration in most textile fibers, which is also those dyed by reactive dyes. Grieve reported more specific contributions of known dyes to the region of the infrared spectra of acrylic fibers [62]. Diffuse reflectance infrared Fourier transform spectroscopy (DRIFTS) has been more successful in characterizing fiber dyes, and the reactive dyes were discriminated on cotton by use of this technique [63].

3. Comparison of Chromatographic Methods for the Identification of Dyed Textile Fibers for Forensic Purposes with a Spectroscopic Technique

Forensic examinations have to give as much evidential information as possible. The identification and comparison studies of microtraces as single dyed fibers for forensic purposes are based on microscopic, spectroscopic, and chromatographic techniques [6] (Figure 10).

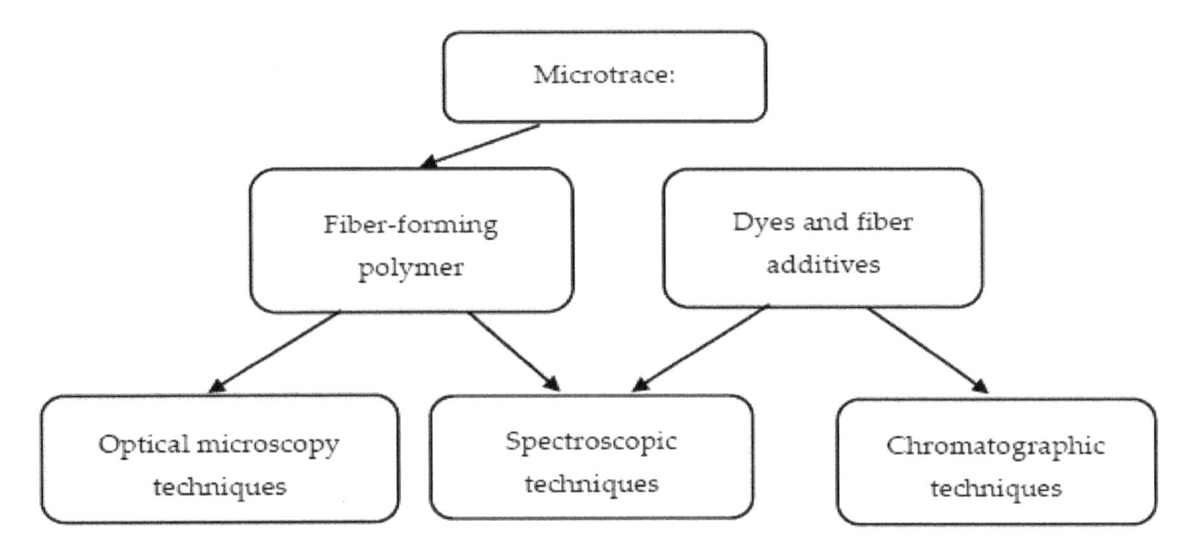

Figure 10. Scheme of forensic examination of textile fibers.

The first step in the forensic laboratory is the investigation of dyed textile fibers using optical microscopy. Optical studies are based on different techniques. A microscopic image of a fiber gives information about its physicochemical structure. Fibers can be investigated mainly by stereoscopic microscopy with reflected light, biological microscopy with transmitted light, polarized light microscopy, and fluorescence microscopy with UV light. For example, scanning electron microscopy can be used to study damage to textile fibers, but as the obtained image is black and white, this technique is not suitable for the study of fiber dyes.

Following the selection of samples and after testing them by microscopy, other methods have to be chosen for further investigations of the fibers, and the choice of such methods usually depends on the laboratory equipment. The most often applied methods are spectroscopic techniques, and the least ones used are chromatographic ones.

All spectroscopic techniques (FTIR, Raman Spectroscopy, and UV-Vis MSP) are nondestructive, which is very important in forensic science. Spectroscopic techniques often give spectra with information about dyes, fiber-forming polymers, or fiber–dye mixtures. The results depend on the applied spectroscopic techniques (Raman Spectroscopy, UV-Vis MSP, or FTIR) and on the analysis conditions.

Chromatographic techniques have a destructive character, but they can give a lot of information about the dyed fibers as evidence. Identification is based on several parameters: retention times, mass spectra, or the selected, characterized *m/z* values of dyes. These parameters allow for the identification and comparative research between the dyes from evidential and comparative materials.

The next parameters that are different for spectroscopic and chromatographic techniques are the time of sample preparation and the time of analysis. Both times are longer for chromatographic than spectroscopic techniques.

In chromatographic techniques, the investigation could be divided into three main steps: (1) extraction of the dye/s from the fiber sample; (2) analysis of the obtained extract by the chromatographic method; (3) analysis of the obtained data. The sample preparation step requires the use of additional reagents and some special equipment, but the chromatographic conditions applied for the identification of individual types of dyes are constant. The identification of dyes is based on chromatographic (retention times) and mass spectrometric (*m/z* values) data.

In spectroscopic techniques, a fiber/a thread is the sample, but the conditions for the analysis must be determined individually.

Summary: the first step-microscopic investigation of the target dyed textile fibers is always the same. After this step, other methods, e.g., both spectroscopic and chromatographic, have to be chosen for further examination of the textile material. This choice of methodology usually depends on the laboratory equipment and such factors as the form and volume of the evidence, the type of textile, the type of dyes, etc. Both spectroscopic and chromatographic methods have advantages and disadvantages regarding the identification of dyed cotton fibers. The results and how they can be interpreted should be discussed within the perspective of previous studies and of working hypotheses. The findings and their implications should be discussed in the broadest context possible. Future investigations may also be proposed.

4. New Challenges in Identifying Cotton Fibers Dyed with Reactive Dyes for Forensic Purposes

The examination of dyed cotton fibers in forensic laboratories (e.g., in the form of microtraces found at a crime scene) usually consists of comparing them with fibers originating from a known source (e.g., the suspect's clothing) or determining their potential source (e.g., knitted or woven fabric) or comparing them with other evidential fibers (e.g., revealed under the victim's fingernails). The developed chromatographic methods, especially with mass spectrometric detection, are very sensitive, and they are very suitable for the comparison of evidential fibers to known fibers, i.e., whether the evidential fiber has the same dye as the known fiber [46]. In order to compare evidential fibers to known ones, it is necessary to establish an HPLC-MS/MS dye analysis database. When presenting the sequence for examining dyed fibers for forensic purposes, it is first necessary to identify the type of fiber, and then, knowing the classes of commonly used dyes for a given type and color of fibers, narrow down the range of possible dyes. Combined with the peak time and molecular and product ions in the database, a comparison between unknown fibers can be achieved.

Many scientists and practitioners are working on the construction of an analysis database of Raman spectroscopy and HPLC-MS/MS for common dyes. However, the development of the textile and dyes industry is so great that new information has to be continuously entered into such a database.

The chromatographic methods described above were shown to be suitable for the analysis of reactive dyes extracted from dyed cotton fibers. The high sensitivity and wide applicability make it possible to implement this methodology in routine case work. However, challenges associated with the compatibility of these chromatographic methods with fiber dye extraction solvent systems remain to be overcome. In addition, it is important to build a database of reactive dye mass spectra for the better detection of unknown fibers.

5. Conclusions

In forensic laboratories, the first step of the identification and comparison of fibers as microtraces is the use of optical microscopy techniques. An important feature in such research is the fiber color. Subsequently, if the colors of the evidential and comparative materials are similar, studies are carried out using UV/VIS MSP and Raman spectrometry, and occasionally by infrared spectrometry, mass spectrometry (MS), high-performance liquid chromatography (HPLC) or by using a combination of the last two techniques. Cotton fibers are natural and they are rather difficult to distinguish on the basis of morphological characteristics. For this reason, their color and consequently dye composition are often their only characteristic features. To our knowledge, this is the first review paper in which (1) the possibility of the application of chromatographic methods for the analysis of cotton fibers dyed with reactive dyes for forensic purposes was presented; (2) a chromatographic approach was compared with a spectroscopic one; and (3) the advantages and limitations of both methodologies were shown. Among chromatographic methods for the identification of reactive dyes for forensic purposes, the most often used is the HPLC-HRMS system, while among spectroscopic ones, UV-Vis MSP dominates. Chromatographic techniques can give much more information about the dyed fibers as evidence (identification is based on retention times, mass spectra or the selected, characterized *m/z* values of dyes) in comparison to spectroscopic ones. On the other hand, spectroscopic techniques are nondestructive, which is very important in forensic science. They often give spectra with information about dyes, fiber-forming polymers, or fiber–dye mixtures. The presentation of the above-mentioned procedures may help to choose the best analytical method for the effective characterization of cotton fibers dyed with reactive dyes. This knowledge could be useful for forensic experts and could help them to avoid mistakes. Moreover, this information/these data could be very interesting also for scientists working in food, cosmetics, or pharmaceutical/medical sciences.

Author Contributions: Writing—original draft preparation, D.Ś.-K.; J.W.-G., J.K.; writing—review and editing, D.Ś.-K.; J.W.-G., J.K.; project administration, P.S.; funding acquisition, P.S. All authors have read and agreed to the published version of the manuscript.

Acknowledgments: Financial support was provided by the Ministry of Science and Higher Education under grant nos.531-T010-D593-20 and BMN 538-8610-B321-19, and by the Institute of Forensic Research Project No. X/K/2019–2021.

References

1. Reichard, E.J.; Bartick, E.G.; Morgan, S.L.; Goodpaster, J.V. Microspectrophotometric analysis of yellow polyester fiber dye loadings with chemometric techniques. *Forensic Chem.* **2017**, *3*, 21–27. [CrossRef]
2. Lepot, L.; De Wael, K.; Gason, F.; Gilbert, B.; Eppe, G.; Malherbe, C. Discrimination of textile dyes in binary mixtures by Raman spectroscopy. *J. Raman Spectrosc.* **2020**, *51*, 717–730. [CrossRef]
3. Góra, P.; Wąs-Gubała, J. Enzymatic extraction of dyes for differentiation of red cotton fibres by TLC coupled with VSC. *Sci. Justice* **2019**, *59*, 425–432. [CrossRef] [PubMed]
4. Grieve, M.; Roux, C.; Wiggins, K.; Champot, C.; Taroni, F. Interpretation of Fibre Evidence Fibres. In *Forensic Examination of Fibres*, 3rd ed.; Robertson, J., Roux, C., Eds.; CRC Press: Boca Raton, FL, USA, 2017; pp. 345–425; ISBN 9781439828649.
5. Athey, S.N.; Adams, J.K.; Erdle, L.M.; Jantunen, L.M.; Helm, P.A.; Finkelstein, S.A.; Diamond, M.L. The Widespread Environmental Footprint of Indigo Denim Microfibers from Blue Jeans. *Environ. Sci. Technol. Lett.* **2020**. [CrossRef]
6. Smigiel-Kaminska, D.; Pospiech, J.; Makowska, J.; Stepnowski, P.; Was-Gubała, J.; Kumirska, J. The identification of polyester fibers dyed with disperse dyes for forensic purposes. *Molecules* **2019**, *24*, 613. [CrossRef] [PubMed]
7. Ahmed, N.S.E.; El-Shishtawy, R.M. The use of new technologies in coloration of textile fibers. *J. Mater. Sci.* **2010**, *45*, 1143–1153. [CrossRef]
8. Benghaya, S.; Mrabet, S.; Elharfi, A. A Review on Classifications, Recent Synthesis and Applications of Textile Dyes. *Inorg. Chem. Rev.* **2020**, *115*, 107891. [CrossRef]
9. Bafana, A.; Devi, S.S.; Chakrabarti, T. Azo dyes: Past, present and the future. *Environ. Rev.* **2011**, *19*, 350–370. [CrossRef]
10. Ferreira, E.S.B.; Hulme, A.N.; McNab, H.; Quye, A. The natural constituents of historical textile dyes. *Chem. Soc. Rev.* **2004**, *33*, 329–336. [CrossRef]
11. Smith, M.; Thompson, K.; Lennard, F. A literature review of analytical techniques for materials characterisation of painted textiles—Part 2: Spectroscopic and chromatographic analytical instrumentation. *J. Inst. Conserv.* **2017**, *40*, 252–266. [CrossRef]
12. Lewis, D.M. Developments in the chemistry of reactive dyes and their application processes. *Coloration Technol.* **2014**, *130*, 382–412. [CrossRef]
13. Blaus, K. Reactive Dyes for Cellulose Fibers. XXX Semianarium Polskich Kolorystów. 2014, pp. 63–87. Available online: https://pdfslide.net/documents/xxx-seminarium-polskich-kolorystow.html (accessed on 27 October 2020).
14. Hoy, S.J. Development and Figures of Merit of Microextraction and Ultra-Performance Liquid Chromatography for Forensic Characterization of Dye Profiles on Trace Acrylic, Nylon, Polyester, and Cotton Textile Fibers. Ph.D. Thesis, University of South Carolina, Portland, OR, USA, 2013.
15. Mahapatra, N.N. *Rective Dyes in Textile Dyes*; Woodhead Publishing India Pvt. Ltd.: Delhi, India, 2016; pp. 175–197; ISBN 9789385059049.
16. Wiggins, K.G. Fibre Dyes Analysis. In *Forensic Examination of Fibres*, 3rd ed.; Robertson, J., Roux, C., Eds.; CRC Press: Boca Raton, FL, USA, 2017; pp. 227, 239–241; ISBN 9781439828649.
17. Chakraborty, J.N. Dyeing with reactive dye. In *Fundamentals and Practices in Colouration of Textiles*; Woodhead Publishing India Pvt. Ltd.: Delhi, India, 2010; pp. 57–75; ISBN 9789380308463.

18. Lewis, D.M. The chemistry of reactive dyes and their application process. In *Handbook of Textile and Industrial Dyeing*; Clark, M., Ed.; Woodhead Publishing: Cambridge, UK, 2011; pp. 303–364; ISBN 9780081016510.
19. Shang, S.M. Process Control in dyeing od textiles. In *Process Control in Textile Manufacturing*; Majumdar, A., Das, A., Eds.; Woodhead Publishing: Cambridge, UK, 2013; pp. 300–338; ISBN 9780857090270.
20. Chattopadhyay, D.P. Chemistry of dyeing. In *Handbook of Textile and Industrial Dyeing*; Clark, M., Ed.; Woodhead Publishing: Cambridge, UK, 2011; pp. 150–183; ISBN 9781845696955.
21. Zollinger, H. Reactive Azo Dyes. In *Color Chemistry: Synthesis, Properties and Application of Organic Dyes and Pigments*, 3rd ed.; Verlag Helvetica Acta, Willej-VCH: Weinheim, Germany, 2003; pp. 225–240; ISBN 9783906390239.
22. Pal, P. Treatment Technology for Textile Water Treatment: Case Studies. In *Industrial Water Treatment Process Technology*; Butterworth—Heinemann: Cambridge, UK, 2017; pp. 243–511; ISBN 9780128103920.
23. Khatri, A.; Peerzada, M.H.; Mohsin, M.; White, M. A review on developments in dyeing cotton fabrics with reactive dyes for reducing effluent pollution. *J. Clean. Prod.* **2015**, *87*, 50–57. [CrossRef]
24. El Harfi, S.; El Harfi, A. Classifications, properties and applications of textile dye: A review. *Appl. J. Environ. Eng. Sci.* **2017**, *3*, 311–320. [CrossRef]
25. Soleimani-Gorgani, A.; Karami, Z. The effect of biodegradable organic acids on the improvement of cotton ink-jet printing and antibacterial activity. *Fibers Polym.* **2016**, *17*, 512–520. [CrossRef]
26. Cantrell, S.; Roux, C.; Maynard, P.; Robertson, J. A textile fibre survey as an aid to the interpretation of fibre evidence in the Sydney region. *Forensic Sci. Int.* **2001**, *123*, 48–53. [CrossRef]
27. Watt, R.; Roux, J.; Robertson, J. The population of coloured textile fibers in domestic washing machine. *Sci. Justice* **2005**, *45*, 75–83. [CrossRef]
28. Grieve, M.C.; Biermann, T.W. The population of coloured textile fibres on outdoor surfaces. *Sci. Justice* **1997**, *37*, 231–239. [CrossRef]
29. Dockery, C.R.; Stefan, A.R.; Nieuwland, A.A.; Roberson, S.N.; Baguley, B.M.; Hendrix, J.E.; Morgan, S.L. Automated extraction of direct, reactive, and vat dyes from cellulosic fibers for forensic analysis by capillary electrophoresis. *Anal. Bioanal. Chem.* **2009**, *394*, 2095–2103. [CrossRef]
30. Morgan, S.L. Validation of Forensic Characterization and Chemical Identification of Dyes Extracted from Milimeter-length Fibers. In *Final Report*; University of South Carolina: Columbia, SC, USA, 2015.
31. Božič, M.; Kokol, V. Ecological alternatives to the reduction and oxidation processes in dyeing with vat and sulphur dyes. *Dyes Pigm.* **2008**, *76*, 299–309. [CrossRef]
32. Home, J.M.; Dudley, R.J. Thin-layer chromatography of dyes extracted from cellulosic fibres. *Forensic Sci. Int.* **1981**, *17*, 71–78. [CrossRef]
33. Sirén, H.; Sulkava, R. Determination of black dyes from cotton and wool fibres by capillary zone electrophoresis with UV detection: Application of marker technique. *J. Chromatogr. A* **1995**, *717*, 149–155. [CrossRef]
34. Xu, X.; Leijenhorst, H.A.L.; Van Den Hoven, P.; De Koeijer, J.A.; Logtenberg, H. Analysis of single textile fibres by sample-induced isotachophoresis—Micellar electrokinetic capillary chromatography. *Sci. Justice* **2001**, *41*, 93–105. [CrossRef]
35. Sultana, N.; Williams, K.; Ankeny, M.; Vinueza, N.R. Degradation studies of CI Reactive Blue 19 on biodegraded cellulosic fabrics via liquid chromatography-photodiode array detection coupled to high resolution mass spectrometry. *Coloration Technol.* **2019**, *135*, 475–483. [CrossRef]
36. Carey, A.; Rodewijk, N.; Xu, X.; Van Der Weerd, J. Identification of dyes on single textile fibers by HPLC-DAD-MS. *Anal. Chem.* **2013**, *85*, 11335–11343. [CrossRef] [PubMed]
37. Schotman, T.G.; Xu, X.; Rodewijk, N.; van der Weerd, J. Application of dye analysis in forensic fibre and textile examination: Case examples. *Forensic Sci. Int.* **2017**, *278*, 338–350. [CrossRef]
38. Bundeskriminalamt. *Collective Work, European Fibres Group, Best Practice Guidelines—Thin Layer Chromatography*; Bundeskriminalamt: Wiesbaden, Germany, 2001.
39. Feng, C.; Sultana, N.; Sui, X.; Chen, Y.; Brooks, E.; Ankeny, M.A.; Vinueza, N.R. High-resolution mass spectrometry analysis of reactive dye derivatives removed from biodegraded dyed cotton by chemical and enzymatic methods. *AATCC J. Res.* **2020**, *7*, 9–18. [CrossRef]
40. Laing, D.K.; Dudley, R.J.; Hartshorne, A.W.; Home, J.M.; Rickard, R.A.; Bennett, D.C. The extraction and classification of dyes from cotton and viscose fibres. *Forensic Sci.* **1991**, *50*, 23–35. [CrossRef]

41. Lewis, S.W. Chapter 11: Analysis of dyes using chromatography. In *Identification of Textile Fibres*; Houck, M., Ed.; Woodhead Publishing: Cambridge, UK, 2009; pp. 203–223; ISBN 9781845692667.
42. Dorrien, D.M. Discrimination of Automobile Carpet Fibers Using Various Analytical Techniques and the Subsequent Creation of a Comprehensive Database. Ph.D. Thesis, University of Central Florida, Orlando, FL, USA, 2006.
43. Zotou, A.; Eleftheriadis, I.; Heli, M.; Pegiadou, S. Ion Ion-pair high performance liquid chromatographic study of the hydrolysis behaviour of reactive fluorotriazinic dyes. *Dyes Pigm.* **2002**, *53*, 267–275. [CrossRef]
44. Chemchame, Y.; Popikov, I.V.; Soufiaoui, M. Study of analytical methods for quantifying unfixed form of bifunctional reactive dyes used in dyeing cellulosic fibers (cotton). *Fibers Polym.* **2010**, *11*, 565–571. [CrossRef]
45. Chemchame, Y.; Popikov, I.V.; Soufiaoui, M. Study on analytical methods for quantifying the non-adsorbed reactive dye forms in an exhausted dyebath. *Coloration Technol.* **2012**, *128*, 169–175. [CrossRef]
46. Hu, C.; Zhu, J.; Mei, H.; Shi, H.; Guo, H.; Zhang, G.; Wang, P.; Lu, L.; Zheng, X. A sensitive HPLC-MS/MS method for the analysis of fiber dyes. *Forensic Chem.* **2018**, *11*, 1–6. [CrossRef]
47. Nayar, S.B.; Freeman, H.S. Hydrolyzed reactive dyes. Part 1: Analyses via fast atom bombardment and electrospray mass spectrometry. *Dyes Pigm.* **2008**, *79*, 89–100. [CrossRef]
48. Chalmers, J.; Edwards, H.; Hargreaves, M. *Infrared and Raman Spectroscopy in Forensic Science*; John Wiley and Sons: Hoboken, NJ, USA, 2012; ISBN 9780470749067.
49. Goodpaster, J.V.; Liszewski, E.A. Forensic analysis of dyed textile fibers. *Anal. Bioanal. Chem.* **2009**, *394*, 2009–2018. [CrossRef] [PubMed]
50. Was-Gubala, J.; Starczak, R. UV-Vis microspectrophotometry as a method of differentiation between cotton fibre evidence coloured with reactive dyes. *Spectrochim. Acta A Mol. Biomol. Spectrosc.* **2015**, *142*, 118–125. [CrossRef] [PubMed]
51. Was-Gubala, J.; Starczak, R. Nondestructive identification of dye mixtures in polyester and cotton fibers using Raman spectroscopy and ultraviolet-visible (UV-Vis) microspectrophotometry. *Appl. Spectrosc.* **2015**, *69*, 296–303. [CrossRef] [PubMed]
52. Buzzini, P.; Massonnet, G. Discrimination of Colored Acrylic, Cotton, and Wool Textile Fibers Using Micro-Raman Spectroscopy. Part 1: In situ Detection and Characterization of Dyes. *J. Forensic Sci.* **2013**, *58*, 1593–1600. [CrossRef]
53. Buzzini, P.; Massonnet, G. The analysis of colored acrylic, cotton, and wool textile fibers using micro-Raman spectroscopy. Part 2: Comparison with the traditional methods of fiber examination. *J. Forensic Sci.* **2015**, *60*, 712–720. [CrossRef]
54. Was-Gubala, J.; Machnowski, W. Application of Raman spectroscopy for differentiation among cotton and viscose fibers dyed with several dye classes. *Spectrosc. Lett.* **2014**, *47*, 527–535. [CrossRef]
55. Puchowicz, D.; Giesz, P.; Kozanecki, M.; Cieślak, M. Surface-enhanced Raman spectroscopy (SERS) in cotton fabrics analysis. *Talanta* **2019**, *195*, 516–524. [CrossRef]
56. Zaffiono, C.; Bruni, S.; Guglielmi, V.; De Luca, E. Fourier-transform surface-enhanced Raman spectroscopy (FT-SERS) applied to the identification of natural dyes in textile fibers: An extractionless approach to the analysis. *J. Raman Spectrosc.* **2014**, *45*, 211–218. [CrossRef]
57. Degano, I.; Ribechini, E.; Modugno, F.; Colombini, M.P. Analytical Methods for the Characterization of Organic Dyes in Artworks and in Historical Textiles. *Appl. Spectrosc. Rev.* **2009**, *44*, 363–410. [CrossRef]
58. Casadio, F.; Leona, M.; Lombardi, J.R.; Van Duyne, R. Identification of Organic Colorants in Fibers, Paints, and Glazes by Surface Enhanced Raman Spectroscopy. *Acc. Chem. Res.* **2010**, *43*, 782–791. [CrossRef] [PubMed]
59. Pozzi, F.; Porcinai, S.; Lombardi, J.R.; Leona, M. Statistical methods and library search approaches for fast and reliable identification of dyes using surface-enhanced Raman spectroscopy (SERS). *Anal. Meth.* **2013**, *5*, 4201–4212. [CrossRef]
60. Sciutto, G.; Prati, S.; Bonacini, I.; Litti, L.; Meneghetti, M.; Mazzeo, R. A new integrated TLC/MU-ATR/SERS advanced approach for the identification of trace amounts of dyes in mixtures. *Anal. Chim. Acta* **2017**, *991*, 104–112. [CrossRef] [PubMed]
61. Kirkbride, K.P. Infrared microspectroscopy of fibres. In *Forensic Examination of Fibres*, 3rd ed.; Robertson, J., Roux, C., Eds.; CRC Press: Boca Raton, FL, USA, 2018; pp. 245–289; ISBN 9781439828649.

62. Grieve, M.C.; Griffin, R.M.E.; Malone, R. An assessment of the value of blue, red, and black cotton fibers as target fibers in forensic science investigations. *Sci. Justice* **1998**, *38*, 27–37. [CrossRef]
63. Kokot, S.; Crawford, K.; Rintoul, L.; Meyer, U. A DRIFTS study of reactive dye states on cotton fabric. *Vib. Spectrosc.* **1997**, *15*, 103–111. [CrossRef]

2

Determination of Cadmium (II) in Aqueous Solutions by In Situ MID-FTIR-PLS Analysis using a Polymer Inclusion Membrane-Based Sensor: First Considerations

René González-Albarrán, Josefina de Gyves and Eduardo Rodríguez de San Miguel *

Departamento de Química Analítica, Facultad de Química, UNAM, Ciudad Universitaria, 04510 Cd. Mx., Mexico; renegalbarran@comunidad.unam.mx (R.G.-A.); degyves@unam.mx (J.d.G.)

* Correspondence: erdsmg@unam.mx

Academic Editor: Ewa Sikorska

Abstract: Environmental monitoring is one of the most dynamically developing branches of chemical analysis. In this area, the use of multidimensional techniques and methods is encouraged to allow reliable determinations of metal ions with portable equipment for in-field applications. In this regard, this study presents, for the first time, the capabilities of a polymer inclusion membrane (PIM) sensor to perform cadmium (II) determination in aqueous solutions by in situ visible (VIS) and Mid- Fourier transform infrared spectroscopy (MID-FTIR) analyses of the polymeric films, using a partial least squares (PLS) chemometric approach. The influence of pH and metal content on cadmium (II) extraction, the characterization of its extraction in terms of the adsorption isotherm, enrichment factor and extraction equilibrium were studied. The PLS chemometric algorithm was applied to the spectral data to establish the relationship between cadmium (II) content in the membrane and the absorption spectra. Furthermore, the developed MID-FTIR method was validated through the determination of the figures of merit (accuracy, linearity, sensitivity, analytical sensitivity, minimum discernible concentration difference, mean selectivity, and limits of detection and quantitation). Results showed reliable calibration curves denoting systems' potentiality. Comparable results were obtained in the analysis of real samples (tap, bottle, and pier water) between the new MID-FTIR-PLS PIM based-sensor and F-AAS.

Keywords: cadmium (II); polymer inclusion membrane; FTIR; chemometrics; PLS

1. Introduction

Heavy metals are persistent toxic metals or metalloids present at low concentrations, in all parts of the environment. Human activity affects the natural geological and biological redistribution of these metals through pollution of air, water, and soil. Therefore, highly accurate and sensitive spectrophotometric and spectrometric analytical techniques are used for their measurement in complex matrices, e.g., atomic absorption (Flame atomic absorption spectroscopy (F-AAS), Graphite furnace atomic absorption spectroscopy (GF-AAS), Hydride generation atomic absorption spectroscopy (HG-AAS), and Cold vapor atomic absorption spectroscopy (CV-AAS)), emission (Inductively coupled plasma optical emission spectroscopy (ICP–OES)), and mass (Inductively coupled plasma mass spectrometry (ICP-MS)) methods. At present, analytics and environmental monitoring are among the most dynamically developing branches of chemical analysis. The general trends in both areas can be classified into two basic groups—(i) development of new methodical procedures, and (ii) new achievements in the construction of measuring instruments (instrumentation) [1]. Examples of the first group developed in the pursuit of obtaining complex information on environmental quality are the introduction of solventless techniques to the analytical practice for sample preparation and multidimensional techniques, while examples of the second are the design of new sensors and detectors for conducting measurements in situ. The general trend in analytical instrumentation toward a smaller size, improved reliability, and easy operation, now makes it possible to count with portable UV-VIS, FTIR, Raman, and NIR spectrometers [2,3]. Successful applications of such technology in process automation [4], chemical reaction monitoring [5], homeland defense [6], food [7,8], controlled drugs [9], artwork [3,10] and hazardous chemical [11] analyses were reported. However, difficulties related to the production of large amount of information and a lack of selectivity and compromised detection capabilities must be handled. As for the first ones, the use of multivariate mathematical and statistical (chemometric) techniques can be profitably exploited. As for the second ones, adequate sample preparation methods might be employed. Due to the growing demand for techniques involving fewer toxic reagents, less time-consuming protocols, with lower limits of detection, facility of sampling and elimination of interferences, simple preparation methods that can be directly coupled with the measurement technique have a great potential in environmental analysis. Due to its easy implementation, simplicity of operation, selectivity, stability, versatility, and minimal power consumption, liquid membrane-based sample preparation and preconcentration techniques have gained growing attention [12,13]. Polymer inclusion membranes (PIM) are a kind of liquid membrane that is adequate for separation and preconcentration [14] of analytes, with the additional advantages of an easy synthesis and the possibility to perform in-situ metal analysis by X-ray [15], VIS [16–18], and fluorescence [19] spectroscopies, including its use in a continuous flow system [20]. Promising applications of PIMs as immunosensors for *Salmonella typhimurium* [21], for the detection of chlorpyrifos, diazinon, and cyprodinil in natural waters samples [22], and for flow injection determination of V(V) [23], and thiocyanate [24] were recently reported. Lately, the increasing interest in these membranes in analytical chemistry was reviewed, as they were adapted to new and novel applications [25].

In this work, a simple multivariate sensor for measuring cadmium (II) in waters employing a PIM was developed and characterized. Direct analyses on the membrane were conducted using VIS and MID-IR spectroscopic techniques. The partial least squares (PLS) chemometric algorithm was applied to quantitatively measure the amount of metal in the membrane. To the best of our knowledge this is the first study in which simultaneous analyses in the VIS and MID-IR spectral regions were performed in a PIM for quantitative reasons, and multivariate regression was applied to such a purpose. The work showed that the results obtained by the validated MID-FTIR-PLS PIM-based sensor compared well to those generated by (F-AAS), thus the new sensor could be quite suitable for on-site analyses with portable equipment.

The FTIR is an analytical technique based on the interaction of IR radiation and a molecule. It is well-known that IR is divided into three regions—near (NIR-IR), mid (MID-IR), and far (F-IR). MID-IR and NIR-IR are non-destructive, fast, repeatable and cost-effective techniques. In the MID-IR region, absorptions are generated by overtones and fundamental vibrations of the –CH, –NH, –OH groups, among others [26]. The MID-IR region can lead to quantitative analysis since the absorbance of the sample is proportional to the number of functional groups. In contrast, NIR-IR is characterized by the presence of broad bands and overtones, making this region less useful for univariate quantitative analysis; nevertheless, the significant differences in positions of functional groups also provide a source of information [27]. A useful way to solve the spectral overlapping and to use all the information contained in the spectrum of complex matrices is the multivariate calibration methods [28], extensively applied to NIR-IR [29–33]. Several works demonstrated that NIR-IR techniques can be applied as a tool for quantitative multivariate analysis, especially with the combination of separation and preconcentration methods for organic compounds [34,35], and metal ions [36–38], where some of them use NIR-IR followed by MID-IR and Raman [39]. Even though it's a drawback, the popularity of NIR is related to the good band assignments, improvements in instrumentation, and progress in statistical and mathematical methods [39]. In some reports, the capabilities of NIR-FTIR and MID-FTIR in the analysis of metals in different matrices were compared, and different advantages and constrains were found between them [27,40,41].

2. Results and Discussion

2.1. Establishment of Liquid-Solid Extraction Conditions

PIM composition was selected according to Aguilar et al. [42], i.e., (23.4 ± 0.2)% CTA, (54.5 ± 0.2)% NPOE, and (21.1 ± 0.4)% (*w*/*w*) Kelex 100. With this composition (97.2 ± 0.4)% of the metal was extracted in 4 h from an initial 1×10^{-4} mol dm^{-3} solution. Extraction experiments at 20, 40, 60, 120, 180, and 240 min showed that after 40 min, equilibrium in the system was reached. Further experiments were then performed using 60 min of extraction time and 1×10^{-4} mol dm^{-3} cadmium (II) solutions. The pH study was evaluated in the range 4 to 9. In Figure 1, the characteristic increase in extraction percentage with pH of acid extractants like 8-hydroxyquinoline, was observed [43]. In the higher pH region, the extraction percentage becomes inversely proportional to the hydrogen ion concentration, due to an increase in the concentration of the dissociated acid form of the extractant, L^-, while in the lower pH region, an opposite behavior was observed due to the predominance of the acid form, HL, which did not favor the extraction. At pH ≥ 8, an extraction percent of at least (97.7 ± 0.2)% was achieved. Using the MEDUSA software [44], the predominance of the free ion species in the medium was evidenced below pH = 8. Above this value, the presence of hydroxide complexes and the $Cd(OH)_2$ precipitate became relevant, and for this reason higher pH values were not further employed.

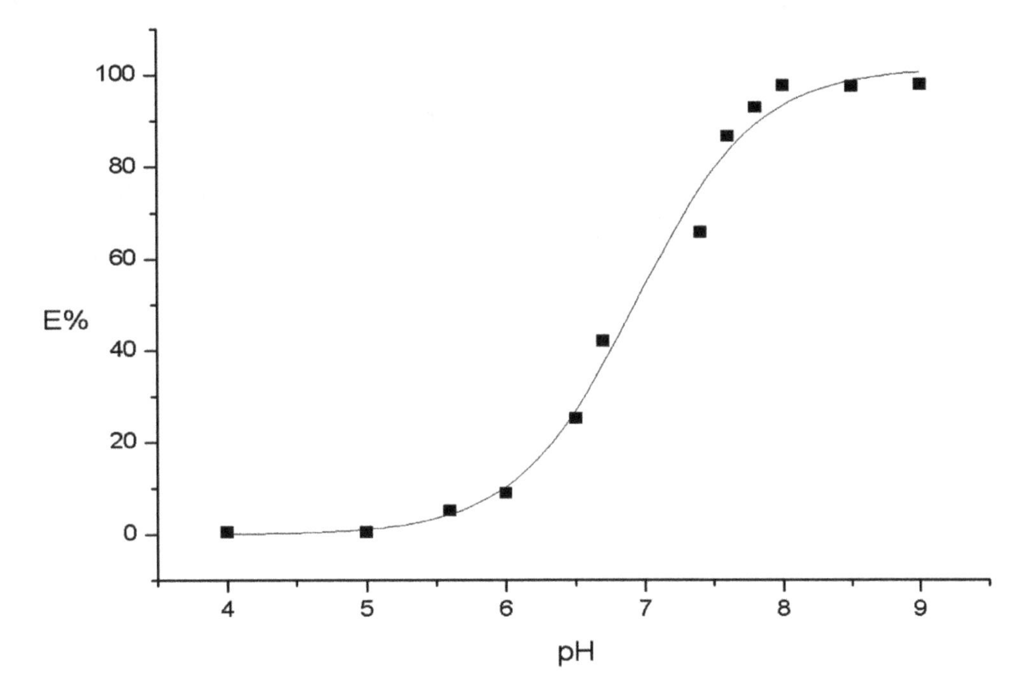

Figure 1. Percent of extraction of cadmium (II) as a function of pH in the PIM. Aqueous phase: 1×10^{-4} mol dm^{-3}, PIM: 23.4% CTA, 54.5% NPOE, and 21.1% Kelex 100 (*w*/*w*).

As for the influence of the initial cadmium (II) content, C_o, metal concentrations were varied in the following sequence: 5×10^{-3}, 1×10^{-3}, 7.5×10^{-4}, 5×10^{-4}, 4×10^{-4}, 2.5×10^{-4}, 1×10^{-4} mol dm^{-3}. Figure 2 shows the remaining aqueous equilibrium metal concentration. It was observed that for high concentrations (7.5×10^{-4}, 1×10^{-3}, 5×10^{-3} mol dm^{-3}), less than 10% of the metal was extracted; in contrast, at about 5×10^{-4} mol dm^{-3}, 50% was extracted, and at 1×10^{-4} mol dm^{-3}, (97.6 ± 0.2)% of the metal was retained in the membrane phase.

It is interesting to note that from 1×10^{-3} mol dm^{-3} and on, the metal was practically not extracted in the system, due to a saturation phenomenon of the extracting phase. Further experiments were then performed, maintaining cadmium (II) concentrations within the range of 5×10^{-4} to 1×10^{-4} mol dm^{-3}. In addition, the characterization of the systems in terms of its absorption capacity was studied.

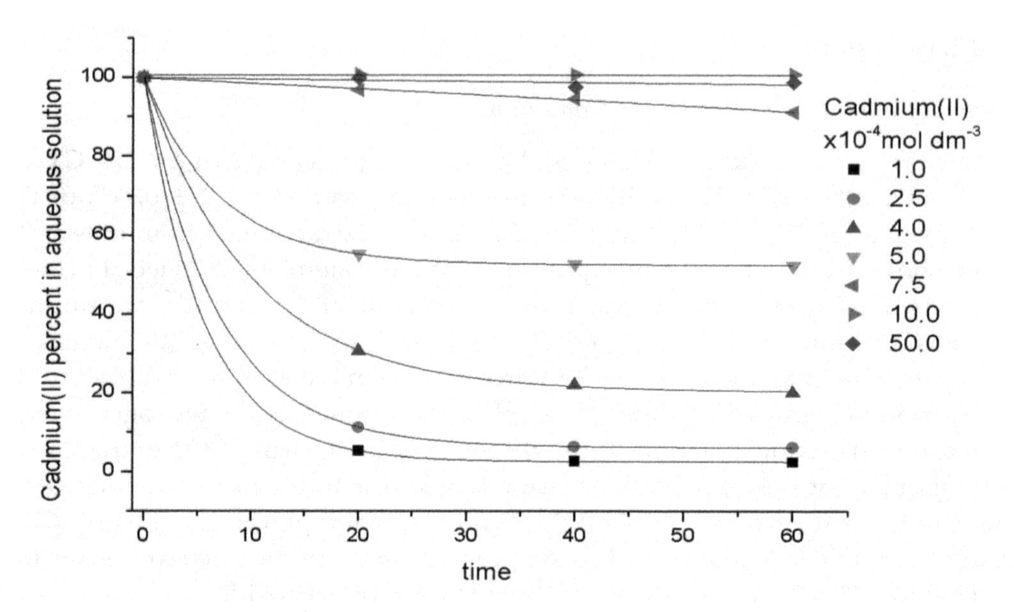

Figure 2. Cadmium (II) concentration profiles in the aqueous phase (pH = 8, 1×10^{-3} mol dm^{-3} TRIS) for different initial concentrations. PIM: 23.4% CTA, 54.5% NPOE, and 21.1% Kelex 100 (*w/w*).

2.2. Adsorption Isotherm

The amount of metal in the membrane phase, q_e [mmol g^{-1}] was plotted with respect to its equilibrium concentration in the aqueous phase, C_e [mmol cm^{-3}] (Figure 3A), for a range of aqueous concentrations from 6.94×10^{-7} to 3.82×10^{-4} mol dm^{-3}. As a Langmuir type isotherm was observed Equation (1), linearization of the data ($C_e/q_e = f(C_e)$) was applied to determine the adsorption constant, K_L (cm^3 mmol^{-1}), and the maximum adsorption capacity, q_{max} (mmol g^{-1}), parameters.

$$q_e = \frac{q_{max} K_L C_e}{1 + K_L C_e} \tag{1}$$

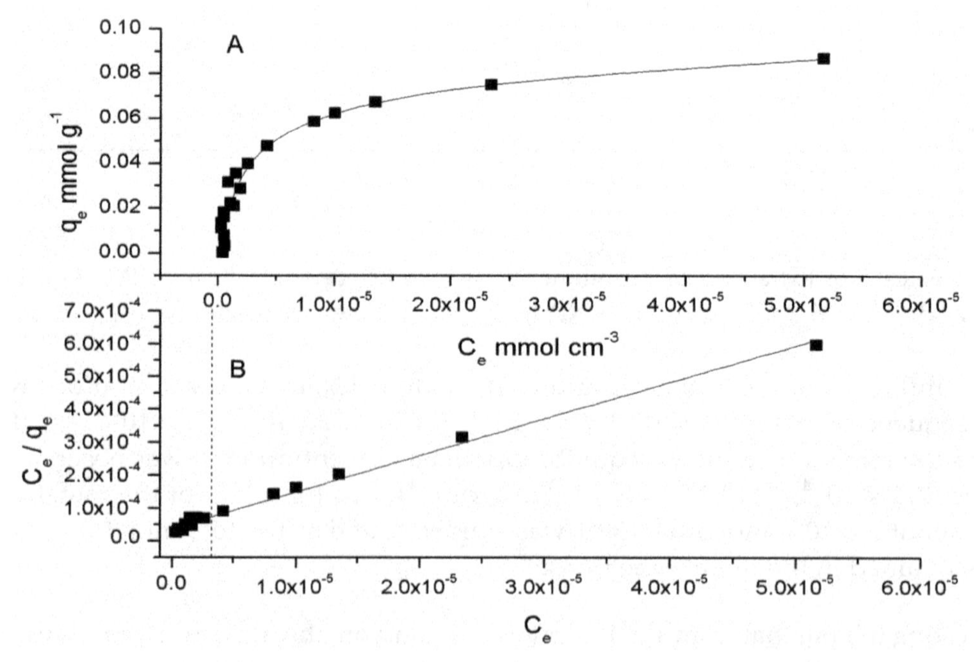

Figure 3. Cadmium (II) adsorption isotherm (**A**) and its linearized form (**B**). Aqueous phase: 6.94×10^{-7}–3.82×10^{-4} mol dm^{-3}, pH = 8.0 (1×10^{-3} mol dm^{-3} TRIS), PIM: 23.4% CTA, 54.5% NPOE, and 21.1% Kelex 100 (*w/w*).

After performing the analysis, a careful inspection of the data pointed toward two different regions, depending on the metal content, high (1.63×10^{-6} to 5.16×10^{-5} (mmol cm^{-3})) and low (1.11×10^{-6} to 9.33×10^{-7} (mmol cm^{-3})), in which the degree of fit of the model was better performed, assuming different model parameters for each region (Figure 3B).

In Table 1, both sets of model parameters are reported. It can be observed that q_{max} decreased with a diminishing C_e, as it probably became independent of the metal content [45], increasing the data dispersion. This behavior was further analyzed through computation of the separation coefficient, R_L, defined by

$$R_L = \frac{1}{1 + K_L C_O} \quad (2)$$

Table 1. Parameters obtained from the linearized form of the absorption isotherms and ANOVA test results.

Parameter	$C_e = 1.11 \times 10^{-6}$ to 9.33×10^{-7}	$C_e = 1.63 \times 10^{-6}$ to 5.16×10^{-5}
K_L [cm^3 mmol^{-1}]	17.028	2.133
q_{max} [mmol g^{-1}]	0.034	0.092
R^2	0.8421	0.9959
R_L	0.005–0.008	0.01–0.03
CV-R^2	0.7539	0.9834
Model sum of squares	2.37×10^{-9}	1.71×10^{-7}
Error sum of squares	7.36×10^{-10}	2.56×10^{-10}
Model mean square	2.37×10^{-9}	1.71×10^{-7}
Error mean square	9.20×10^{-11}	6.41×10^{-11}
F-value	25.81	2665.23
p-value	9.53×10^{-4}	8.42×10^{-7}

Values of $R_L > 1$ were indicative of a non-favorable adsorption; $R_L = 1$ indicates a linear adsorption, while $0 < R_L < 1$ were observed in favorable adsorption. An irreversible adsorption was present in such a case where $R_L = 0$ [46], in the high concentration range of the studied PIM system $0 < R_L < 1$, pointing out a favorable adsorption [47]. On the contrary, in the low range, $R_L \approx 0$, denoted an irreversible adsorption [46]. This observation agreed well with the two distinct regions observed in the adsorption isotherm. Comparing q_{max} to other cadmium (II) sorbents, Fan et al. reported $q_{max} = 0.545$ mmol g^{-1} for *P. Simplicissimum* [48], while Chakravarty et al. reported $q_{max} = 0.0948$ mmol g^{-1} for *Areca catechu* [49], and Singh et al. reported $q_{max} = 9.43 \times 10^{-4}$ mmol g^{-1} for *Trichoderma viridae* [50]. Some inorganic sorbents like activated alumina CNT nanoclusters, oxidized CNTs, and Fe_3O_4@TA showed maximum cadmium (II) adsorption capacities of 229.9, 11.01, and 286 mg g^{-1} [51] (equivalent to 2.04, 0.098, and 2.54 mmol g^{-1}, respectively). These results showed that a wide interval of maximum sorption capacities for the metal could be obtained, depending on the type of sorbent, pH, temperature, ionic force, among other factors. The obtained q_{max} values then lies within the reported ranges.

2.3. Enrichment Factor

The enrichment factor, E, defined by

$$E = \frac{[Cd(II)]_{membrane}}{[Cd(II)]_{initial}} \quad (3)$$

is a measure of the preconcentration efficiency of the system. When plotting $[Cd(II)]_{membrane}$ (mmol g^{-1}) as a function of $[Cd(II)]_{initial}$ (mmol dm^{-3}) within the interval 6.94×10^{-7} a 3.82×10^{-4} mol dm^{-3} a linear relationship was observed, denoting a constant value of this parameter. From the slope, a value of $E = 29.2$ was determined. This result guaranteed the application of the sorption in the PIM as an adequate preconcentration method. The initial amount of cadmium (II) in the aqueous phase could be predicted from the amount of cadmium (II) in the membrane phase and the constant value of E in the metal concentration range, in which this constant value was attained. From the comparison of the E value with other membrane-based cadmium (II) preconcentration methods, the PIM once again presented an acceptable intermediate value, as Castro et al. reported $E = 17.9$, with the use of liquid membranes containing 2-APHB in toluene as extractant [52], while Peng et al. reported $E = 387$ using hollow fibers with dithizone dissolved in a mixture of 1-octanol and oleic acid [53]. Evidently, the E value should be dependent on the type of sorbent, cadmium (II) content, pH, temperature, ionic force, among other factors.

2.4. Stoichiometry of the Extracted Complex

Experiments in which Kelex 100 concentration was varied (1.9, 2.8, 3.6, 4.6, 7.1, 9.3, 10.9, and 12.3 *w*/*w*%), maintaining constant amounts of CTA and NPOE, were performed to determine the stoichiometry of the extracted complex through conventional graphical slope analysis. According to Aguilar et al. [42], cadmium (II) extraction with Kelex 100 proceeded through the reaction

$$Cd^{2+} + n\overline{HL} + NO_3^- \rightarrow \overline{CdH_{n-1}L_nNO_3} + H^+ \quad (4)$$

in which $\overline{HL}$ stands for the extractant, $\overline{CdH_{n-1}L_nNO_3}$ for the extracted species, the bar denotes species in the membrane phase, and $n = 1$ and 2, depending on the nature of the ionic medium. From the extraction equilibrium constant, K_{ext}, defined by

$$K_{ext} = \frac{\left[\overline{CdH_{n-1}L_nNO_3}\right][H^+]}{[Cd^{2+}]\left[\overline{HL}\right]^n\left[NO_3^-\right]} \quad (5)$$

it is possible to write

$$logD = logK_{ext} + pH + \log\left[NO_3^-\right] + nlog\overline{[HL]} \quad (6)$$

once the distribution coefficient, D, is considered

$$D = \frac{\left[\overline{Cd(II)}\right]}{[Cd(II)]} \quad (7)$$

where $\left[\overline{Cd(II)}\right]$ and $[Cd(II)]$ are total membrane and aqueous phases equilibrium concentrations, respectively.

From the plot $logD = f(\overline{[HL]})_{pH}$ a value of $n \approx 2$. Such results perfectly agree with that reported by Aguilar et al. [42] for the extraction of the analyte with Kelex 100 in a solvent extraction system, using kerosene as solvent in nitrate medium. The determined extraction constant is $logK_{ext} = 0.02$.

2.5. Multivariate Regression Analysis

2.5.1. PLS modeling of VIS and FTIR data

As it was observed that VIS and FTIR information varied with cadmium concentration (Figure 4A,B), the obtained spectral data were submitted to PLS regression analysis to correlate the two data matrices, the X matrix (the VIS and FTIR spectra) and Y matrix (the property, i.e., cadmium content).

The employed concentration range was selected so that a Langmuir-type absorption of the metal ion by the PIM and a favorable preconcentration factor were attained. In the beginning, the complete spectral range was employed. However, from the analysis of the regression coefficients and model parameters, an improvement in the regression parameters was observed when the FTIR range was restricted to 700–410 cm^{-1}, and, consequently, further processing was performed using this interval. In the first instance, it was verified that all samples were representative of the same population, using a Hotelling T^2 test, in conjunction with an F-residual plot. Once no outliers were detected, the analyses results were interpreted. From Figure 5A it was observed that 96% of variability in the VIS spectra was explained using 2 factors; similarly, in the case of FTIR spectra (Figure 5B), the variability explained by the two first factors almost reached 100% for the spectral data.

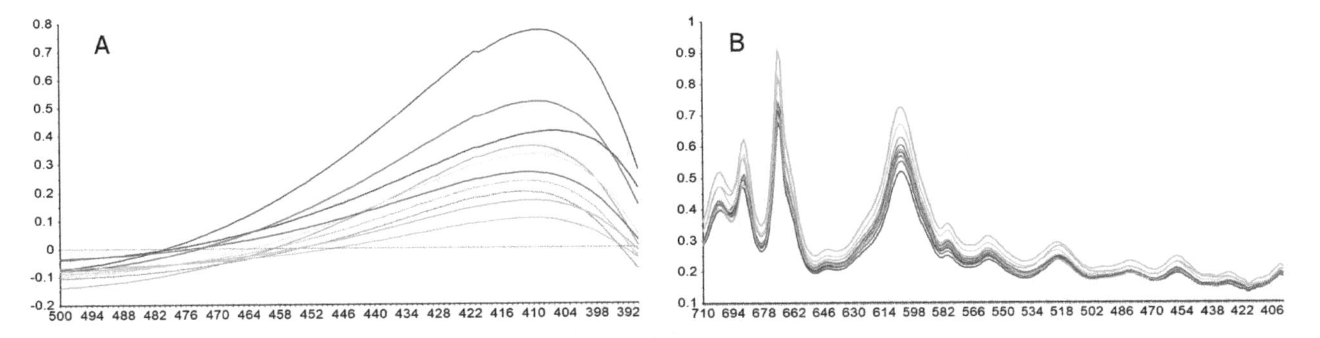

Figure 4. VIS (500–390 nm) (**A**) and MID–FTIR (710–400 cm^{-1}) (**B**) spectra of PIMs after equilibration with different cadmium (II) initial concentrations in the aqueous phase (1.63×10^{-6}–3.82×10^{-4} mol dm^{-3}) at pH = 8.0 (1×10^{-3} mol dm^{-3} TRIS), PIM: 23.4% CTA, 54.5% NPOE, and 21.1% Kelex 100 (*w/w*).

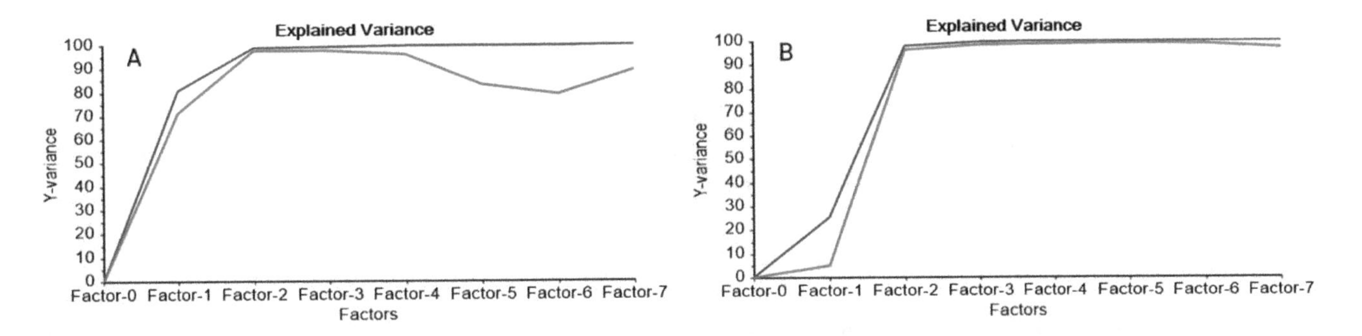

Figure 5. Percent of spectral variance explained by the VIS (**A**) and MID-FTIR (**B**) models as a function of the number of latent variables for calibration (blue) and cross-validation (red).

While the model for VIS data required just two latent variables, the model for FTIR data incorporated a third one, as selected according to a leave-one-out cross-validation process. This result was a direct consequence of the differences in complexity between the VIS and FTIR spectra. The accuracy of the models was quantitatively measured through the RMSEC, the RMSECV, and the slope, intercept, and determination coefficient (R^2) from the reference vs. predicted values of the property during calibration and cross-validation (Figure 6A,B and Table 2).

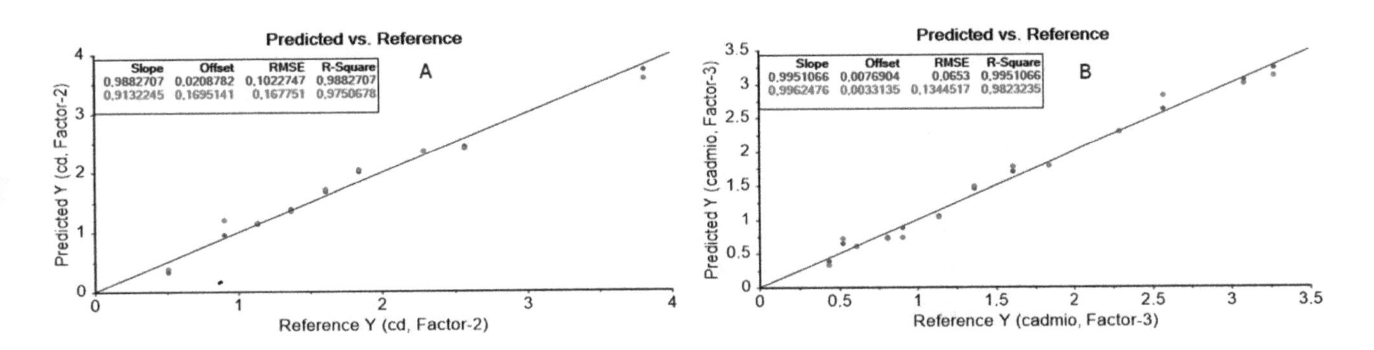

Figure 6. Cadmium (II) observed vs. predicted results for VIS (**A**), and MID–FTIR (**B**) data showing the fitting parameters during calibration (blue) and cross-validation (red). The reference line with zero intercept and with a slope of one is included.

Table 2. Statistical parameters associated with the accuracy of the developed methods.

Parameters	VIS Calibration	VIS Validation	MID-FTIR Calibration	MID-FTIR Validation
RMSEC	0.1022		0.0653	
RMSECV		0.1677		0.1344
Slope	0.9882	0.9132	0.9951	0.9962
Intercept	0.0208	0.1695	0.0076	0.0033
R^2	0.9882	0.9750	0.9951	0.9823

As observed good performance was accomplished by the models and no systematic variations were detected based on the slope (b_1) and intercept values (b_0) of the regression equations, i.e., the joined F-test for both statistical parameters gave no significant differences at 95% confidence between these values and the expected ones for the slope ($\beta_1 = 1$) and intercept ($\beta_0 = 0$) (p-values were 0.9590 and 0.3703 for VIS data for calibration and cross-validation, respectively, and 0.9733 and 0.2731 for FTIR data for calibration and cross-validation, respectively) according to the statistics:

$$F = \frac{(\beta_0 - b_0)^2 + 2\overline{x}(\beta_0 - b_0)(\beta_1 - b_1) + \left(\sum x_i^2/n\right)(\beta_1 - b_1)^2}{2S_e^2/n} \quad (8)$$

where S_e^2 error mean square; n, number of data points; $\overline{x}$, mean of the reference values; and $\sum x_i^2/n$, mean sum of squares of the reference values. The high values of the determination coefficients give information about the goodness of fit of the models, as this parameter is a statistical measure of how well the regression line approximates the real data points. In addition, the CV-determination coefficients showed a good predictive ability. The agreement between model predictions and ideal behavior is clearly seen in Figure 6A,B from the closeness of the data with the ideal reference line. The regression coefficients (Figure 7A,B) showed similarities with spectral data, which was simpler with VIS than with FTIR data. While the VIS regression coefficient showed a curve profile with a maximal contribution typical of the 8-hydroxyquinoline-Cd(II) complex [54], the FTIR regression coefficient profile included Kelex 100 characteristic IR vibrational bands at about 687 cm^{-1} and 459 cm^{-1}, related to C–H out-of-plane bending and to C–O in-plane bending, respectively [55], as expected from the coordination properties of the extractant through its oxygen group.

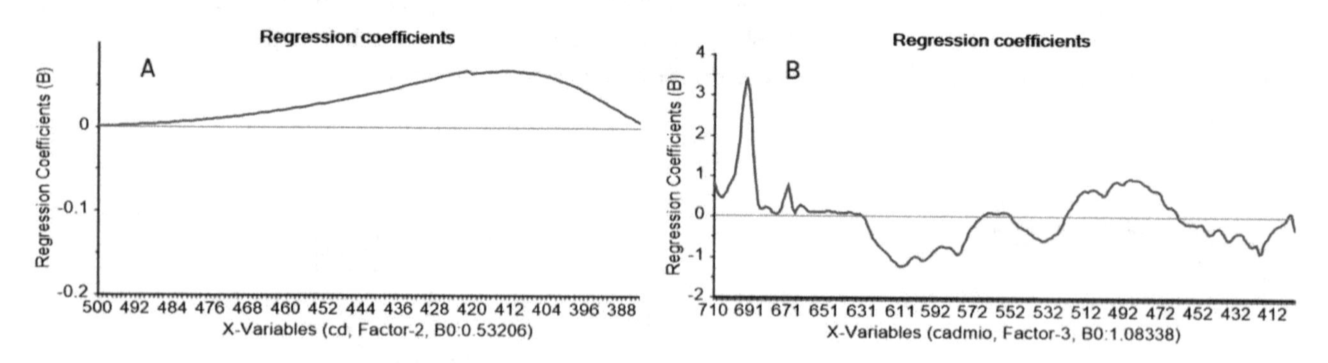

Figure 7. Regression coefficients of the VIS (**A**) and MID-FTIR (**B**) analyses as a function of wavelength (500–390 nm) or wave number (710–400 cm^{-1}), respectively.

2.5.2. Figures of Merit of the MID-FTIR-PLS PIM-Based Sensor

At this point it is important to mention that both models gave very good results, being slightly better for FTIR than for the VIS data, as indicated by the higher values of R^2 and lower values of RMSEC, RMSECV. Based on these observations, the MID-FTIR model was further characterized to extend its application to complex natural waters. Due to the specific spectroscopic signals between the metal ion and the extractant, a minimal or no effect caused by the impurities and the suspended particles is expected, and an accurate analysis could be performed. The unnecessary use of a chromophore agents is an additional advantage of the method. In Table 3, the figures of merit of the FTIR-developed model are presented (linearity, evaluated from $RMSEC = \sqrt{\frac{\sum_{i=1}^{n}(y_i-\hat{y}_i)^2}{n-1}}$, cross validation *RMSE* (*RMSECV*), determination coefficient (R^2), cross-validation R^2 (CV-R^2), slope, and intercept of the cadmium (II) observed vs. predicted results; sensitivity, as $sen = ||s_k^*|| = \frac{1}{||b||}$; analytical sensitivity as $\gamma = \frac{sen}{|\delta x|}$; minimum discernible concentration difference, $\gamma^{-1} = \frac{|\delta x|}{sen}$; and limits of detection and quantitation ($LD = 3.3\delta x \frac{1}{sen}$ and $LQ = 10\delta x \frac{1}{sen}$, respectively), where y_i and $\hat{y}_i$ are the estimated and reference values, respectively, of the I, sample, n the total number of samples, $||s_k||$ stands for the norm of the sensitivity coefficients of the spectra containing the analyte k at unit concentration and $||s_k^*||$ for that corresponding to its *NAS*, $NAS_i = (x_i \cdot b) \cdot \left(b^T \cdot b\right)^{-1} \cdot b^T$ where x_i is a sample spectrum after preprocessing and b is a column vector of the PLS regression coefficients, $||b||$ is the norm of the vector of regression coefficients of the calibration model, and δx is the instrumental noise [56–58]. Overall, good performance characteristics were observed. The low value of multivariate selectivity (4.03%) was anticipated according to the employed experimental conditions, as this parameter was a measure of the fraction of the spectrum that was related to the cadmium content in the PIM, and it should be considered that the Kelex 100: cadmium(II) ratio in the PIM was very high (approximately, a ratio of 21).

Table 3. Analytical figures of merit for the developed MID-FTIR method.

Figure of Merit	Value
RMSEE	0.0653
RMSECV	0.1344
R^2	0.9951
CV-R^2	0.9823
slope	0.9951
intercept	0.0076
range	(0.45–3.27) × 10^{-4} mol dm^{-3}
sen	0.0164 mol^{-1} dm^3
γ	7.26 mol dm^{-3}
γ^{-1}	0.14 mol^{-1} dm^3
Mean selectivity	0.0403
LD	0.45 × 10^{-4} mol dm^{-3}
LQ	1.37 × 10^{-4} mol dm^{-3}

2.5.3. Application of the MID-FTIR-PLS PIM-Based Sensor

As the model's accuracy was dependent on the presence of interferences, the application of the model to three different natural waters (tap, bottle, and pier water) representing complex matrices due to the presence of different ions (i.e., calcium, magnesium, sodium, potassium, chloride, nitrate, sulfate, bicarbonate, fluoride, among others [59,60]) and particulates (e.g., dissolved organic compounds) was evaluated. Samples were spiked with different analyte concentrations and the results of the FTIR method was compared with F-AAS analysis. As observed from Table 4, the non-specific character of 8-hydroxyquinoline, i.e., Kelex 100, which could form complexes with Na (I), Ca (II), and Mg (II), among other ions [61,62], was perfectly compensated by the Cd (II)-Kelex 100 complex specific information contained within the analyzed FTIR spectral region and the pH value selected for the analysis. The joined F-test for the slope and intercept values of the regression equation between the reference vs. the determined values gave no significant differences at the 95% confidence between these values and the expected ones (p-value = 0.0774). Consequently, comparable results were obtained between both methods, even in the analysis of a challenging medium like pier water.

Table 4. Results of the analysis of cadmium (II) in real water samples spiked with the analyte.

Sample	Reference Value (F-AAS) $\times 10^4$ mol dm^{-3}	Determined Value (MID-FTIR) $\times 10^4$ mol dm^{-3}
Tap	3.10 ± 0.10	3.25 ± 0.25
	1.00 ± 0.10	1.35 ± 0.22
	0.56 ± 0.10	0.52 ± 0.35
Bottle	0.87 ± 0.10	1.08 ± 0.20
	0.51 ± 0.10	0.52 ± 0.30
Pier	2.75 ± 0.10	3.08 ± 0.25
	0.96 ± 0.10	1.07 ± 0.20
	0.56 ± 0.10	0.47 ± 0.30

One major advantage of the developed MID-FTIR-PLS PIM-based method was that it did not require the presence in the membrane of a chemical reagent with special properties, either a chromophore species that is able to complex the metal ion, i.e., acting as ionophore [19], or a mixture of an ionophore and a chromophore in the same PIM [17], or a fluorescent reagent [18]. Consequently, there was no need to optimize the PIM composition for chromophore/ionophore/support compatibility [25], so that in practice the methodology is transferable to any PIM system reported up till now. Future work will be addressed toward extending the range of application of the methodology to—(i) a lower analyte concentration range by a careful selection of the dielectric nature of the medium and the dipole moment of the bond associated with IR vibrations of the extracted complex (variation in PIM component's composition and nature); the larger the dipole moment change and the smaller the position change of the atoms (i.e., of bond lengths or bond angles), the higher the band intensities [63]; and (ii) different analytes, either alone or in a mixture, to fit the purpose of environmental monitoring. In this regard, taking into account a similar behavior of Kelex 100 to its parent structure 8-hydroxyquinoline ($\log K_{extraction}$: Cu^{2+} (1.77) > Ni^{2+} (−2.18) > Zn^{2+} (−2.41) > Co^{2+} (−3.7) > Cd^{2+} (−5.29) > Mg^{2+} (−15.13) > Ca^{2+} (−17.89) [64], it is evident that the influence of the presence of other heavy metal ions is a challenge to handle. Chemometric selectivity based on specific absorption bands for the different metal ions might represent a promising alternative to be investigated in future applications. Overall, this article showed the potentiality of the proposed methodology and allowed a proof of the concept for the target purpose.

3. Materials and Methods

In PIM preparation, Kelex 100 (Sherex Chemical Co. Inc. Dublin, Ohio, USA), cellulose triacetate (CTA, Honeywell Fluka, Charlotte, N.C., USA), and 2-nitrophenyl octyl ether (NPOE, ≥99.0% Honeywell Fluka, Charlotte, N.C., USA) were used as extractants, support, and plasticizer, respectively, using dichloromethane (Merck, Kenilworth, N.J., USA) as a casting solvent. Working solutions were prepared from tetrahydrated cadmium (II) nitrate (≥99.0% Fluka) while a 1000 mg/L Sigma-Aldrich AAS standard solution was diluted using deionized water, for the preparation of the standards for F-AAS determinations. Tris(hydroxymethyl)aminomethane (TRIS, 99.8% Aldrich, pH 7–9), 4-morpholino ethanesulfonic acid (MES, 99.5% Sigma, pH 5.5–6.7), sodium acetate (99% Aldrich Chem. Co., St. Louis, MO, USA) / acetic acid (99.7%, Sigma-Aldrich) buffer solution (pH 3.8–5.8), and hydrochloric acid (37% Sigma-Aldrich Chem. Co., St. Louis, MO, USA) were employed to adjust the pH of the aqueous solutions.

Extraction experiments were carried out with a model 75 Wrist ActionTM shaker (Burrell Scientific Inc, Pittsburgh, Pa., USA). The spectrometers Perkin Elmer 3100, Perkin Elmer Lambda 2 and Perkin Elmer Spectrum GX (Waltham, Mass., USA) were used for F-AAS, VIS, and FTIR determinations, respectively. A Metrohm 620 pH-meter (Herisau, Switzerland) was employed for pH measurement and adjustment. A Fowler IP54 micrometer (Fowler High Precision, Newton, Mass., USA) was used for measuring PIM thickness. The Unscrambler 10.5.1 software (Camo Analytics, Oslo, Norway) was employed for PLS analyses.

PIMs were prepared by dissolving the weighted amounts of CTA, NPOE, and Kelex 100 in dichloromethane. The mixture was stirred for 1 h on a magnetic plate with a stirring bar. The solution was then casted in a 5 cm diameter Petri dish and rested for 24 h for solvent evaporation. Finally, the membrane was carefully peeled off and the whole piece was used in all experiments. PIMs were transparent films with an average thickness of (46 ± 11) μm, average weight of (112.3 ± 0.0056) mg and an average diameter of (4.89 ± 0.0322) cm.

As for the biphasic solid-liquid extraction experiments, the membranes were introduced in 50 mL polypropylene Falcon tubes, in the presence of 30 mL of aqueous solution, with cadmium (II) at fixed concentrations. The tubes were shaken for regular time intervals and the aliquots of 400 μL were taken and diluted to 2 mL before F-AAS analysis, using the conditions recommended by the manufacturer (λ 228.8 nm, 7 nm slit, air/acetylene flame). Experiments were performed on a duplicate basis with an average RSD of 5%. No analyte elution step was required as direct analysis of the PIMs was performed. This was opposite to the traditional three-phases configuration (feed, membrane, strip) usually employed for metal ion removal. The employed set-up was commonly found when the PIMs were used for sensing in chemical analysis [25].

Chemometric analyses were conducted using a training set of 15 different concentrations (in duplicate), ranging from 6.94×10^{-7} to 3.82×10^{-4} mol dm^{-3}. The concentration of the calibration standards was determined by F-AAS (Perkin Elmer 3100, Waltham, Mass., USA). The same PIMs were analyzed by VIS and MID-FTIR spectroscopies. VIS spectra were recorded in transmission mode by triplicates, in the range 500–390 nm, after sandwiching the membrane between two Petri dishes to avoid wrinkles and movement. FTIR spectra were recorded in transmission mode by triplicates in the range of 4000–400 cm^{-1}. The PIM was mounted in the transmission accessory of the equipment and scanned 45 times to record the spectrum. The six spectra for each concentration in VIS and IR modes were then averaged for multivariate analysis, using the spectra calculator of the software. The best results were obtained after mean-centering the spectra and, in the case of the FTIR data, applying the unit vector normalization. Chemometric analyses were validated through cross-validation procedures. An in-house made MATHLAB program was used for the figures of merit determination, using the outputs of the Unscrambler 10.5.1 (Camo Analytics, Oslo, Norway) software. The applicability of the method was tested by spiking with reference concentrations bottle, tap, and pier (Cuemanco, Xochimilco, Mexico) water, after filtration of the samples.

4. Conclusions

The results showed that sorption coupled with direct spectroscopic analysis, using a PLS chemometric approach in a PIM, constituted a potential sensor for metal ion determination in waters with results comparable to F-AAS. On-site analysis with portable equipment was anticipated through the proof of the concept of the methodology in the case of the measurement of cadmium (II) in aqueous solutions, using the commercial extractant Kelex 100 immobilized in the membrane. MID-FTIR showed to be an adequate technique for cadmium (II) analysis in the membrane, once the metal was preconcentrated, which was barely used for quantitative purposes in this type of application. The specificity of bonding between the metal and the extractant, together with the optimization of the uptake conditions, allowed the selectivity of the system toward competing ions, while the chemometric treatment of the spectral data allowed the selectivity of the metal signal toward the organic components conforming the membrane. A wide range of future applications is anticipated to target different metal ions with specific extractants immobilized in PIMs, as no chromophores or fluorescent reagents are needed in this novel type of application. The simplicity of the system is expected to optimize the PIM composition for chromophore/ionophore/support compatibility, and in its transferability, as all organic extractants present active functional groups in the MID-IR region.

Author Contributions: Draft preparation, experimentation, and formal analysis, R.G.-A.; writing—review and editing, J.d.G.; conceptualization, writing—original and supervision, E.R.d.S.M. All authors have read and agreed to the published version of the manuscript.

Acknowledgments: Q. Nadia Marcela Munguía Acevedo and Q.F.B. María Guadalupe Espejel Maya are thanked for their help in technical services. R. González-Albarrán acknowledges the CONACYT scholarship.

Abbreviations

CTA	cellulose triacetate
CV-AAS	Cold vapor atomic absorption spectroscopy
CV	cross-validation
E	Enrichment factor
F-AAS	Flame atomic absorption spectroscopy
F-IR	Far infrared spectroscopy
FTIR	Fourier-transform infrared spectroscopy
GF-AAS	Electrothermal or graphite furnace atomic absorption spectroscopy
HG-AAS	Hydride generation atomic absorption spectroscopy
ICP-MS	Inductively coupled plasma mass spectrometry
ICP-OES	Inductively coupled plasma optical emission spectroscopy
IR	infrared spectroscopy
Kelex 100	8-hydroxyquinoline derivative
LD	limit of detection
LQ	limit of quantitation
MEDUSA	Make Equilibrium Diagrams Using Sophisticated Algorithms software
MES	4-morpholino ethanesulfonic acid
MID-IR	Mid infrared spectroscopy
NAS	Net analyte signal
NIR-IR	Near infrared spectroscopy
NPOE	2-nitrophenyl octyl ether
PIM	polymer inclusion membrane
PLS	partial least squares
RMSEC	root mean square error of calibration
RMSECV	root mean square error of cross-validation
RSD	Relative standard deviation
TRIS	Tris(hydroxymethyl)aminomethane
VIS	visible spectroscopy

References

1. Namieśnik, J. Modern Trends in Monitoring and Analysis of Environmental Pollutants. *Pol. J. Environ. Stud.* **2001**, *10*, 127–140.
2. Capitán-Vallvey, L.F.; Palma, A.J. Recent developments in handheld and portable optosensing—A review. *Anal. Chim. Acta* **2011**, *696*, 27–46. [CrossRef]
3. Vandenabeele, P.; Castro, K.; Hargreaves, M.; Moens, L.; Madariaga, J.M.; Edwards, H.G.M. Comparative study of mobile Raman instrumentation for art analysis. *Anal. Chim. Acta* **2007**, *588*, 108–116. [CrossRef]
4. Malinen, J.; Känsäkoski, M.; Rikola, R.; Eddison, C.G. LED-based NIR spectrometer module for hand-held and process analyser applications. *Sens. Actuator B-Chem.* **1998**, *51*, 220–226. [CrossRef]
5. Osticioli, I.; Ciofini, D.; Mencaglia, A.A.; Siano, S. Automated characterization of varnishes photo-degradation using portable T-controlled Raman spectroscopy. *Spectrochim. Acta A* **2017**, *172*, 182–188. [CrossRef]
6. López-López, M.; García-Ruiz, C. Infrared and Raman spectroscopy techniques applied to identification of explosives. *TrAC* **2014**, *54*, 36–44. [CrossRef]
7. Ayvaz, H.; Rodriguez-Saona, L.E. Application of handheld and portable spectrometers for screening acrylamide content in commercial potato chips. *Food Chem.* **2015**, *174*, 154–162. [CrossRef] [PubMed]
8. Ayvaz, H.; Sierra-Cadavid, A.; Aykas, D.P.; Mulqueeney, B.; Sullivan, S.; Rodriguez-Saona, L.E. Monitoring multicomponent quality traits in tomato juice using portable mid-infrared (MIR) spectroscopy and multivariate analysis. *Food Control.* **2016**, *66*, 79–86. [CrossRef]
9. Tsujikawa, K.; Miyaguchi, H.; Kanamori, T.; Iwata, Y.T.; Yoshida, T.; Inoue, H. Development of an on-site screening system for amphetamine-type stimulant tablets with a portable attenuated total reflection Fourier transform infrared spectrometer. *Anal. Chim. Acta* **2008**, *608*, 95–103. [CrossRef]

10. Saviello, D.; Toniolo, L.; Goidanich, S.; Casadio, F. Non-invasive identification of plastic materials in museum collections with portable FTIR reflectance spectroscopy: Reference database and practical applications. *Microchem J.* **2016**, *124*, 868–877. [CrossRef]
11. Gredilla, A.; Fdez-Ortiz de Vallejuelo, S.; Elejoste, N.; de Diego, A.; Madariaga, J.M. Non-destructive Spectroscopy combined with chemometrics as a tool for Green Chemical Analysis of environmental samples: A review. *TrAC* **2016**, *76*, 30–39. [CrossRef]
12. López-López, J.A.; Mendiguchía, C.; Pinto, J.J.; Moreno, C. Liquid membranes for quantification and speciation of trace metals in natural waters. *TrAC* **2010**, *29*, 645–653. [CrossRef]
13. Carasek, E.; Merib, J. Membrane -based microextraction techniques in analytical chemistry: A review. *Anal. Chim. Acta* **2015**, *880*, 8–25. [CrossRef] [PubMed]
14. Almeida, A.M.I.G.S.; Cattrall, R.W.; Kolev, S.D. Recent trends in extraction and transport of metal ions using polymer inclusion membranes (PIMs). *J. Membrane Sci.* **2012**, *415*, 9–23. [CrossRef]
15. Fontàs, C.; Queralt, I.; Hidalgo, M. Novel and selective procedure for Cr (VI) determination by X-ray fluorescence analysis after membrane concentration. *Spectrochim. Acta B* **2006**, *61*, 407–413. [CrossRef]
16. Tharakeswar, Y.; Kalyan, Y.; Gangadhar, B.; Kumar, K.S.; Naidu, G.R. Optical Chemical Sensor for Screening Cadmium (II) in Natural Waters. *J. Sensor Technol.* **2012**, *2*, 68–74. [CrossRef]
17. Ngarisan, N.I.; Zanariah, C.W.; Ngah, C.W.; Ahmad, M.; Kuswandi, B. Optimization of polymer inclusion membranes (PIMs) preparation for immobilization of Chrome Azurol S for optical sensing of aluminum (III). *Sens. Actuator B-Chem.* **2014**, *203*, 465–470. [CrossRef]
18. Mohd Suah, F.B.; Ahmad, M.; Heng, L.Y. Highly sensitive fluorescence optode for aluminium(III) based on non-plasticized polymer inclusion membrane. *Sens. Actuators B-Chem.* **2014**, *201*, 490–495. [CrossRef]
19. Mohd Suah, F.B.; Ahmad, M.; Heng, L.Y. A novel polymer inclusion membrane based optode for sensitive determination of Al^{3+}. *Spectrochim Acta A* **2015**, *144*, 81–87. [CrossRef]
20. Denna, M.C.F.J.; Camitan, R.A.B.; Yabut, D.A.O.; Rivera, B.A.; dlC Coo, L. Determination of Cu (II) in environmental water samples using polymer inclusion membrane-TAC optode in a continuous flow system. *Sens. Actuators B-Chem.* **2018**, *260*, 445–451. [CrossRef]
21. Silva, N.F.D.; Magalhães, J.M.C.S.; Barroso, M.F.; Oliva-Teles, T.; Freire, C.; Delerue-Matos, C. In situ formation of gold nanoparticles in polymer inclusion membrane: Application as platform in a label-free potentiometric immunosensor for Salmonella typhimurium detection. *Talanta* **2019**, *194*, 134–142. [CrossRef] [PubMed]
22. Vera, R.; Insa, S.; Fontàs, C.; Anticó, E. A new extraction phase based on a polymer inclusion membrane for the detection of chlorpyrifos, diazinon and cyprodinil in natural water samples. *Talanta* **2018**, *185*, 291–298. [CrossRef] [PubMed]
23. Reza Yaftian, M.; Almeida, M.I.G.S.; Cattrall, R.W.; Kolev, S.D. Flow injection spectrophotometric determination of V(V) involving on-line separation using a poly(vinylidene fluoride-cohexafluoropropylene)-based polymer inclusion membrane. *Talanta* **2018**, *181*, 385–391. [CrossRef]
24. Ohshima, T.; Kagaya, S.; Gemmei-Ide, M.; Cattrall, R.W.; Kolev, S.D. The use of a polymer inclusion membrane as a sorbent for online preconcentration in the flow injection determination of thiocyanate impurity in ammonium sulfate fertilizer. *Talanta* **2014**, *129*, 560–564. [CrossRef]
25. Almeida, M.I.G.S.; Cattrall, R.W.; Kolev, S.D. Polymer inclusion membranes (PIMs) in chemical analysis—A review. *Anal. Chim. Acta* **2017**, *987*, 1–14. [CrossRef]
26. Bellon Maurel, V.; McBratney, A. Near-Infrared (NIR) and Mid-Infrared (MIR) Spectroscopic Techniques for Assessing the Amount of Carbon Stock in Soils—Critical Review and Research Perspectives. *Soil Biol. Biochem.* **2011**, *43*, 1398–1410. [CrossRef]
27. Türker, S.; Huck, C. A Review of Mid-Infrared and Near-Infrared Imaging: Principles, Concepts and Applications in Plant Tissue Analysis. *Molecules* **2017**, *22*, 168. [CrossRef]
28. Ibañez, G.; Escandar, G. Luminescence Sensors Applied to Water Analysis of Organic Pollutants—An Update. *Sensors* **2011**, *11*, 11081–11102. [CrossRef] [PubMed]
29. Liu, Z.; Cai, W.; Shao, X. A weighted multiscale regression for multivariate calibration of near infrared spectra. *Analyst* **2009**, *134*, 261–266. [CrossRef] [PubMed]
30. Shao, X.; Bian, X.; Liu, J.; Zhang, M.; Cai, W. Multivariate calibration methods in near infrared spectroscopic analysis. *Anal. Methods* **2010**, *2*, 1662–1666. [CrossRef]
31. Xu, H.; Cai, W.; Shao, X. Weighted partial least squares regression by variable grouping strategy for multivariate calibration of near infrared spectra. *Anal. Methods* **2010**, *2*, 289–294. [CrossRef]

32. Shao, X.; Zhang, M.; Cai, W. Multivariate calibration of near-infrared spectra by using influential variables. *Anal. Methods* **2012**, *4*, 467–473. [CrossRef]
33. Li, X.; Cai, W.; Shao, X. Correcting Multivariate Calibration Model for near Infrared Spectral Analysis without Using Standard Samples. *J. Near Infrared Spectrosc.* **2015**, *23*, 285–291. [CrossRef]
34. Pérez, R.L.; Escandar, G.M. Experimental and chemometric strategies for the development of Green Analytical Chemistry (GAC) spectroscopic methods for the determination of organic pollutants in natural waters. *Sustain. Chem. Pharm.* **2016**, *4*, 1–12. [CrossRef]
35. Zhang, Y.; Hao, Y.; Caia, W.; Shao, X. Simultaneous determination of phenol and p-nitrophenol in wastewater using near-infrared diffuse reflectance spectroscopy with adsorption preconcentration. *Anal. Methods* **2011**, *3*, 703–708. [CrossRef]
36. Sheng, N.; Cai, W.; Shao, X. An approach by using near-infrared diffuse reflectance spectroscopy and resin adsorption for the determination of copper, cobalt and nickel ions in dilute solution. *Talanta* **2009**, *79*, 339–343. [CrossRef]
37. Li, J.; Zhang, Y.; Cai, W.; Shao, X. Simultaneous determination of mercury, lead and cadmium ions in water using near-infrared spectroscopy with preconcentration by thiol-functionalized magnesium phyllosilicate clay. *Talanta* **2011**, *84*, 679–683. [CrossRef]
38. Chen, G.; Mei, Y.; Tao, W.; Zhang, C.; Tang, H.; Iqbal, J.; Du, Y. Micro near infrared spectroscopy (MicroNIRS) based on on-line enrichment: Determination of trace copper in water using glycidyl methacrylate-based monolithic material. *Anal. Chim. Acta* **2010**, *670*, 39–43. https://doi.org/10.1016/j.aca.2010.04.064. *Anal. Chim. Acta* **2011**, *698*, 84. [CrossRef]
39. Moros, J.; Garrigues, S.; Guardia, M.d.l. Vibrational spectroscopy provides a green tool for multi-component analysis. *TrAC* **2010**, *29*, 578–591. [CrossRef]
40. Shi, T.; Chen, Y.; Liu, Y.; Wu, G. Visible and near-infrared reflectance spectroscopy—An alternative for monitoring soil contamination by heavy metals. *J. Hazard. Mater.* **2014**, *265*, 166–176. [CrossRef]
41. Wang, J.; Cui, L.; Gao, W.; Shi, T.; Chen, Y.; Gao, Y. Prediction of low heavy metal concentrations in agricultural soils using visible and near-infrared reflectance spectroscopy. *Geoderma* **2014**, *216*, 1–9. [CrossRef]
42. Aguilar, J.C.; Sánchez-Castellanos, M.; Rodríguez de San Miguel, E.; de Gyves, J. Cd (II) and Pb (II) extraction and transport modeling in SLM and PIM systems using Kelex 100 as carrier. *J. Membrane Sci.* **2001**, *190*, 107–118. [CrossRef]
43. Rydberg, J.; Choppin, G.R.; Musikas, C.; Sekine, T. Solvent Extraction Equilibria. In *Solvent Extraction and Practice*; Rydberg, J., Cox, M., Musikas, C., Choppin, G.R., Eds.; CRC Press: Boca Raton, FL, USA, 2004.
44. Puigdomenech, I. Make Equilibrium Diagrams Using Sophisticated Algorithms (MEDUSA) software, Royal Institute of Technology (KTH), Stockholm. 2015. Available online: https://www.kth.se/che/medusa/downloads-1.386254 (accessed on 20 June 2020).
45. Worch, E. *Adsorption Technology in Water Treatment. Fundamentals, Processes, and Modeling*; De Gruyter: Dresden, Germany, 2012. [CrossRef]
46. Foo, K.Y.; Hameed, B.H. Insights into the modeling of adsorption isotherm systems. *Chem. Eng. J.* **2010**, *156*, 2–10. [CrossRef]
47. Zheng, H.; Liu, D.; Zheng, Y.; Liang, S.; Liu, Z. Sorption isotherm and kinetic modeling of aniline on Cr-bentonite. *J. Hazard. Mater.* **2009**, *167*, 141–147. [CrossRef] [PubMed]
48. Fan, T.; Liu, Y.; Feng, B.; Zeng, G.; Yang, C.; Zhou, M.; Tan, Z.; Wang, X. Biosorption of cadmium (II), zinc (II) and lead(II) by Penicillium simplicissimum: Isotherms, kinetics and thermodynamics. *J. Hazard Mater.* **2008**, *160*, 655–661. [CrossRef] [PubMed]
49. Chakravarty, P.; Sarma, N.S.; Sarma, H.P. Biosorption of cadmium (II) from aqueous solution using heartwood powder of Areca catechu. *Chem. Eng. J.* **2010**, *162*, 949–955. [CrossRef]
50. Singh, R.; Chadetrik, R.; Kumar, R.; Bishnoi, K.; Bhatia, D.; Kumar, A.; Bishnoi, N.R.; Singh, N. Biosorption optimization of lead (II), cadmium (II) and copper (II) using response surface methodology and applicability in isotherms and thermodynamics modeling. *J. Hazard. Mater.* **2010**, *174*, 623–634. [CrossRef]
51. Malik, L.A.; Bashir, A.; Qureashi, A.; Pandith, A.H. Detection and removal of heavy metal ions: A review. *Environ. Chem. Lett.* **2019**, *17*, 1495–1521. [CrossRef]
52. Granado-Castro 1, M.D.; Galindo-Riaño, M.D.; García-Vargas, M. Separation and preconcentration of cadmium ions in natural water using a liquid membrane system with 2-acetylpyridine benzoylhydrazone as carrier by flame atomic absorption spectrometry. *Spectrochim. Acta B* **2004**, *59*, 577–583. [CrossRef]

53. Peng, J.-f.; Liu, R.; Liu, J.-f.; He, b.; Hu, X.-l.; Jiang, G.-b. Ultrasensitive determination of cadmium in seawater by hollow fiber supported liquid membrane extraction coupled with graphite furnace atomic absorption spectrometry. *Spectrochim. Acta B* **2007**, *62*, 499–503. [CrossRef]
54. Shar, G.A.; Soomro, G.A. 8-Hydroxyquinoline as a complexing reagent for the determination of Cd (II) in micellar medium. *J. Chem. Soc. Pakistan* **2005**, *27*, 471–475.
55. Marchon, B.; Bokobza, L.; Coté, G. Vibrational study of 8-quinolinol and 7-(4-ethyl-1-methyloctyl)-8-quinolinol (Kelex 100), two representative members of an important chelating agent family. *Spectrochim. Acta A* **1986**, *42*, 537–542. [CrossRef]
56. Olivieri, A.C.; Faber, N.M.; Ferré, J.; Boqué, R.; Kalivas, J.H.; Mark, H. Uncertainty estimation and figures of merit for multivariate calibration (IUPAC Technical Report). *Pure Appl. Chem.* **2006**, *78*, 633–661. [CrossRef]
57. de Carvalho Rocha, W.F.; Nogueira, R.; Vaz, B.G. Validation of model of multivariate calibration: An application to the determination of biodiesel blend levels in diesel by near-infrared spectroscopy. *J. Chemometr.* **2012**, *26*, 456–461. [CrossRef]
58. Braga, J.W.B.; Trevizan, L.C.; Nunes, L.C.; Rufini, I.A.; Santos, D.; Krug, F.J., Jr. Comparison of univariate and multivariate calibration for the determination of micronutrients in pellets of plant materials by laser induced breakdown spectrometry. *Spectrochim. Acta B* **2010**, *65*, 66–74. [CrossRef]
59. Rosborg, I.; Nihlgård, B.; Gerhardsson, L.; Gernersson, M.-L.; Ohlin, R.; Olsson, T. Concentrations of Inorganic Elements in Bottled Waters on the Swedish Market. *Environ. Geochem. Health.* **2005**, *27*, 217–227. [CrossRef] [PubMed]
60. Güler, C.; Alpaslan, M. Mineral content of 70 bottled water brands sold on the Turkish market: Assessment of their compliance with current regulations. *J. Food Compos. Anal.* **2009**, *22*, 728–737. [CrossRef]
61. Zhan, L.; Ren, X.; Jiang, D.; Lu, A.; Yuan, J. Photolumininescence and UV-Vis absorption spectra characteristics of metal complexes of 8-hydroxyquinoline. *Guang pu xue guang pu fen* **1997**, *17*, 12–15.
62. Stary, J. *Critical Evaluation of Equilibrium Constants Involving 8-hydroxyquinoline and Its Metal Chelates. Critical Evaluation of Equilibrium Constants in Solution: Part. B: Equilibrium Constants of Liquid–Liquid Distribution Systems*; Pergamon: Oxford, England, 1979. [CrossRef]
63. Skrabal, P.M. *Spectroscopy-An Interdisciplinary Integral Description of Spectroscopy from UV to NMR*; Vdf Hochschulverlag AG an der ETH Zürich: Zurich, Switerland, 2012.
64. Stary, J.; Freiser, H. 8-Hydroxyquinolines. In *Chelating Extractants*; Stary, J., Freiser, H., Eds.; Pergamon Press: Oxford, UK, 1978.

3

Monitoring Virgin Olive Oil Shelf-Life by Fluorescence Spectroscopy and Sensory Characteristics: A Multidimensional Study Carried out under Simulated Market Conditions

Ana Lobo-Prieto [1], Noelia Tena [2], Ramón Aparicio-Ruiz [2], Diego L. García-González [1,*] and Ewa Sikorska [3]

[1] Instituto de la Grasa (CSIC), Campus Universidad Pablo de Olavide-Edificio 46, Ctra. de Utrera, Km. 1, 41013 Sevilla, Spain; ana.lobo@ig.csic.es

[2] Department of Analytical Chemistry, Universidad de Sevilla, C/Prof. García González 2, 41012 Sevilla, Spain; ntena@us.es (N.T.); aparicioruiz@us.es (R.A.-R.)

[3] Institute of Quality Science, The Poznan University of Economics and Business, al. Niepodleglosci 10, 61-875 Poznan, Poland; ewa.sikorska@ue.poznan.pl

* Correspondence: dlgarcia@ig.csic.es

Abstract: The control of virgin olive oil (VOO) freshness requires new tools that reflect the diverse chemical changes that take place during the market period. Fluorescence spectroscopy is one of the techniques that has been suggested for controlling virgin olive oil (VOO) freshness during its shelf-life. However, a complete interpretation of fluorescence spectra requires analyzing multiple parameters (chemical, physical–chemical, and sensory) to evaluate the pace of fluorescence spectral changes under moderate conditions with respect to other changes impacting on VOO quality. In this work, four VOOs were analyzed every month with excitation–emission fluorescence spectra. The same samples were characterized with the concentration of fluorophores (phenols, tocopherols, chlorophyll pigments), physical–chemical parameters (peroxide value, K_{232}, K_{270}, free acidity), and sensory attributes (medians of defects and of the fruity attribute). From the six components extracted with parallel factor analysis (PARAFAC), two components were assigned to chlorophyll pigments and those assigned to tocopherols, phenols, and oxidation products were selected for their ability to discriminate between fresh and aged oils. Thus, the component assigned to oxidation products correlated with K_{270} in the range 0.80–0.93, while the component assigned to tocopherols–phenols correlated with the fruity attribute in the range 0.52–0.90. The sensory analysis of the samples revealed that the changes of these PARAFAC components occurred at the same time as, or even before, the changes of the sensory characteristics.

Keywords: fluorescence spectroscopy; virgin olive oil; shelf-life; PARAFAC; sensory assessment; quality

---

1. Introduction

The healthy and organoleptic properties of virgin olive oil (VOO) make this product highly valued by consumers. Furthermore, due to the current preference of consumers for less-processed products, VOO is already consumed in greater quantity and for more countries than ten years ago [1]. Since VOO is only produced during a few months per year, it must be stored and carefully handled to guarantee the supply during the entire year. Several studies have focused on tracking the chemical alteration of VOO during its storage under different conditions [2–4]. The oxidation process causes the loss of its antioxidant compounds and the reduction of its sensory and healthy properties. These changes may even lead to a downgrading of the category (e.g., from "extra virgin olive oil" to "virgin olive oil"), with the resulting reduction of the product value and consumer acceptability [5]. For that reason, quality control of VOO during its shelf-life is a current concern in the olive oil sector. The regulatory bodies have established several individual parameters to determine the quality and oxidation state of VOO. However, the complexity of the degradation process, where many parameters are involved and can influenced by each other, make necessary the development of new analytical tools that are able to assess the quality state of VOO since a multiparametric perspective.

Light and temperature, even under moderate conditions, have a great influence on the degradation of VOO [6–10]. Several studies have been focused on the effect of light and temperature upon VOO shelf-life [11–13]. These studies have highlighted the strong effect of light on the VOO oxidation stability; nevertheless, the response of VOO to moderate light and temperature is highly conditioned by the chemical composition of the oil [14]. During its sale and distribution, VOO stability can be considerably affected in different manners by these and others variables, which may lead to some disparities between the results of control testing (quality parameters, Rancimat method, etc.) and "the best before date" declared on the label. This problem has caused a deep concern in the regulatory and control bodies, which have established some guidelines for an optimum storage of olive oil [15,16] in order to prevent a rapid degradation of the oils. However, they do not establish a maximum period of storage or specify how VOO degradation could be controlled during the storage. Thus, currently, the analytical tools to ensure the VOO freshness and quality during the storage are not clear enough in their interpretation. The lack of harmonized analytical tools for this purpose means that most producers define the "shelf-life" of each VOO batch following their own criteria. In consequence, some discrepancies are sometimes found between the actual quality of the product in a supermarket and the quality declared on the label and expected by consumers when they purchase it.

Although there are several methods available to estimate the VOO stability, such as the oil stability index or active oxygen method, they use experimental conditions that are different from those found in actual storage (e.g., a temperature of 100 °C or more is applied). These differences in conditions modify the kinetics of the oxidation process and its effect upon VOO when it is stored under moderate conditions [14,17]. The control of the degradation process involves the monitoring of many parameters (peroxide value, free acidity, ultraviolet absorbance, organoleptic assessment, phenol content, etc.) with different time-trends in the course of the storage and informing about a particular aspect of quality, which makes it difficult to interpret quality with an overall perspective.

The need for controlling a high number of parameters during the different steps of the food chain, including storage, turns spectroscopy into an adequate technique capable of providing global information of the quality state of foods. Spectroscopic techniques have been extensively applied in food analysis since they allow a rapid and efficient measurement of a large variety of chemical parameters in food matrices [18]. Due to the usefulness and simplicity of measurements with these techniques, they are presented as an effective alternative to the classical analytical methods [19]. Currently, their applications in the analysis of different edible vegetable oils is rapidly growing.

Particularly, total luminescence spectroscopy has been implemented in food analysis and it permits the characterization of samples regarding different quality and authenticity issues [20]. Thus, it is presented as a useful and accurate technique to get information about the fluorescence compounds in vegetable oils, this technique being able to detect lower concentrations than absorption spectroscopy [21]. Total luminescence spectroscopy provides the total intensity profile associated with the fluorescent compounds present in a sample in a determined excitation and emission range of wavelengths. The obtained excitation–emission matrix (EEM) is a three-dimensional spectrum or contour map, and contains signals from all fluorophores that are present in the sample. The analysis of the EEMs with multivariate methods allows information to be extracted about the different fluorescent compounds found in a food sample simultaneously. This characteristic makes total luminescence spectroscopy an adequate analytical technique to study virgin olive oil (VOO), which is a complex food system that contains several fluorescent compounds, such as phenols, tocopherols, and pheophytins.

Several authors have proposed using fluorescence spectroscopy to characterize VOO [22–26]. Other authors have applied this technique to detect frauds such as blends of olive oils with other vegetable oils [27,28]. Furthermore, fluorescence spectroscopy has also been proposed to discriminate oils with different geographical provenances [29]. The tracking of the different fluorophores during the VOO storage provides global information about the chemical changes taking place, and consequently, provides knowledge of how the oxidation process progresses depending on the VOO chemical composition and the storage conditions.

This study proposes the total luminescence spectroscopy combined with (parallel factor) PARAFAC analysis as an appropriate technique, which is able to monitor the changes of VOO produced during the storage under moderate conditions from a multidimensional perspective. Four monovarietal VOOs from three different cultivars were stored during 21 months under conditions close to the real ones, in order to study their fluorescence characteristics, and, at the same time, the chemical quality parameters and sensory attributes. The objective of this study was to verify if the results obtained from fluorescence spectroscopy and chemometrics could provide real information of the changes occurring in VOO during its storage, and the ability of this method to distinguish between fresh and aged oils. Furthermore, due to the relevance of the sensory quality of VOO, the results obtained by an accredited panel are included to have complete information about the quality of the oils.

2. Materials and Methods

2.1. Samples

Four virgin olive oils (VOOs) from Picual, Hojiblanca, and Arbequina (2 oils) cultivars were used in this study. These three cultivars were selected because of their distribution and for being predominant in a particular region. The four VOOs were directly provided by Spanish producers and they were taken from the vertical centrifuge at the oil mill, in order to guarantee the freshness of the oils. Subsequently, the VOOs were filtered to remove moisture, and bottled. The filtration in the laboratory was carried out in the dark to avoid photooxidation, using folded filter paper (filter paper 600, Dorsan Living Filtration, Barcelona, Spain). After the filtration, the samples were randomly subjected to moisture analysis according to ISO 662 to check the filtration efficiency. Three randomly selected portions of each oils were analyzed in duplicate and the moisture contents were below 0.1% *m*/*m* in all cases. The samples were named as follows: VOO1, Hojiblanca; VOO2, Arbequina-A; VOO3, Picual; VOO4, Arbequina-B.

2.2. Storage Experiment

VOOs were stored for 21 months in a compartment where the conditions that are given in a supermarket were simulated. Each VOO was packaged in 22 transparent PET (polyethylene terephthalate) bottles of 500 mL (one per month of storage plus one for the fresh sample), and they were hermetically sealed. The bottles were kept under a light intensity of ≈1000 lx in 12 h light/dark cycles, while the temperature and humidity were controlled. In this storage experiment, the temperature, daily controlled, varied between 16.3 and 29.7 °C, and the humidity varied between 21% and 70%. During the storage experiment, one bottle was opened and analyzed every month. The oil remained in the bottle after the analyses were discarded. Therefore, only oils from a freshly opened bottle were used for the analyses.

2.3. Quality Parameters

The quality parameters were analyzed in the fresh samples (time zero) to determine the VOO category before starting the storage. During the storage experiment, they were also analyzed each month in order to monitor their changes. These parameters were the peroxide value (PV), free fatty acid content (FFA), and extinction coefficients from ultra-violet absorbance (K_{270} and K_{232}), which were measured by applying the International Olive Council methods [30–32].

2.4. Sensory Assessment

The sensory characteristics of the VOO samples were determined by the panel of Instituto de la Grasa [33] applying the standard COI/T.20/Doc. No 15/Rev.10 2018 [34]. The panelists evaluated the median of the fruity attribute (Mf) and defect (Md) for the four VOOs subjected to the storage experiment. The sensory assessment results were generated every month. Thus, it provided chronological information about the sensory characteristics' changes of the oils during the storage period, which made it possible to identify changes in the quality category of the oils and in their sensory characteristics.

2.5. Phenol Content

The phenol composition were determined by applying the method described by Mateos et al. [35], slightly modified by Aparicio–Ruiz et al. [36]. An amount of 2.5 g of the sample was solved in 6 mL of hexane, and *p*-hydroxyphenylacetic (0.12 mg/mL) and *o*-coumaric (0.01 mg/mL) were added as internal standards. The isolation of the phenolic fraction was carried out with methanol by solid phase extraction using diol-bonded phase cartridges. After that, the concentrated phenolic fraction was injected in the HPLC system (Agilent Technologies 1200, Waghaeusel–Wiesental, Germany), equipped with a diode array detector. The column was a Lichrospher 100RP-18 column (4.0 i.d. × 250 mm; 5 μm, particle size) (Darmstadt, Germany) kept at 30 °C. The flow rate of 1.0 mL/min was used and the gradient elution was performed using a mixture of water/ortho-phosphoric acid (99.5:0.5 *v*/*v*) (solvent A) and methanol/acetonitrile (50:50 *v*/*v*) (solvent B). The change in solvent gradient was programed as follows: From 95% (A) and 5% (B) to 70% (A) and 30% (B) in 25 min; 65% (A) and 35% (B) in 10 min; 60% (A) and 40% (B) in 5 min; 30% (A) and 70% (B) in 10 min and 100% (B) in 5 min, followed by 5 min of maintenance. The chromatographic signals were obtained at 235, 280, and 335 nm. Figure S1 shows an example of a chromatogram obtained in the analysis and Table S2 shows the phenolic compounds identified in the VOO samples. The quantification of the phenols was carried out following the procedure described by Mateos et al. [35]. Quantification of phenols, lignans, and cinnamic acid was carried out at 280 nm using *p*-hydroxyphenylacetic acid as internal standard, whereas the quantification of flavones was at 335 nm using *o*-coumaric acid as internal standard.

2.6. α-Tocopherol Content

The method ISO 9936:2016 [37] was applied for the determination of α-tocopherol. The sample (0.1 g) was dissolved with 10 mL of hexane. From this solution, 20 μL was injected into the HPLC Agilent Technologies 1200 (Waghaeusel–Wiesental, Germany), equipped with a fluorescence detector. The column used was a silica gel column Superspher®RP-18 (4 i.d. × 250 mm length, 5μm particle size) purchased from Merck (Darmstadt, Germany). The identification of α-tocopherol was carried out with λ_{ex} = 290 nm and λ_{em} = 330 nm. Figure S2 shows an example of a chromatogram obtained in this analysis. The quantification of α-tocopherol was carried out by calibration curve of a stock solution of α-tocopherol (Sigma–Aldrich–Fluka, Darmstadt, Germany). The preparation of the stock solution and the development of its calibration curve was carried out following ISO 9936:2016 [37]. The real concentration of the stock solution was determined by its maximum absorbance in a wavelength range between 270 and 310 nm using a UV VIS spectrometer Thermo Scientific GENESYS 10s (Waltham, MA, USA) and 10-mm path length cell, Hellma Analytics (Müllheim, Germany).

2.7. Pigment Analysis

The determination of the degradation products of chlorophyll a, such as pheophytin a and pyropheophytin a, were measured using the method ISO 29841:2012 [37]. The fraction of chlorophyll pigment was extracted by solid phase extraction using silica cartridge 1000 mg/6 mL, 55 μm, 700 nm (Supelco, Bellefonte, PA, USA). The analysis of pigments was carried out using an HPLC system LaChrom Elite de Hitachi (Tokyo, Japan) with a diode array detector. The column used was a Lichrospher RP18 HPLC column, 250 mm length, 4.0 mm internal diameter, filled with reversed-phase particles size 5 μm (Merck, Darmstadt, Germany). The peak identification was carried out using the standard of pheophytin a and pyropheophytin a, which were obtained from a dissolution of ethyl ether and chlorophyll a, from spinach (Sigma–Aldrich, Darmstadt, Germany), following the procedure explained by Sierves and Hynninen [38] in the case of pheophytin a, and Schwartz et al. [39] in the case of pyropheophytin a. Figure S3 shows an example of a chromatogram of the degradation products of chlorophyll a obtained in the analysis.

2.8. Fluorescence Measurements

The fluorescence spectra were obtained with an AqualogTM (Horiba, Montpellier, France) spectrofluorometer. A xenon lamp was used as an excitation source. Before the measurements, the instrument performance was checked using a standard procedure. The excitation and emission slit widths were 5 nm. The gain of a charge-coupled device (CCD) detector was set to a low range. The corrected three-dimensional spectra were obtained by measuring the emission spectra from 250 to 830 nm with an average increment of 4.66 nm repeatedly, at excitation wavelengths from 240 to 800 nm, spaced by 5 nm intervals. Right-angle geometry was used for analyzing the oil samples diluted in n-hexane (3% *v*/*v*) in a 10-mm fused-quartz cuvette. This low concentration was chosen to avoid spectral distortions. Additionally, the inner filter effect was corrected based on the simultaneous absorbance measurements, using AqualogTM built-in software.

2.9. Statistical Analysis

The excitation–emission matrices in the excitation range of 280–800 nm and emission range of 300–830 nm (EEMs) were used for the statistical analysis. The EEMs of 22 samples per each of the four oils were arranged in a three-dimensional structure with a size of 88 × 114 × 105 (number of samples x number of emission wavelengths x number of excitation wavelengths). The entire data set was analyzed using PARAFAC, which is able to break down the EEMs into the contributions of the individual fluorescent components. The Rayleigh scattering bands were removed with a Rayleigh-masking algorithm. Core consistency diagnostics (CONCORDIA) and the explained variance were used to find the optimal number of components in the PARAFAC model [40,41]. The PARAFAC analysis was carried out with SOLO v.8.7.1 software (Eigenvector Research Inc., Wenatchee, WA, USA).

Principal component analysis (PCA) was performed with the chemical parameters (quality parameters, phenols, and α-tocopherol content) analyzed in the stored samples and the fluorescence components extracted by PARAFAC analysis. The loading plot and the scores plot were studied in order to identify the relationship between the PARAFAC components and the chemical parameters, and to characterize the stored VOOs.

Stepwise linear discriminant analysis (SLDA) was performed in fresh (0–5 months) and aged (16–21 months) samples using the fluorescence components obtained by PARAFAC, in order to identify the PARAFAC components that were able to discriminate between these two kinds of samples. Significance discrimination was accepted when $p < 0.05$.

The multivariate analyses were carried out using the STATISTICA 8 package (Statsoft, Tulsa, OK, USA).

3. Results and Discussion

3.1. Physical–Chemical Characterization

Peroxide value, free acidity (free fatty acids or FFA), extinction coefficients (K_{270} and K_{232}), and the total concentration of phenols, α-tocopherol, and pigments derived from chlorophyll a (pheophytin a and pyropheophytin a) were analyzed in the four fresh VOOs previous to the storage ("time zero") in order to characterize the VOOs at the moment of bottling. Table 1 shows the results for peroxide value, free acidity, and K_{270} and K_{232} for each VOO during the storage experiment. Furthermore, Table 2 shows the concentrations of phenols, α-tocopherol, pheophytin a, and pyropheophytin a for each VOO during the entire experiment. The values of peroxide value, FFA, K_{270}, and K_{232} in the fresh samples ("time zero") revealed that the four VOOs belonged to the "extra virgin olive oil" category, according to the European Commission (EC) regulation [15]. VOO2 showed the highest value of peroxide value, K_{270}, and K_{232}, which pointed out a certain degree of alteration despite the fact that all the VOOs were fresh and they were directly taken from the vertical centrifuge.

Regarding the concentration of phenols, VOO3 and VOO4 were characterized with the highest concentrations, 564.82 and 451.25 mg/kg, respectively. VOO1 and VOO2 showed lower concentrations of total phenols (Table 2), 246.71 and 338.90 mg/kg respectively. The method used for this determination was successfully used in previous works [35,36]. This method allowed for a good separation of several kinds of phenols in a single chromatographic run.

The initial concentration of α-tocopherol was similar between VOO2 and VOO3, with a value of 272.28 and 256.91 mg/kg. The other two VOOs showed a lower concentration, 212.62 in VOO1 and 192.94 mg/kg in VOO4 (Table 2). The study of minor compounds, such as phenolic and tocopherols compounds, can provide information about how stable the oil is under oxidation, since they interrupt the propagation chain of lipid oxidation [42–44].

The tracking of pheophytin a and pyropheophytin a concentrations in VOO over time have been used in the monitoring of its stability and its loss of freshness [45]. The highest concentration of pheophytin a in the fresh samples (Table 2) was found in VOO3 with a value of 23.43 mg/kg. The highest concentration of pyropheophytin a in the fresh oils was found in VOO4 with a value of 0.11 mg/kg.

As soon as the storage experiment started, all the quality parameters evolved immediately (Tables 1 and 2). The quality indexes (peroxide value, free acidity, K_{270} and K_{232}) showed their maximum values in the last month of storage, although they were within the "extra virgin olive oil" category according to the limits stated in European regulation [15], except for K_{270} (Table 1). This parameter surpassed the limit established for the "extra virgin olive oil" category in the first months of storage (2–4 month) for the four VOOs. VOO1 and VOO3 showed the highest K_{270} values, while VOO2 and VOO4 showed the highest value of K_{232}. The rest of the parameters, phenols, α-tocopherol, and pheophytin a, decreased their concentration from the beginning of the storage (Table 2). Pyropheophytin a concentration also increased from the beginning of the storage, but it later decreased until the end of the experiment. In this case, these compounds had fluorescence properties and their changes were reflected in the fluorescence spectra.

3.2. *Changes of Fluorescence Excitation–Emission Matrices during the Storage*

The time-trend of the main fluorescence compounds present in the stored VOOs were studied by excitation–emission fluorescence spectroscopy. All these data were studied simultaneously during the whole storage experiment, in order to obtain a multiparametric perspective of the degradation process of VOO under moderate conditions.

Figure 1a shows the contour maps of excitation–emission matrices (EEMs) of the four stored VOOs before starting the storage (fresh oils). The EEMs exhibited general features as other authors reported in previous works [4,46]. Thus, their EEMs showed two groups of bands observed in all the oils (Figure 1a). A group of bands was found in the emission wavelengths range of 600–700 nm. The most intense band in this group was found at the excitation/emission maximum ($\lambda_{ex}/\lambda_{em}$) of 408/678 nm in all the samples studied. According to previous works, this band is associated with the presence of chlorophyll pigments, mainly pheophytin a [2,4,19,47]. The second group of bands was identified at 250–350 nm of emission wavelengths, the excitation/emission maximum ($\lambda_{ex}/\lambda_{em}$) being found at 293/322 nm in all the oils. This band corresponded simultaneously to tocopherols and phenols, as it was extensively reported in previous works [19,21,24,26].

Table 1. Quality parameters (peroxide value, free acidity or FFA, K_{270} and K_{232}) are shown for the four fresh virgin olive oils (VOOs) before starting the storage ("time zero") and in every month during the entire storage time. According to European commission regulation [15]: Limits for extra virgin olive oil: Peroxide value $\leq$ 20 meq O_2/kg, FFA $\leq$ 0.8%, $K_{270} \leq 0.22$, $K_{232} \leq 2.50$. Limits for virgin olive oil: Peroxide value $\leq$ 20 meq O_2/kg, FFA $\leq$ 2.0%, $K_{270} \leq 0.25$, $K_{232} \leq 2.60$.

	Peroxide Value (meq O_2/kg)				FFA (% m/m Oleic Acid)				K_{270} (Absorbance Units)				K_{232} (Absorbance Units)			
Months of Storage	VOO1	VOO2	VOO3	VOO4	VOO1	VOO2	VOO3	VOO4	VOO1	VOO2	VOO3	VOO4	VOO1	VOO2	VOO3	VOO4
0	4.30	5.13	3.63	4.82	0.15	0.21	0.20	0.20	0.06	0.18	0.04	0.18	1.53	1.87	1.73	1.84
1	7.54	5.37	4.12	4.94	0.16	0.22	0.21	0.22	0.21	0.18	0.17	0.20	1.95	1.87	1.78	1.96
2	7.77	5.17	4.85	4.74	0.16	0.21	0.21	0.22	0.24	0.19	0.18	0.21	1.93	1.83	1.82	1.97
3	7.56	5.27	5.45	5.30	0.18	0.22	0.21	0.21	0.23	0.20	0.22	0.22	1.88	1.82	1.82	1.96
4	7.46	5.86	5.48	6.22	0.18	0.22	0.21	0.21	0.25	0.22	0.22	0.21	1.92	1.89	1.82	1.95
5	7.69	5.98	5.63	5.83	0.19	0.23	0.21	0.21	0.26	0.23	0.22	0.23	1.90	1.85	1.77	1.95
6	7.63	5.77	6.18	6.17	0.19	0.23	0.21	0.21	0.27	0.23	0.24	0.23	1.91	1.97	1.80	1.96
7	8.31	5.68	5.93	6.11	0.19	0.23	0.22	0.22	0.27	0.23	0.25	0.23	1.91	1.90	1.80	1.97
8	8.38	5.98	5.99	7.11	0.20	0.24	0.22	0.23	0.27	0.24	0.28	0.24	1.93	1.90	1.82	1.98
9	9.19	7.30	6.24	8.12	0.21	0.26	0.23	0.24	0.28	0.24	0.27	0.23	1.97	1.90	1.84	1.98
10	10.04	7.48	6.74	9.28	0.21	0.27	0.23	0.23	0.29	0.24	0.28	0.24	1.97	1.88	1.83	1.98
11	10.60	8.36	6.66	9.73	0.22	0.28	0.22	0.24	0.30	0.26	0.29	0.24	1.97	1.92	1.83	1.99
12	10.69	8.39	7.27	10.31	0.21	0.27	0.23	0.24	0.32	0.26	0.33	0.24	1.94	1.93	1.84	1.99
13	11.64	8.86	7.72	10.56	0.22	0.28	0.22	0.25	0.31	0.26	0.31	0.25	1.97	1.97	1.84	2.00
14	11.80	9.47	7.45	10.33	0.21	0.28	0.23	0.25	0.31	0.26	0.28	0.25	1.97	1.99	1.85	2.03
15	11.25	9.32	7.32	10.70	0.22	0.29	0.23	0.26	0.32	0.26	0.30	0.26	1.98	2.12	1.85	2.04
16	11.94	9.43	7.41	10.74	0.21	0.29	0.24	0.27	0.32	0.26	0.30	0.26	2.02	2.05	1.86	2.07
17	11.97	9.78	7.60	10.61	0.22	0.30	0.23	0.27	0.31	0.26	0.31	0.26	1.99	2.06	1.85	2.07
18	11.98	9.82	8.01	10.88	0.23	0.29	0.23	0.27	0.32	0.26	0.33	0.27	2.04	2.06	1.90	2.07
19	12.54	10.48	8.05	11.06	0.23	0.30	0.23	0.27	0.33	0.26	0.32	0.27	1.99	2.07	1.89	2.07
20	13.06	10.33	8.89	11.75	0.22	0.31	0.23	0.28	0.33	0.26	0.32	0.27	1.98	2.02	1.90	2.07
21	13.59	10.64	8.84	12.38	0.23	0.30	0.23	0.29	0.33	0.28	0.33	0.28	2.04	2.23	1.90	2.08

Table 2. Concentrations of phenols, α-tocopherol, and pigments derived from chlorophyll a (pheophytin a and pyropheophytin a) are shown for the four fresh virgin olive oils (VOOs) before starting the storage ("time zero") and in every month during the entire storage time.

Months of Storage	Total Phenols (mg/kg)				α-Tocopherol (mg/kg)				Pheophytin a (mg/kg)				Pyropheophytin a (mg/kg)			
	VOO1	VOO2	VOO3	VOO4	VOO1	VOO2	VOO3	VOO4	VOO1	VOO2	VOO3	VOO4	VOO1	VOO2	VOO3	VOO4
0	246.71	338.90	564.82	451.25	212.62	272.28	256.91	192.94	7.06	3.02	23.43	4.61	0.03	0.04	0.07	0.11
1	239.51	337.08	558.45	422.98	190.47	260.76	210.03	157.49	7.68	3.00	22.98	4.60	0.14	0.04	0.39	0.26
2	238.82	338.66	547.93	399.56	160.05	220.01	134.21	125.84	5.17	1.30	22.59	0.53	0.33	0.05	0.63	0.12
3	231.70	333.44	521.74	381.09	147.63	194.34	134.78	122.78	2.60	0.43	21.32	0.33	0.37	0.05	0.95	0.16
4	218.11	325.57	499.02	373.03	140.11	187.77	140.81	120.65	0.49	0.12	16.73	0.20	0.22	0.04	1.18	0.09
5	209.46	316.30	486.27	360.98	136.22	178.83	138.86	115.21	0.47	0.11	16.83	0.19	0.20	0.04	2.74	0.07
6	204.68	308.34	474.05	348.81	133.77	162.18	131.86	115.52	0.18	0.10	17.17	0.12	0.08	0.03	2.13	0.06
7	198.97	298.84	459.09	340.11	130.48	154.93	137.04	110.22	0.03	0.03	5.36	0.03	0.03	0.02	1.23	0.02
8	190.74	292.30	441.80	324.71	131.55	147.99	129.99	92.26	0.03	0.05	2.22	0.03	0.05	0.02	1.22	0.02
9	183.06	280.30	417.35	315.62	129.12	136.40	129.09	85.48	0.02	0.04	2.39	0.05	0.03	0.02	1.72	0.02
10	168.05	271.11	405.58	312.19	125.98	123.43	124.91	81.60	0.03	0.02	0.44	0.03	0.02	0.02	1.58	0.01
11	155.06	261.88	392.51	287.31	122.78	113.15	116.48	79.05	0.02	0.04	0.62	0.02	0.04	0.02	1.51	0.02
12	149.66	253.01	366.32	273.11	122.30	112.77	111.11	75.12	0.02	0.03	0.27	0.02	0.02	0.02	0.54	0.02
13	144.38	247.86	351.22	260.54	121.81	109.56	93.18	73.43	0.02	0.04	0.13	0.02	0.02	0.01	0.32	0.02
14	138.29	240.03	339.06	253.55	122.06	107.97	92.08	70.76	0.01	nd	0.09	nd	0.02	0.02	0.22	nd
15	131.43	231.23	325.29	235.88	120.36	103.17	90.44	60.51	nd	nd	nd	nd	nd	0.03	nd	nd
16	128.41	222.52	326.64	225.60	119.64	99.73	92.84	54.64	nd	nd	nd	nd	nd	0.03	nd	nd
17	124.19	217.34	314.18	225.25	114.55	97.04	93.07	59.10	nd	nd	nd	nd	nd	0.03	nd	nd
18	118.84	209.22	290.96	223.78	105.07	96.47	91.47	53.67	nd	nd	nd	nd	nd	0.03	nd	nd
19	112.45	203.57	277.17	205.86	106.30	96.94	84.28	50.20	nd	nd	nd	nd	nd	0.03	nd	nd
20	108.06	199.34	265.78	205.83	102.30	91.28	84.92	22.42	nd	nd	nd	nd	nd	0.03	nd	nd
21	106.87	193.95	252.69	205.76	102.39	87.47	85.69	20.38	nd	nd	nd	nd	nd	0.03	nd	nd

Note: nd; not detected.

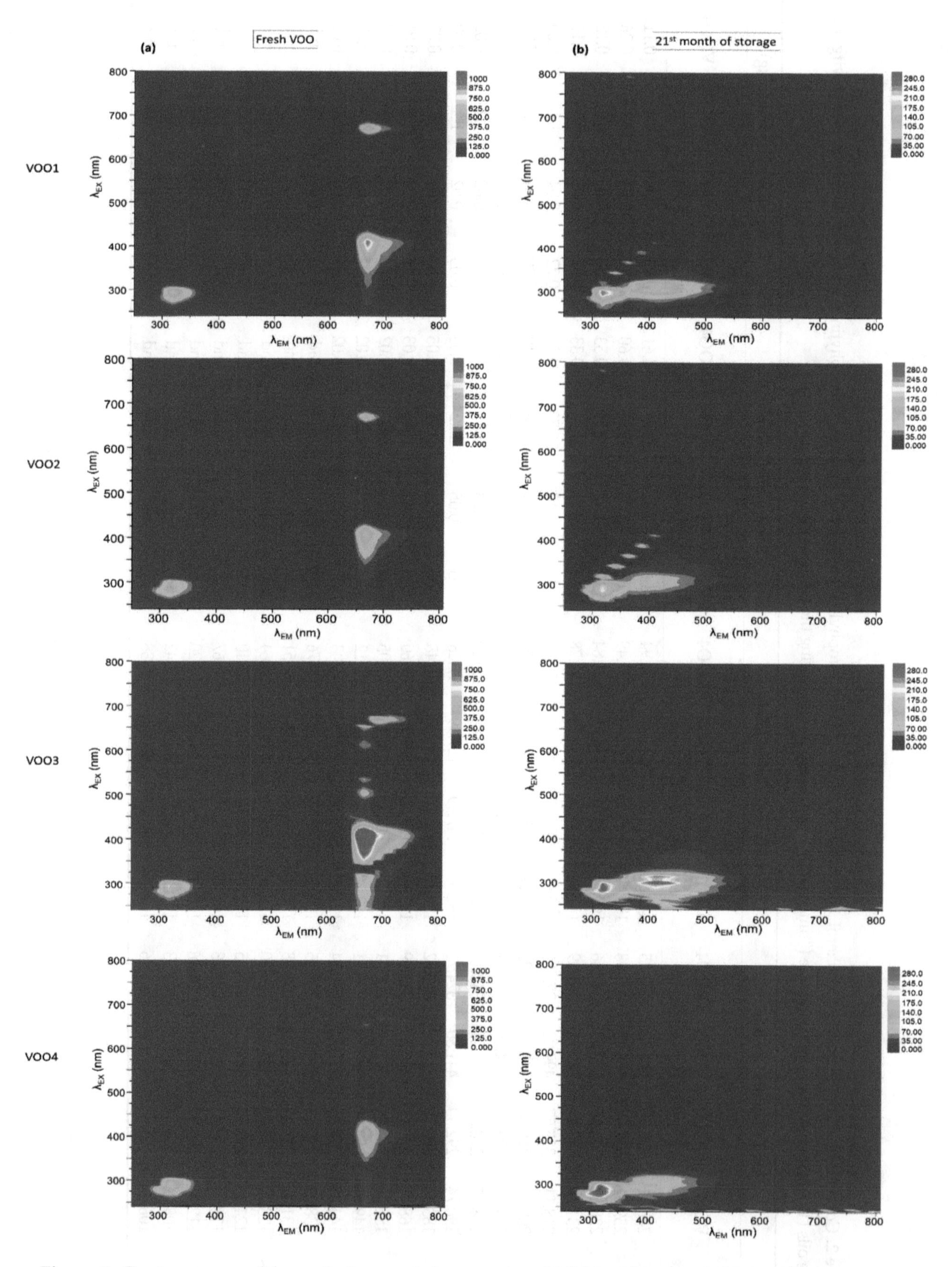

Figure 1. Contour maps of the excitation–emission matrices (EEMs) of the stored VOOs in two moments of their storage under moderate conditions: (**a**) Before starting the storage (fresh sample) and (**b**) at the end of the storage (twenty-first month of storage).

Figure 1b shows the changes of the contour maps of the EEMs of the VOOs during the storage under moderate conditions. Particularly, this figure shows the EEMs at the last months of the storage (21 months). The two groups of bands associated with the fresh oils (Figure 1a) decreased progressively during the storage. Figure 2a,b display the time-trends of the bands associated with pigments ($\lambda_{ex}/\lambda_{em}$ 408/678 nm) and tocopherols and phenols ($\lambda_{ex}/\lambda_{em}$ 293/322 nm) during the 21 months of storage. These bands decreased during the storage in the four VOOs due to the degradation reactions that were taking place [4].

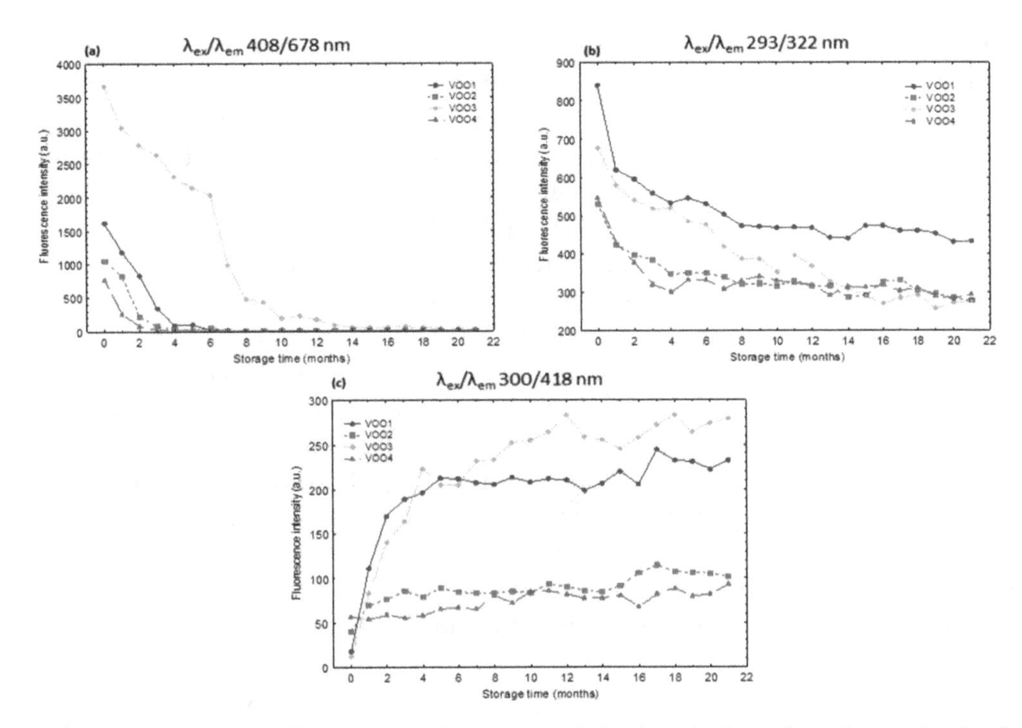

Figure 2. Time-trend of the fluorescence intensity of the bands found at the excitation/emission maxima of (**a**) 408/678, (**b**) 293/322, and (**c**) 300/418 nm for the virgin olive oils during the storage under moderate conditions.

The intensity of the band assigned to the pigments abruptly decreased during the first months of storage under moderate conditions (Figure 2a). Thus, this band completely disappeared in the fourth month of storage in VOO2 and VOO4, in the fifth month in VOO1, and in the eighteenth month in VOO3. The fluorescence intensity of this band for VOO3 was at least double compared to the other three oils before the storage (Figure 2a). This difference explained that this band required a longer time (18 months) to be undetected in the fluorescence spectra of VOO3. This high intensity matched with the high concentration of pheophytin a determined by HPLC in this oil: 23.43 mg/kg in VOO3, while the values for the rest of the oils were 7.06 in VOO1, 3.02 in VOO2, and 3.31 mg/kg in VOO4 (Table 2). Furthermore, the time-trend of the pheophytin a concentration determined by HPLC during the storage was similar to that of the fluorescence band assigned to pigments. Thus, the concentration values of pheophytin a also decreased in all of the cases during the storage (Table 2). These concentrations reached values close to zero ($\leq$0.03 mg/kg) after 7 months of storage of the samples, except for VOO3, in which such a reduction was observed after 15 months (Table 2). The time-trend similarities between pheophytin a concentration (HPLC data) and the intensity of this band ($\lambda_{ex}/\lambda_{em}$ 408/678 nm) were supported by the high correlation coefficients between these two variables, which were 0.98 in VOO1, VOO2, and VOO3, and 0.79 in VOO4. The positive relationship between pheophytin a and the fluorescence band at $\lambda_{ex}/\lambda_{em}$ 408/678 nm was previously reported by several authors [2,4,29]. On the contrary, no relationship was found between the pyropheophytin a concentration and the fluorescence intensity of this band. In fact, the correlation coefficients in this case were 0.45 or lower in the four VOOs.

The intensity of the band assigned to tocopherols and phenols ($\lambda_{ex}/\lambda_{em}$ 293/322 nm) also decreased during storage. Unlike the band assigned to pigments, the intensity of this band never decreased to values close to zero, although it was reduced by approximately up to 50% of its initial values at the end of the experiment. However, as the band assigned to pigments, this band also underwent the highest decrease in the first five months of storage. Regarding the chemical analysis by HPLC, the concentrations of α-tocopherol and phenols also decreased during the storage experiment (Table 2). On the one hand, the α-tocopherol concentration underwent a reduction of their initial values of 51.84% for VOO1, 66.65% for VOO3, 67.87% for VOO2, and 89.44% for VOO4 at the end of the storage (Table 2). Nevertheless, they showed the highest decreases during the first five months of storage, which was also well represented by this fluorescence band. In fact, the correlation coefficients between HPLC results and the spectral intensity of this band in the whole storage experiment were 0.93 for VOO1 and VOO2, 0.94 for VOO3, and 0.79 for VOO4. On the other hand, the concentration of total phenols determined by HPLC revealed a decrease with respect to their initial concentration of 42.77% for VOO2, 54.40% for VOO4, 55.26% for VOO3, and 56.68% for VOO1 (Table 2). The correlation coefficients between the HPLC results and the intensity of this fluorescence band were 0.74 for VOO4, 0.79 for VOO1, 0.85 for VOO2, and 0.89 for VOO3. The individual contribution of tocopherols and phenols has been studied by synchronous fluorescence spectroscopy [4,24] and by using a vitamin E standard and a VOO phenol extract [48]. Some works used this band to develop models to estimate the concentration of tocopherols in vegetables oils [24,49] or even to classify oils according to their concentration of phenols [26].

In addition to the aforementioned bands, a new fluorescence band appeared at the intermediate-wavelength emission region ($\lambda_{ex}/\lambda_{em}$ 300–319/418 nm) during the storage. This band was previously reported and attributed to oil oxidation products by other authors [25,46,48,50]. Figure 2c shows the time-trend of the fluorescence intensity of this band in the VOOs during the storage time. In VOO4 and VOO2, this band was already observed at low intensity in the fresh oils, while in the rest of the VOOs it was barely detected. This agreed with the initial oxidation status of the samples according to the K_{270}, K_{232}, and peroxide value, which identified VOO2 and VOO4 as the most oxidized oils (Table 1). Furthermore, the fluorescence intensity of this band increased during the storage (Figure 2c), reaching its maximum in the last month of the storage. However, the time-trend of the fluorescence intensity of this band showed two different behaviors. Thus, VOO1 and VOO3 showed an abrupt increase of the fluorescence intensity of this band during the first five months, while it increased at a lower rate after this moment. VOO2 and VOO4, the two Arbequina oils, showed a different time-trend consisting in a continuous increment of the band intensity during the whole period and at lower rate compared to VOO1 and VOO3. During the storage experiment, all the quality parameters related to oxidation products (K_{270}, K_{232}, and peroxide value) also showed an increase (Table 1), which also reached their maximum value in the last month (twenty-first month) of storage. The highest relation of the intensity of this band with respect to the quality parameters previously mentioned was found for K_{270}, which showed correlation coefficients of 0.70 for VOO2, 0.79 for VOO4, 0.85 for VOO1, and 0.92 for VOO3. The time-trend of K_{270} (Table 1) and the intensity of this fluorescence band (Figure 2c) showed that VOO1 and VOO3 were the most oxidized samples at the end of the experiment.

3.3. Multivariate Analysis of VOO Excitation–Emission Fluorescence Spectra

Multivariate exploratory methods were used to study the fluorescent compounds of the sample set. The 88 EEMs (22 EEMs per 4 VOOs, one per month during the 21 months of storage and the EEM of the fresh oil) were analyzed by the PARAFAC algorithm. The number of PARAFAC components was six, which was selected according to the core consistency (CONCORDIA = 87%) and the inspections of the residuals and the loadings (variance explained = 97.97%) [40,41]. Figure 3 shows the PARAFAC excitation and emission profiles for the 6 extracted components. The scores of PARAFAC components for each of the four studied oils are presented in Figure S4 (Supplementary Material). In order to carry out the analysis of the PARAFAC results, the six selected components were assigned to the fluorescent compounds. Firstly, the emission profiles of component 1 ($\lambda_{ex}/\lambda_{em}$ 408/678 nm) and component 3 ($\lambda_{ex}/\lambda_{em}$ 408/668 nm) were assigned to the chlorophyll pigments [2,4]. The existence of two components for chlorophyll pigments could be related with the fact that the emission wavelengths (maximum intensity) were different depending on the chlorophyll derivative and they varied in the range (658–672 nm) [51]. However, it was difficult to assign each component to one specific derivative. Secondly, the emission profiles of component 2 ($\lambda_{ex}/\lambda_{em}$ 293/322 nm) and component 5 ($\lambda_{ex}/\lambda_{em}$ 280/314 nm) were both assigned to tocopherols and phenols [21,26,29]. Previous research works dealing with the study of the individual contribution of tocopherols and phenols by means of synchronous fluorescence spectroscopy reported that tocopherols were related with higher excitation/emission wavelengths compared with phenols [21]. Therefore, it could be thought that component 2 would be more related with tocopherols and component 5 with phenols, although a mixture of contribution from both kinds of compounds was expected. Finally, the emission profiles of component 4 ($\lambda_{ex}/\lambda_{em}$ 300/418 nm) and component 6 ($\lambda_{ex}/\lambda_{em}$ 340/450 nm) were attributed to oxidized compounds by several studies [46,52–54].

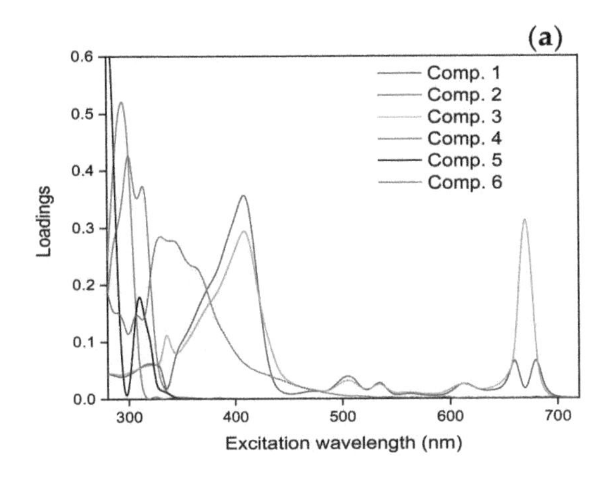

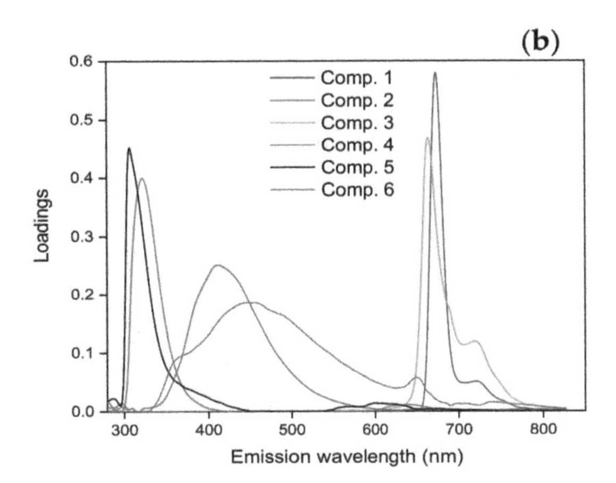

Figure 3. Parallel factor analysis (PARAFAC) excitation (**a**) and emission (**b**) profiles of the entire sample set (four monovarietal samples during the storage under moderate conditions) for the six components: Component 1 ($\lambda_{ex}/\lambda_{em}$ 408/678 nm), component 2 ($\lambda_{ex}/\lambda_{em}$ 293/322 nm), component 3 ($\lambda_{ex}/\lambda_{em}$ 408/668 nm), component 4 ($\lambda_{ex}/\lambda_{em}$ 300/418 nm), component 5 ($\lambda_{ex}/\lambda_{em}$ 280/314 nm), and component 6 ($\lambda_{ex}/\lambda_{em}$ 340/450 nm).

A principal component analysis (PCA) was applied to the 6 PARAFAC components and the chemical and physical–chemical parameters (Tables 1 and 2) to observe the distribution of the samples according to the storage time with a multivariate perspective. Figure 4 shows the loading (Figure 4a) and score (Figure 4b) plots obtained for the two first principal components (PC1 and PC2) of the PCA.

Figure 4a revealed that the physical-chemical parameters were distributed in two groups according to PC1. This distribution divided the parameters between those related to freshness markers (pheophytins and pyropheophytins) and antioxidant compounds (phenols and α-tocopherol), plotted in the negative side of PC1, and those related to oxidative and quality indexes of the oil (K_{232}, K_{270}, free acidity, and peroxide value) were placed in the positive side of PC1. The score plot presented in Figure 4b shows a sequential shift of the samples collected every month along the PC1 axis. The freshest oils were located in the left quadrant, which matched with the quadrant where pheophytins, pyropheophytins, phenols, and α-tocopherol were placed (Figure 4a). However, as storage progressed, the VOOs were plotted in the right quadrant, where the K_{232}, K_{270}, free acidity, and peroxide value were located. Due to the fact that the time is a continuous variable and the chemical changes between consecutive months are moderate in the collected samples, no clear groups were discriminated between samples. Nevertheless, the distribution of the stored oil samples along the PC2 (Figure 4b) was able to distinguish between Arbequina oils (VOO2 and VOO4) and the other two cultivars, Picual (VOO3) and Hojiblanca (VOO1).

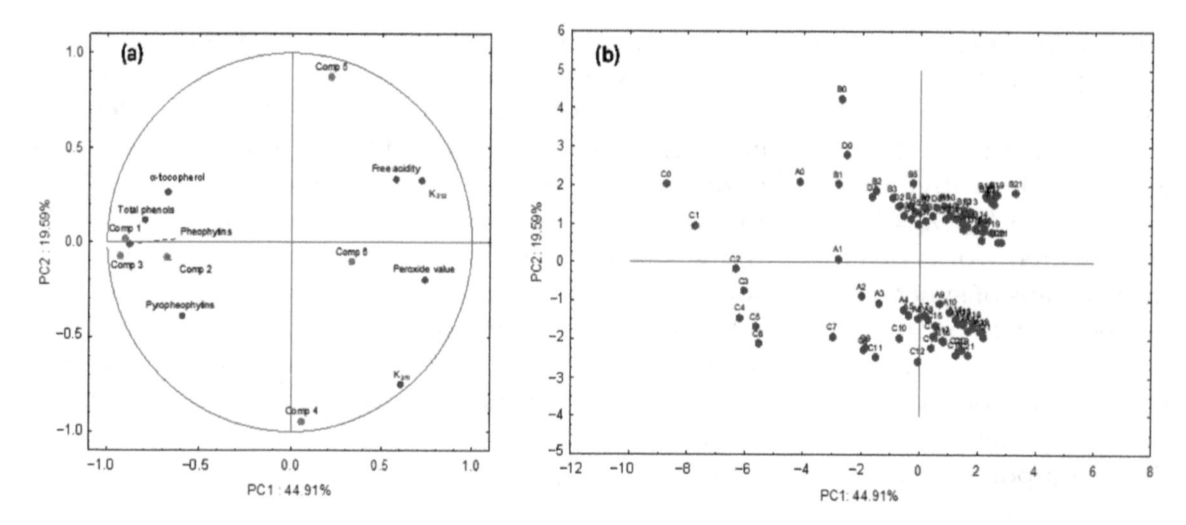

Figure 4. Principal component analysis (PCA) of all of physical–chemical parameters analyzed and the components extracted by PARAFAC analysis from the entire data set: Loading plot (**a**) and score plots (**b**) obtained of the two first principal components (PC1 and PC2). Codes: A, VOO1; B, VOO2; C, VOO3; D, VOO4. The numbers after the codes mean the months of the storage when the samples were collected.

The PCA was also used for studying the relationship between the physical–chemical parameters (peroxide value, K_{232}, K_{270}, free acidity, phenols, α-tocopherol, pheophytin, and pyropheophytin contents), which characterize the oils, and the fluorescence components obtained by PARAFAC. The PARAFAC components 1 and 3 (chlorophyll pigments), and 2 (tocopherols with contribution of phenols) were plotted near their related chemical parameters (pheophytins, pyropheophytins, phenols, and α-tocopherol) (Figure 4a). However, PARAFAC component 5, whose excitation and emission wavelengths are mainly assigned to phenols in the literature [21,26], was plotted far from this chemical parameter (Figure 4a). The high diversity of phenols and their different fluorescent characteristics [21,46] may partially explain the lack of correlation of component 5 with total phenol content. On the other hand, PARAFAC components 4 and 6 were located in the right quadrant, the same quadrant where the peroxide value, free acidity, K_{232}, and K_{270} were plotted. The position of these two PARAFAC components in the loading plot (Figure 4a) indicated that they were related to oxidation products. The relation of the emission region (400–600 nm) of the components 4 and 6 with the oxidation products were reported in previous studies [4,50,53].

PARAFAC components 4 ($\lambda_{ex}/\lambda_{em}$ 300/418 nm) and 6 ($\lambda_{ex}/\lambda_{em}$ 340/450 nm) were represented in a 2D plot, which is shown in Figure 5, to analyze the differences between the four VOOs according to their oxidation state in the course of the storage. The intensity of both components revealed a different oxidation state of the VOOs at the beginning of the storage. Despite the intensity of both components changing over time, the differences of the oxidation state between oils were maintained according to these two components. Thus, four distinguishable groups were observed in the 2D plot associated with the four VOOs. The most remarkable increase of the component intensities was observed in the first months of storage (approximately 0–5 months) (Figure 5). The correlation study of components 4 and 6 with respect to the oxidation indexes (K_{232}, K_{270}, peroxide value) revealed that the best correlation coefficients were found for component 4 and K_{270} ($R = 0.89$, 0.80, 0.93, and 0.81 for oils VOO1, VOO2, VOO3, and VOO4, respectively).

A further study was carried out with a stepwise linear discriminant analysis (SLDA) using all PARAFAC components in order to select which one was the most efficient at discriminating between fresh and aged oils. For this aim, only the samples at the beginning (0–5 months) and at the end (16–21 months) of the experiment were considered as the two classes to be distinguished in the classification model. The selection of the classifying variables (PARAFAC components) included in the model was carried out through F-to-enter and F-to-remove values [55]. The procedure selected components 2 and 4 from the initial six PARAFAC components to build the classification model. These two variables, associated with tocopherols with the contribution of phenols (component 2) and oxidation products (component 4), provided complementary information about the chemical changes that are taking place during storage.

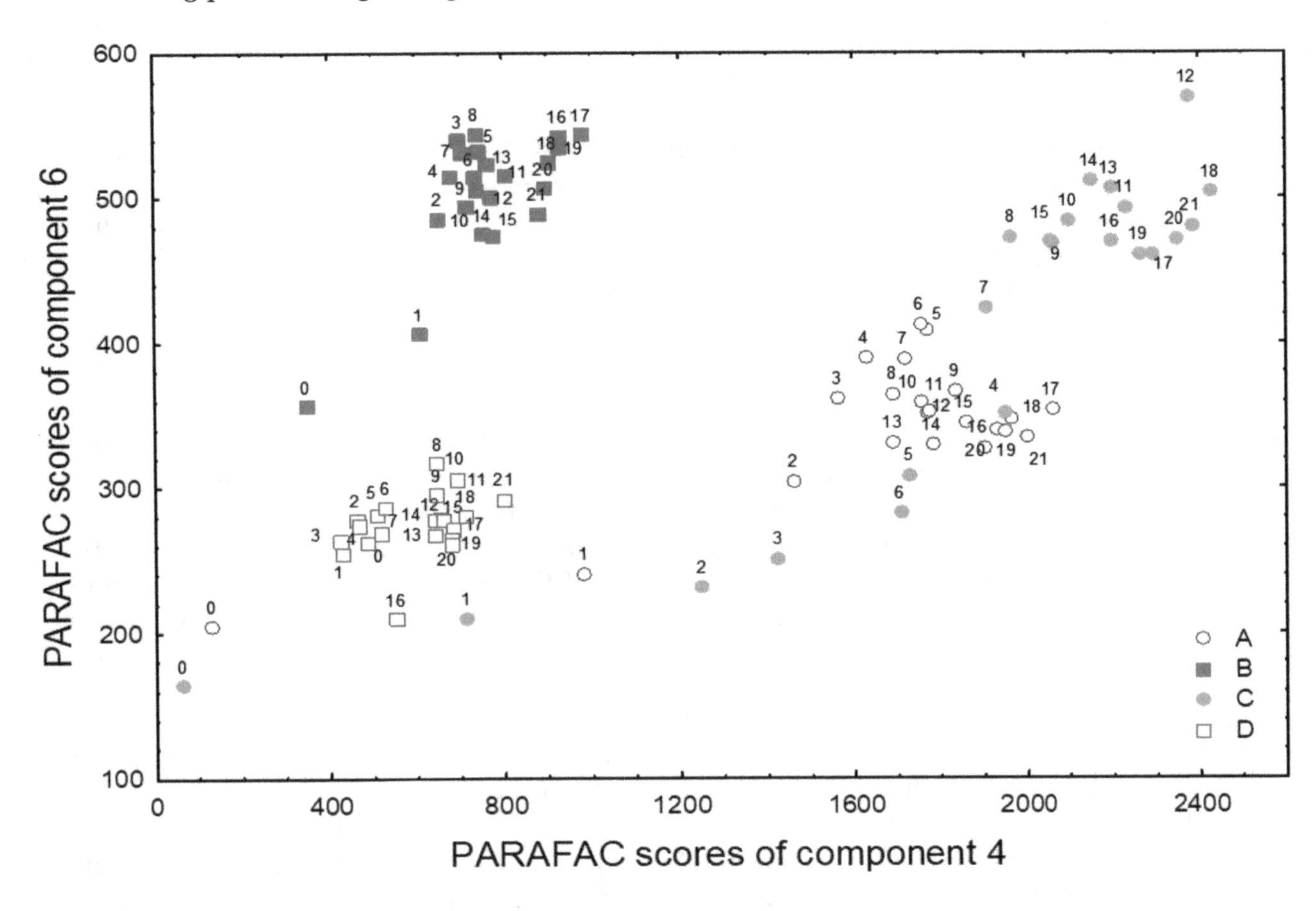

Figure 5. 2D plot of components 4 ($\lambda_{ex}/\lambda_{em}$ 300/418 nm) and 6 ($\lambda_{ex}/\lambda_{em}$ 340/450 nm) for all VOOs during the entire storage time. Codes: A, VOO1 (blue circle); B, VOO2 (red square); C, VOO3 (green circle); D, VOO4 (pink square). The numbers mean the months of the storage when the samples were collected.

3.4. Sensory Quality Changes in the Samples Analyzed by Fluorescence Spectroscopy

The sensory assessment of the fresh oils determined that all VOOs were within "extra virgin olive oil" category according to European regulation [15], except VOO4. This oil was categorized within the "virgin olive oil" category, due to a winey-vinegary defect (Md = 2.1) detected by the panelists before starting the storage. The panelists classified the fresh oils according to their medians of the fruity attribute in the following order: VOO1 (Mf = 4.7) > VOO3 (Mf = 3.8) > VOO2 (Mf = 3.5) > VOO4 (Mf = 3.0).

During storage under moderate conditions, the panelists detected some changes in the flavor of the oils. As it was highlighted in a previous publication, these changes are explained by the changes in the volatile composition during the storage [13]. These changes were enough to lead a change in their categories at different time during the storage. Table S1 (Supplementary Material) shows the changes produced in the median of the defect and fruity attribute during the storage experiment. The median of the fruity attribute decreased in the four oils during the storage experiment. VOO2 showed the fastest decrease, so it was reduced by 57.15% of its initial value during the first five months of storage. Furthermore, the increment of its median of the defect (Md = 1.0) in the fifth month of storage resulted in a downgrading of category to "virgin olive oil", due to a detection of a winey-vinegary defect at this time. VOO3 changed to the "virgin olive oil" category in the mid-term of the storage (tenth storage month) when a winey-vinegary defect (Md = 2.6) was detected by the panelists. VOO1 was the oil that remained unchanged in its category longer, changing to "virgin olive oil" category in the fifteenth month of storage, because an incipient rancid defect (Md = 2.5) was detected. VOO4 was the only oil that reached the "lampante virgin olive oil" category during the storage; the median of the defect reached a value of 3.5 in the eighteenth month.

The study of the physical–chemical parameters during the VOO storage revealed that the category downgrading was due to the increment of K_{270} and the changes in the median of fruity and defect values (Table 1 and Table S1). However, the downgrading of category according to K_{270} occurred in the first 4 months of storage, while the detection of sensory defects occurred at different moments (between 5 and 18 months) depending on the oil. This fact revealed the complexity of quality changes during the storage in which each physical–chemical parameter informed a different aspect of quality. Any analytical method being proposed to control virgin olive oil degradation should consider this complexity. In particular, sensory quality needs special attention due to the discrepancies sometimes found in the sensory assessment results, which is considered by the regulatory bodies as a main problem in the quality assessment of VOO. Furthermore, the sensory quality is the characteristic most appreciated by consumers; therefore, its control during the commercialization of VOO should be extremely important. Although fluorescence spectroscopy determined compounds that were not related to sensory defects, it is important to know if the changes determined by this method occurred before or after sensory defects were clearly detected by panelists and consequently a category downgrading took place. Thus, this information is necessary for a correct interpretation of the results.

Figure 6 shows the median values of the defect and fruity attribute during the storage, together with the time-trend of the PARAFAC components 2 and 4, which were the components previously selected by SLDA. The median of the fruity attribute decreased at the same rate as component 2 (associated with tocopherols and phenols), the correlation coefficients between both variables being 0.70, 0.88, 0.90, 0.52 for VOO1, VOO2, VOO3, and VOO4, respectively. Only the latter showed a correlation coefficient lower than 0.70, probably due to the fact that this oil was already within the "virgin olive oil" category and its median of fruity attribute was the lowest among the studied oils (Mf = 3.0).

In the case of the median of the defect, this variable was related to component 4, associated with oxidation products. In this case, the changes in the median of the defect were marked, unlike the median of fruity attribute, increasing from zero to approximately 2, while the change of component 4 intensity was continuous during the storage. It was also important to note that the increase of the median of the defect could be due to the detection of rancidity, associated with oxidation products, but also to the detection of some fermentative defects, such as the winey-vinegary defect, already existing in the fresh samples and masked by the fruitiness. That explains the low correlation coefficients between the median of the defect and component 4 for the four oils (0.46, 0.67, 0.66, and 0.34 for VOO1, VOO2, VOO3, and VOO4, respectively). In these oils, the increase of the intensity of component 4 up to the plateau was observed before the abrupt change in the median of the defect.

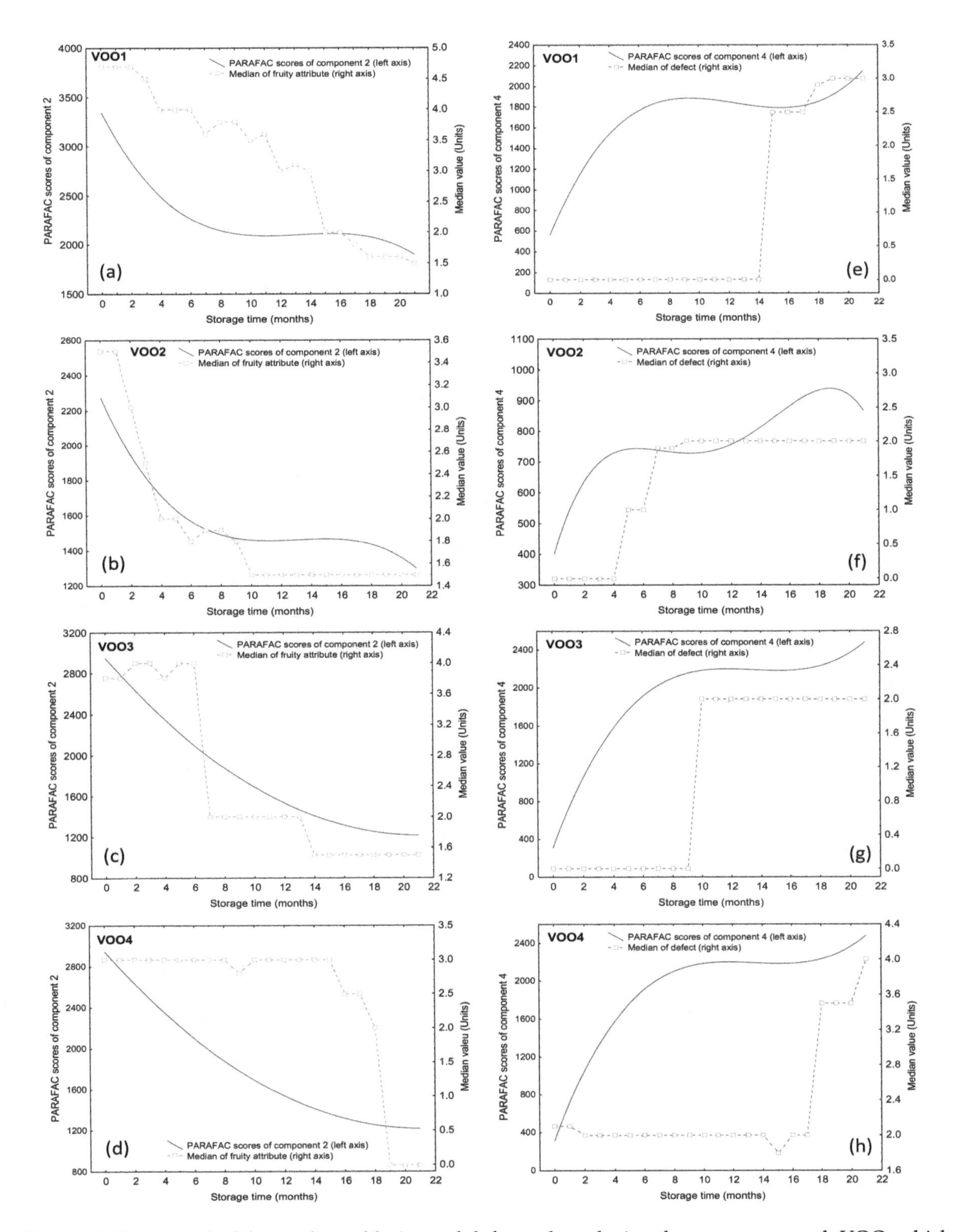

Figure 6. Time-trend of the median of fruity and defect values during the storage per each VOO, which is represented together with the time-trend of the PARAFAC components 2 (**a–d**) and 4 (**e–h**) in a double-y graph, respectively.

4. Conclusions

This study verified the ability of fluorescence spectroscopy to monitor the chemical changes of virgin olive oils during storage under moderate conditions and assessed the relationship with the different quality parameters. In this study, 4 monovarietal VOOs were examined in their stability. The samples were obtained by different producers. The quality and the stability of the samples were influenced by many variables (the state of olive ripeness, the method of extraction, the geographical provenance, and the agricultural practices, among many other factors). This explains that the varietal influence was not so evident for some parameters. For example, the two Arbequina oils presented different phenol concentrations and changes in the sensory characteristics.

The main correlations between the bands identified in the excitation–emission matrices and the quality parameters were found in: The intensity of the band assigned to pigments ($\lambda_{ex}/\lambda_{em}$ 408/678 nm) with the concentration of pheophytin a, the intensity of the band assigned to tocopherols and phenols ($\lambda_{ex}/\lambda_{em}$ 293/322 nm) with the concentration of α-tocopherol, and the intensity of the band assigned to oxidation products ($\lambda_{ex}/\lambda_{em}$ 300/418 nm) with K_{270}. Excitation–emission fluorescence spectroscopy combined with PARAFAC analysis could give information about fluorescent compounds that contribute to the fluorescence emission of VOO, thereby providing a degradation map of the oil. The components extracted by PARAFAC were associated with certain groups of compounds and therefore the observed changes could be interpreted according to the related quality parameters determined in the same oils. Thus, a study of all this information permitted a correct interpretation of the spectra. PARAFAC components 2 and 4 were selected as the best components to distinguish between fresh and aged oils. Both components provided complementary information since they informed on the content of tocopherols with contribution of phenols (component 2) and oxidation products (component 4). Due to the importance of the VOO sensory characteristics for consumer acceptation, and considering that one of the main reasons for downgrading the oils to a lower quality category is the detection of sensory defects in aged oils, the sensory evaluation of the samples was also studied in relation to components 2 and 4 in order to have information of all kinds of degradations. In both components, the changes in their intensity were observed at the same time, or even earlier, than the changes in the medians of the fruity attribute and the defect were determined. This observation could be used as a basis for future studies centered on the interpretation of fluorescent spectra for a practical application in aging control of oils. The challenges ahead should be focused on establishing rules for an easy interpretation of the fluorescence spectra by producers for a daily routine analysis, and also verifying if these rules are dependent on the cultivars. Fluorescence spectroscopy is still scarcely distributed in the labs of olive oil companies, although this technique is affordable and it does not require special training.

Supplementary Materials:
Table S1: Sensory assessment results (medians of the fruity attribute and defect) during the storage experiment for each VOOs. Table S2: Phenolic compounds identified in the VOO samples subjected to storage under moderate conditions. The compounds are grouped according to the excitation wavelength chromatogram where they were registered. Figure S1: An example of chromatogram obtained from the phenol analysis. The chromatograms correspond to VOO1 (fresh oil). The codes are shown in Table S3. Figure S2: An example of chromatogram obtained from the α-tocopherol analysis. The chromatogram corresponds to VOO1 (fresh oil). Figure S3: An example of chromatogram of the degradation products of chlorophyll a obtained in the analysis. The chromatogram corresponds to VOO3 (fresh oil). Figure S4: PARAFAC scores of the sample set (different monovarietal samples during the storage under moderate conditions). Component 1 ($\lambda_{ex}/\lambda_{em}$ 408/678 nm), component 2 ($\lambda_{ex}/\lambda_{em}$ 293/322 nm), component 3 ($\lambda_{ex}/\lambda_{em}$ 408/668 nm), component 4 ($\lambda_{ex}/\lambda_{em}$ 300/418 nm), component 5 ($\lambda_{ex}/\lambda_{em}$ 280/314 nm), and component 6 ($\lambda_{ex}/\lambda_{em}$ 340/450 nm).

Author Contributions: Conceptualization, D.L.G.-G. and E.S.; methodology, E.S., D.L.G.-G., N.T., and R.A.-R.; validation, N.T. and R.A.-R.; formal analysis, A.L.-P., D.L.G.-G., and E.S.; investigation, N.T. and E.S.; resources, N.T., D.L.G.-G., and E.S.; writing—original draft preparation, A.L.-P.; writing—review and editing, N.T., D.L.G.-G., A.L.-P., E.S., and R.A.-R.; visualization, N.T.; supervision, D.L.G.-G. and E.S.; project administration, D.L.G.-G.; funding acquisition, D.L.G.-G. All authors have read and agreed to the published version of the manuscript.

References

1. International Olive Council. *World Olive Oil and Table Olive Figures. World Olive Oil Consumption*; International Olive Council: Madrid, Spain, 2019.

2. Aparicio-Ruiz, R.; Tena, N.; Romero, I.; Aparicio, R.; García-González, D.L.; Morales, M.T. Predicting extra virgin olive oil freshness during storage by fluorescence spectroscopy. *Grasas Aceites* **2017**, *68*, 1–9. [CrossRef]
3. Hernández-Sánchez, N.; Lleó, L.; Ammari, F.; Cuadrado, T.R.; Roger, J.M. Fast fluorescence spectroscopy methodology to monitor the evolution of extra virgin olive oils under illumination. *Food Bioprocess. Technol.* **2017**, *10*, 949–961. [CrossRef]
4. Sikorska, E.; Khmelinskii, I.; Sikorski, M.; Caponio, F.; Bilancia, M.T.; Pasqualone, A.; Gomes, T. Fluorescence spectroscopy in monitoring of extra virgin olive oil during storage. *Int. J. Food Sci. Technol.* **2008**, *43*, 52–61. [CrossRef]
5. Morales, M.T.; Aparicio-Ruiz, R.; Aparicio, R. Chromatographic methodologies: Compounds for olive oil odor issues. In *Handbook of Olive Oil: Analysis and Properties*; Aparicio, R., Harwood, J., Eds.; Springer: Boston, MA, USA, 2013; pp. 261–309.
6. Gonçalves, T.R.; Rosa, L.N.; Torquato, A.S.; da Silva, L.F.O.; Março, P.H.; Gomes, S.T.M.; Matsushita, M.; Valderrama, P. Assessment of brazilian monovarietal olive oil in two different package systems by using data fusion and chemometrics. *Food Anal. Methods* **2019**, *13*, 86–96. [CrossRef]
7. Krichene, D.; Salvador, M.D.; Fregapane, G. Stability of virgin olive oil phenolic compounds during long-term storage (18 months) at temperatures of 5–50 °C. *J. Agric. Food Chem.* **2015**, *63*, 6779–6786. [CrossRef] [PubMed]
8. Lolis, A.; Badeka, A.V.; Kontominas, M.G. Effect of bag-in-box packaging material on quality characteristics of extra virgin olive oil stored under household and abuse temperature conditions. *Food Packag. Shelf Life* **2019**, *21*, 100368. [CrossRef]
9. Méndez, A.I.; Falqué, E. Effect of storage time and container type on the quality of extra-virgin olive oil. *Food Control.* **2007**, *18*, 521–529. [CrossRef]
10. Tena, N.; Lobo-Prieto, A.; Aparicio, R.; García-González, D.L. Storage and preservation of fats and oils. In *Encyclopedia of Food Security and Sustainability*; Ferranti, P., Berry, E., Jock, A., Eds.; Elsevier: Amsterdam, The Netherlands, 2018; pp. 605–618.
11. Trypidis, D.; García-González, D.L.; Lobo-Prieto, A.; Nenadis, N.; Tsimidou, M.Z.; Tena, N. Real time monitoring of the combined effect of chlorophyll content and light filtering packaging on virgin olive oil photo-stability using mesh cell-FTIR spectroscopy. *Food Chem.* **2019**, *295*, 94–100. [CrossRef]
12. Tena, N.; Aparicio, R.; García-González, D.L. Photooxidation effect in liquid lipid matrices: Answers from an innovative FTIR spectroscopy strategy with "mesh Cell" incubation. *J. Agric. Food Chem.* **2018**, *66*, 3541–3549. [CrossRef]
13. Lobo-Prieto, A.; Tena, N.; Aparicio-Ruiz, R.; Morales, M.T.; García-González, D.L. Tracking sensory characteristics of virgin olive oils during storage: Interpretation of their changes from a multiparametric perspective. *Molecules* **2020**, *25*, 1686. [CrossRef]
14. Tena, N.; Aparicio, R.; García-González, D.L. Virgin olive oil stability study by mesh cell-FTIR spectroscopy. *Talanta* **2017**, *167*, 453–461. [CrossRef] [PubMed]
15. European Commission. Commission regulation (EEC) No 2568/91 of 11 July 1991 on the characteristics of olive oil and olive-residue oil and on the relevant methods of analysis. *Offic. J. L* **1991**, *248*, 1–83, updates.
16. International Olive Council. *Standard COI/BPS/Doc. No1. Best Practice Guidelines for the Storage of Olive Oils and Olive-Pomace Oils for Human Consumption*; International Olive Council: Madrid, Spain, 2018.
17. Velasco, J.; Dobarganes, C. Oxidative stability of virgin olive oil. *Eur. J. Lipid Sci. Technol.* **2002**, *104*, 661–676. [CrossRef]
18. García-González, D.L.; Baeten, V.; Fernández Pierna, J.A.; Tena, N. Infrared, raman, and fluorescence spectroscopy: Methodologies and applications. In *Handbook of Olive Oil: Analysis and Properties*; Aparicio, R., Harwood, J., Eds.; Springer: Boston, MA, USA, 2013; pp. 336–383. ISBN 9781461477778.
19. Sikorska, E.; Khmelinskii, I.; Sikorski, M. Vibrational and electronic spectroscopy and chemometrics in analysis of edible oils. In *Methods in Food Analysis*; Cruz, R.M.S., Khmelinskii, I., Vieira, M., Eds.; CRC Press: Boca Raton, FL, USA, 2014; pp. 201–234.
20. Christensen, J.; Nørgaard, L.; Bro, R.; Engelsen, S.B. Multivariate autofluorescence of intact food systems. *Chem. Rev.* **2006**, *106*, 1979–1994. [CrossRef] [PubMed]
21. Sikorska, E.; Khmelinskii, I.; Sikorski, M. Analysis of olive oils by fluorescence Spectroscopy: Methods and applications. In *Olive Oil—Constituents, Quality, Health Properties and Bioconversions*; Boskou, D., Ed.; IntechOpen: London, UK, 2012; pp. 63–88.

22. Ammari, F.; Cordella, C.B.Y.; Boughanmi, N.; Rutledge, D.N. Independent components analysis applied to 3D-front-face fluorescence spectra of edible oils to study the antioxidant effect of Nigella sativa L. extract on the thermal stability of heated oils. *Chemom. Intell. Lab. Syst.* **2012**, *113*, 32–42. [CrossRef]
23. Cabrera-Bañegil, M.; Martín-Vertedor, D.; Boselli, E.; Durán-Merás, I. Control of olive cultivar irrigation by front-face fluorescence excitation-emission matrices in combination with PARAFAC. *J. Food Compos. Anal.* **2018**, *69*, 189–196. [CrossRef]
24. Sikorska, E.; Gliszczyńska-Świgło, A.; Khmelinskii, I.; Sikorski, M. Synchronous fluorescence spectroscopy of edible vegetable oils. Quantification of tocopherols. *J. Agric. Food Chem.* **2005**, *53*, 6988–6994. [CrossRef]
25. Sikorska, E.; Romaniuk, A.; Khmelinskii, I.V.; Herance, R.; Bourdelande, J.L.; Sikorski, M.; Koziol, J. Characterization of edible oils using total luminescence spectroscopy. *J. Fluoresc.* **2004**, *14*, 25–35. [CrossRef]
26. Squeo, G.; Caponio, F.; Paradiso, V.M.; Summo, C.; Pasqualone, A.; Khmelinskii, I.; Sikorska, E. Evaluation of total phenolic content in virgin olive oil using fluorescence excitation–emission spectroscopy coupled with chemometrics. *J. Sci. Food Agric.* **2019**, *99*, 2513–2520. [CrossRef]
27. Sayago, A.; Morales, M.T.; Aparicio, R. Detection of hazelnut oil in virgin olive oil by a spectrofluorimetric method. *Eur. Food Res. Technol.* **2004**, *218*, 480–483. [CrossRef]
28. Sayago, A.; García-González, D.L.; Morales, M.T.; Aparicio, R. Detection of the presence of refined hazelnut oil in refined olive oil by fluorescence spectroscopy. *J. Agric. Food Chem.* **2007**, *55*, 2068–2071. [CrossRef] [PubMed]
29. Dupuy, N.; Le Dréau, Y.; Ollivier, D.; Artaud, J.; Pinatel, C.; Kister, J. Origin of french virgin olive oil registered designation of origins predicted by chemometric analysis of synchronous excitation-emission fluorescence spectra. *J. Agric. Food Chem.* **2005**, *53*, 9361–9368. [CrossRef] [PubMed]
30. International Olive Council. *Standard. COI/T.20/Doc. No 35/Rev.1. Determination of Peroxide Value*; International Olive Council: Madrid, Spain, 2017.
31. International Olive Council. *Standard. COI/T.20/Doc. No 34/Rev. 1. Determination of Free Fatty Acids, Cold Method*; International Olive Council: Madrid, Spain, 2017.
32. International Olive Council. *Standard. COI/T.20/Doc. No 19/Rev. 5. Spectrophotometric Investigation in the Ultraviolet*; International Olive Council: Madrid, Spain, 2019.
33. Asociación Española de Normalización. UNE-EN ISO/IEC 17025. In *Requisitos Generales Para la Competencia de los Laboratorios de Ensayo y Calibración*; Asociación Española de Normalización: Madrid, Spain, 2017.
34. International Olive Council. *Sensory Analysis of Olive Oil. Method for the Organoleptic Assessment of Virgin Olive Oil. COI/T.20/Doc. No 15/Rev. 10*; International Olive Council: Madrid, Spain, 2018.
35. Mateos, R.; Espartero, J.L.; Trujillo, M.; Ríos, J.J.; León-Camacho, M.; Alcudia, F.; Cert, A. Determination of phenols, flavones, and lignans in virgin olive oils by solid-phase extraction and high-performance liquid chromatography with diode array ultraviolet detection. *J. Agric. Food Chem.* **2001**, *49*, 2185–2192. [CrossRef] [PubMed]
36. Aparicio-Ruiz, R.; García-González, D.L.; Oliver-Pozo, C.; Tena, N.; Morales, M.T.; Aparicio, R. Phenolic profile of virgin olive oils with and without sensory defects: Oils with non-oxidative defects exhibit a considerable concentration of phenols. *Eur. J. Lipid Sci. Technol.* **2016**, *118*, 299–307. [CrossRef]
37. International Organization for Standarization. Standard ISO 29841:2014/A1:2016. In *Vegetable Fats and Oils—Determination of the Degradation Products of Chlorophylls a and a' (Pheophytins a, a' and Pyropheophytins)*; International Organization for Standarization: Geneva, Switzerland, 2016.
38. Sievers, G.; Hynninen, P.H. Thin-layer chromatography of chlorophylls and their derivatives on cellulose layers. *J. Chromatogr.* **1977**, *134*, 359–364. [CrossRef]
39. Schwartz, S.J.; Woo, S.L.; von Elbe, J.H. High-Performance Liquid Chromatography of chlorophylls and their derivatives in fresh and processed spinach. *J. Agric. Food Chem.* **1981**, *29*, 533–535. [CrossRef]
40. Andersen, C.M.; Bro, R. Practical aspects of PARAFAC modeling of fluorescence excitation-emission data. *J. Chemom.* **2003**, *17*, 200–215. [CrossRef]
41. Bro, R.; Kiers, H.A.L. A new efficient method for determining the number of components in PARAFAC models. *J. Chemom.* **2003**, *17*, 274–286. [CrossRef]
42. Bendini, A.; Cerretani, L.; Carrasco-Pancorbo, A.; Gómez-Caravaca, A.M.; Segura-Carretero, A.; Fernández-Gutiérrez, A.; Lercker, G. Phenolic molecules in virgin olive oils: A survey of their sensory properties, health effects, antioxidant activity and analytical methods. An overview of the last decade. *Molecules* **2007**, *12*, 1679–1719. [CrossRef]

43. Psomiadou, E.; Tsimidou, M. Stability of virgin olive oil. 1. Autoxidation studies. *J. Agric. Food Chem.* **2002**, *50*, 716–721. [CrossRef]
44. Velasco, J.; Andersen, M.L.; Skibsted, L.H. Evaluation of oxidative stability of vegetable oils by monitoring the tendency to radical formation. A comparison of electron spin resonance spectroscopy with the Rancimat method and differential scanning calorimetry. *Food Chem.* **2004**, *85*, 623–632. [CrossRef]
45. Aparicio-Ruiz, R.; Roca, M.; Gandul-Rojas, B. Mathematical model to predict the formation of pyropheophytin a in virgin olive oil during storage. *J. Agric. Food Chem.* **2012**, *60*, 7040–7049. [CrossRef] [PubMed]
46. Tena, N.; Aparicio, R.; García-González, D.L. Chemical changes of thermoxidized virgin olive oil determined by excitation-emission fluorescence spectroscopy (EEFS). *Food Res. Int.* **2012**, *45*, 103–108. [CrossRef]
47. Galeano Díaz, T.; Durán Merás, I.; Correa, C.A.; Roldán, B.; Rodríguez Cáceres, M.I. Simultaneous fluorometric determination of chlorophylls a and b and pheophytins a and b in olive oil by partial least-squares calibration. *J. Agric. Food Chem.* **2003**, *51*, 6934–6940. [CrossRef] [PubMed]
48. Cheikhousman, R.; Zude, M.; Bouveresse, D.J.R.; Léger, C.L.; Rutledge, D.N.; Birlouez-Aragon, I. Fluorescence spectroscopy for monitoring deterioration of extra virgin olive oil during heating. *Anal. Bioanal. Chem.* **2005**, *382*, 1438–1443. [CrossRef] [PubMed]
49. Baltazar, P.; Hernández-Sánchez, N.; Diezma, B.; Lleó, L. Development of rapid extra virgin olive oil quality assessment procedures based on spectroscopic techniques. *Agronomy* **2020**, *10*, 41. [CrossRef]
50. Giungato, P.; Aveni, M.; Rana, F.; Notarnicola, L. Modifications induced by extra virgin olive oil frying processes. *Ind. Aliment.* **2006**, *45*, 148–154.
51. Lozano, V.A.; Muñoz de la Peña, A.; Durán-Merás, I.; Espinosa Mansilla, A.; Escandar, G.M. Four-way multivariate calibration using ultra-fast high-performance liquid chromatography with fluorescence excitation-emission detection. Application to the direct analysis of chlorophylls a and b and pheophytins a and b in olive oils. *Chemom. Intell. Lab. Syst.* **2013**, *125*, 121–131. [CrossRef]
52. Díaz, G.; Pega, J.; Primrose, D.; Sancho, A.M.; Nanni, M. Effect of light exposure on functional compounds of monovarietal extra virgin olive oils and oil mixes during early storage as evaluated by fluorescence spectra. *Food Anal. Methods* **2019**, *12*, 2709–2718. [CrossRef]
53. Guzmán, E.; Baeten, V.; Pierna, J.A.F.; García-Mesa, J.A. Evaluation of the overall quality of olive oil using fluorescence spectroscopy. *Food Chem.* **2015**, *173*, 927–934. [CrossRef]
54. Mishra, P.; Lleó, L.; Cuadrado, T.; Ruiz-Altisent, M.; Hernández-Sánchez, N. Monitoring oxidation changes in commercial extra virgin olive oils with fluorescence spectroscopy-based prototype. *Eur. Food Res. Technol.* **2018**, *244*, 565–575. [CrossRef]
55. Aparicio, R.; García-González, D.L. Olive oil characterization and traceability. In *Handbook of Olive Oil: Analysis and Properties*; Aparico, R., Harwood, J., Eds.; Springer: Boston, MA, USA, 2013; pp. 431–472. ISBN 9781461477778.

4

Comparison and Identification for Rhizomes and Leaves of *Paris yunnanensis* based on Fourier Transform Mid-Infrared Spectroscopy Combined with Chemometrics

Yi-Fei Pei [1,2], Qing-Zhi Zhang [2], Zhi-Tian Zuo [1,*] and Yuan-Zhong Wang [1,*]

1 Institute of Medicinal Plants, Yunnan Academy of Agricultural Sciences, Kunming 650200, China; feifei950222@gmail.com
2 College of Traditional Chinese Medicine, Yunnan University of Traditional Chinese Medicine, Kunming 650500, China; ynkzqz@126.com
* Correspondence: yaaszztian@126.com (Z.-T.Z.); boletus@126.com (Y.-Z.W.)

Academic Editor: Marcello Locatelli

Abstract: *Paris polyphylla*, as a traditional herb with long history, has been widely used to treat diseases in multiple nationalities of China. Nevertheless, the quality of *P. yunnanensis* fluctuates among from different geographical origins, so that a fast and accurate classification method was necessary for establishment. In our study, the geographical origin identification of 462 *P. yunnanensis* rhizome and leaf samples from Kunming, Yuxi, Chuxiong, Dali, Lijiang, and Honghe were analyzed by Fourier transform mid infrared (FT-MIR) spectra, combined with partial least squares discriminant analysis (PLS-DA), random forest (RF), and hierarchical cluster analysis (HCA) methods. The obvious cluster tendency of rhizomes and leaves FT-MIR spectra was displayed by principal component analysis (PCA). The distribution of the variable importance for the projection (VIP) was more uniform than the important variables obtained by RF, while PLS-DA models obtained higher classification abilities. Hence, a PLS-DA model was more suitably used to classify the different geographical origins of *P. yunnanensis* than the RF model. Additionally, the clustering results of different geographical origins obtained by HCA dendrograms also proved the chemical information difference between rhizomes and leaves. The identification performances of PLS-DA and the RF models of leaves FT-MIR matrixes were better than those of rhizomes datasets. In addition, the model classification abilities of combination datasets were higher than the individual matrixes of rhizomes and leaves spectra. Our study provides a reference to the rational utilization of resources, as well as a fast and accurate identification research for *P. yunnanensis* samples.

Keywords: *Paris polyphylla* Smith var. *yunnanensis*; multivariate analysis; chemometrics; Fourier transform infrared

---

1. Introduction

The perennial herb plant *Paris* is a genus in the Liliaceae family. *Paris* is one of more than 2000 medicinal plants described in the Chinese Pharmacopoeia (2015 edition), and it has utmost important medicinal effects on treating diseases, including snake bite and insect sting, innominate toxin swelling, and various inflammatory and traumatic injuries with ancient history in China. In addition, *Paris* is also used as an ethnobotanical medicinal herb in Nepal and India, which export *Paris* raw materials every year to China to meet the Chinese traditional medicine (TCM) market demand [1]. *Paris* medicinal plants sold in today's TCM markets were both of wild and cultivated types, with the number of wild *Paris* gradually decreasing, with a long-term growth cycle, immoderate harvesting, and huge commercial activities [2]. Additionally, amongst almost 28 species and varieties of *Paris*, only *Paris polyphylla* Smith var. *chinensis* (Franch.) Hara (*P. chinensis*) and *P. polyphylla* var. *yunnanensis* (Franch.) Hand. -Mazz (*P. yunnanensis*) are officially described by the Chinese Pharmacopoeia (2015 edition), which further restricted the number of *Paris* medicinal plants [3–5]. Hence, substitutes with similar medicinal effects and chemical compounds are considered for selection from the closely related species of *P. yunnanensis* and *P. chinensis*, and other parts of the plants, such as stems and leaves.

A serious problem is that many number of leaves of *Paris* medicinal plants were abandoned every year, with the rhizomes being unable to meet the market demand. Thus, the use of *P. yunnanensis* and *P. chinensis* leaves as substitutes for the primary choice was to be considered. Currently, Qin et al. have reviewed the feasibility for whether renewable above-ground parts (leaves and stems) of *P. yunnanensis* could be used as an alternative source to rhizomes [6]. They concluded that the above-ground parts can be the substitute source for the rhizomes of *P. yunnanensis*, in that similar pharmacological properties, including antimicrobial, hemostatic, cytotoxic, and other effects. A variety of quality of Paridis Rhizomes in TCM markets may affect the quality of Chinese patent medicines based on *P. yunnanensis* rhizomes. On these basis, it is necessary and meaningful to quickly assess the quality of *P. yunnanensis* rhizomes and leaves.

Yunnan possesses complex climatic conditions, which means that the quality of TCM plants varies with different climatic conditions of different geographical origins in Yunnan. A variety of analytic techniques have been applied to determine the active chemical components and fingerprints to assess the quality of *P. yunnanensis* samples, including ultra-high performance liquid chromatography-mass spectrometry (UHPLC-MS) [7,8], ultraviolet-visible (UV-Vis) [9], high performance liquid chromatography (HPLC) [10], and Fourier transform mid infrared (FT-MIR) [8,11,12], and so on. Up to now, chemometrics has been widely applied to herbal medicines and plant spectral analyses [13,14]. For example, principal component analysis (PCA) often was used to research Chinese herbal medicines of multiple tissues and geographical origins [15,16]. Partial least squares discriminant analysis (PLS-DA) and random forest (RF) have been gradually applied to the field of traditional Chinese herbs in recent years, such as *Panax notoginseng*, *Dendrubium officinale*, etc. [17,18]. Our previous studies have demonstrated that all of these techniques have obtained better identification abilities for *P. yunnanensis* from different geographical origins. Compared with chromatography, the better classification abilities, more convenience, and time-saving techniques were displayed using spectroscopy techniques. To date, combined with various analytical techniques, chemometrics methods have been successfully applied to assess *P. yunnanensis* samples with better classification and identification abilities, including support vector machine [19], RF [11,12], hierarchical cluster analysis (HCA) [10,12,20,21], PLS-DA [9,12,22], and PCA [9,12,22]. However, they failed to analyze other parts of *P. yunnanensis* to fast assess their quality, as well as comparing and combining rhizomes and leaves to identify *P. yunnanensis* from a variety of geographical origins. Hence, the purpose of our study is to assess the quality of *P. yunnanensis* medicinal materials by determining their rhizomes and leaves in FT-MIR spectra, combined with chemometrics.

In this study, to further obtain better, faster, and reliable identification methods for *P. yunnanensis* raw materials from different geographical origins, we investigated *P. yunnanensis* samples from six regions from Yunnan Province by FT-MIR spectroscopy, combined with four chemometrics methods, including PCA, PLS-DA, RF, and HCA. The influence on the fast-quality assessment effects of different parts, including leaves and rhizomes of *P. yunnanensis* were compared. The results may demonstrate the importance of the leaves of *P. yunnanensis*, and they can provide direction for the future development of *P. yunnanensis* medicinal plants.

2. Results and Discussion

2.1. Comparison Analysis between Rhizomes and Leaves

The raw and SD FT-MIR spectra of rhizomes and leaves of *P. yunnanensis* samples from six geographical regions are showed in Figure 1. The peaks height, character, and position among different geographical regions are showed in Figure 1. The peaks height, character, and position among different geographical origins samples are similarly shown in Figure 1a. Characteristic peaks appeared at ~3328 cm^{-1}, and were assigned to O–H absorption, at ~2726, 1414, and 1370 cm^{-1} to methylene and methyl stretching, and bending vibration. Absorption at ~1742 cm^{-1} was endorsed to C=O stretching vibration, at ~1650 cm^{-1} it was attributed to C=C and C=O stretching vibration, which may be attributed to oils, saccharides, steroid saponins, and flavonoids. Besides, absorption at ~1244 cm^{-1} was assigned to C–O stretching vibration, while ~1151, 1078, and 1020 cm^{-1} were endorsed to C–C, C–O stretching vibration and C–OH bending vibration, as well as the main attribute to saccharides and glycosides. Absorption at ~929 cm^{-1} was assigned to the sugar skeleton. These attributes for characteristic peaks were in accordance with studies by Sun et al. and Yang et al. [23,24]. Absorption at ~2855, 1547, 1340, 862, 765, 708, 611, and 580 cm^{-1} were also showed in these FT-MIR spectra. Absorption at ~1650 cm^{-1} and ~1020 cm^{-1} were the key peaks among all absorption peaks of the raw FT-MIR spectra of rhizomes. Additionally, many details of spectral information were shown by standard normal variate–second derivative (SNV-SD) FT-MIR rhizomes spectra in Figure 1c. In detail, among the peaks regions of 1200–900 cm^{-1}, the peaks absorptions were at 1173, 1135, 1093, 1065, 1050, 1035, 996, 976, and 950 cm^{-1}, which are not shown in the raw FT-MIR spectra of rhizomes.

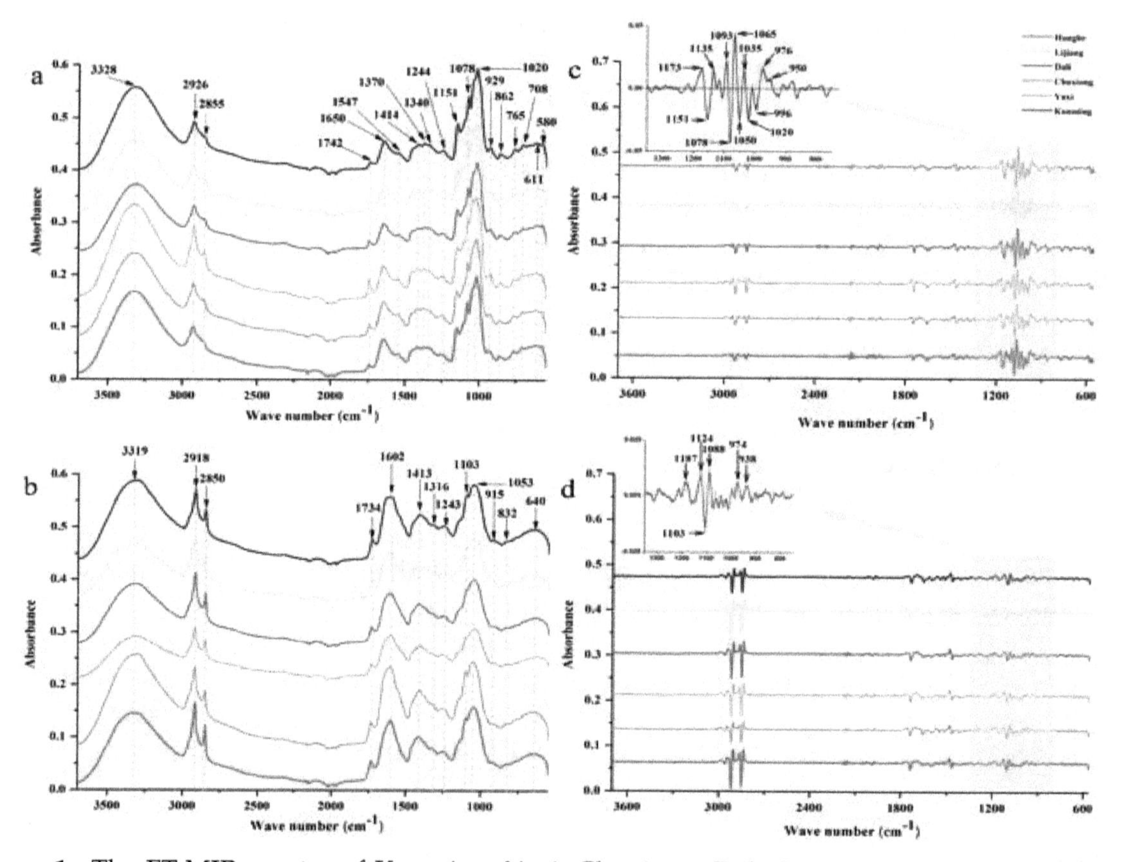

Figure 1. The FT-MIR spectra of Kunming, Yuxi, Chuxiong, Dali, Lijiang, and Honghe, Yunnan: (**a**) the raw spectra of rhizomes, (**b**) the raw spectra of leaves, (**c**) the best preprocessing spectra of rhizomes, (**d**) the best preprocessing spectra of leaves.

The raw FT-MIR spectra of leaves showed different peak heights, characters and positions and numbers of the characteristic peaks for those of rhizomes, which are shown in Figure 1b. Compared with the raw rhizomes FT-MIR spectra, absorption for the raw leaves spectra exhibited a red-shift at

1750–1290 cm^{-1}, and a blue-shift at 1290–950 cm^{-1}. In other words, various differences of chemical information was reflected by the raw rhizomes and leaf FT-MIR spectra. Similar to the raw rhizome FT-MIR spectra, the absorption was mainly attributed to oils, saccharides, steroid saponins, flavonoids saccharides, and glycosides. Namely, absorption at 1602 cm^{-1} and 1053 cm^{-1} are the two key peaks of the raw leaf FT-MIR spectra. Similarly, certain details from the spectral information are shown in SNV-SD leaf FT-MIR spectra in Figure 1d. In detail, among peaks regions of 1200–900 cm^{-1}, the peak absorptions at 1187, 1124, 1088, 974, and 938 cm^{-1} are proven, which are not shown in the FT-MIR spectra of raw leaves.

The PCA score plot and loading plot based on the total FT-MIR spectra are shown in Figure 2. Besides, 72.9% and 17% FT-MIR spectra information were exhibited by PC 1 and PC 2, respectively. Two parts (rhizomes and leaves) were well separated by the first two principal components (PCs) in the PCA score plot. Absorption at 1300–550 cm^{-1} by PC 1 contributed to a higher importance than that of PC 2. In other words, the bands of this region are more important to PC 2.

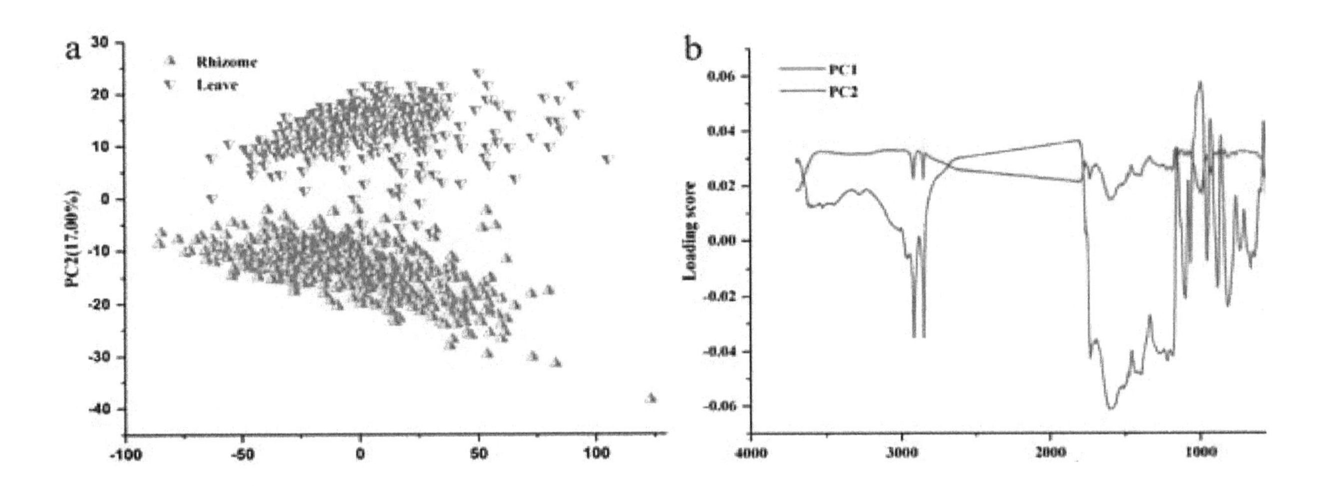

Figure 2. Principal component analysis (PCA) result based on Fourier transform mid infrared (FT-MIR) spectra: (**a**) Score plot, (**b**) Loading plot.

2.2. *Origin Traceability Based on Chemometrics*

2.2.1. Using Rhizome FT-MIR Spectra Datasets

Raw FT-MIR rhizomes spectra were pretreated by SNV, standard normal variate-first-derivative (SNV-FD), SNV-SD, and SD preprocessing methods, and to select the best pretreatment method. All parameters for these pretreatment methods are shown in Table S1. Comparing parameters to the raw FT-MIR spectra, all parameters are better after preprocessing. Among them, SNV-SD was defined as the optimal preprocessing method for the fundamental for the larger values of cumulative interpretation ability (R^2), cumulative prediction ability (Q^2), and accuracy of the calibration set, as well as the lower values of the root mean square error of estimation (RMSEE) and the root mean square error of cross-validation (RMSECV). Despite SD obtaining a better accuracy, SNV-SD obtained a lower RMSEE, RMSECV, and latent variables (LVs). In our following study for rhizomes, models established by raw and the best preprocessing (SNV-SD) FT-MIR spectra data will be compared.

The variable importance for the projection (VIP) scores for values greater than 1 of the raw rhizome FT-MIR data are shown in Figure 3a. The regions of 1750–1500 cm^{-1} and 1200–750 cm^{-1} are important variables regions for differentiating six geographical origins of *P. yunnanensis* by FT-MIR spectra. The bonds at 1750–1500 cm^{-1} are mainly attributed to oils, saccharides, steroid saponins, and flavonoids compounds. Besides, the bands at 1200–750 cm^{-1} are mainly endorsed to saccharides and glycosides compounds. The two key peaks of raw rhizome FT-MIR spectra were contained in these two bands. What's more, there were also some peaks that were not clearly identified, and these peaks are equally important for the identification of *P. yunnanensis* samples from different origins. On the basis of the SNV-SD rhizome FT-MIR data, the VIP scores for values greater than 1 are shown in Figure 3b. The degrees of important variables regions from 1750–750 cm^{-1} seem to be similar in importance for the differentiation of six geographical origins of *P. yunnanensis* by FT-MIR spectra. It was further demonstrated that each peak was important for distinguishing *P. yunnanensis* samples from different geographical origins.

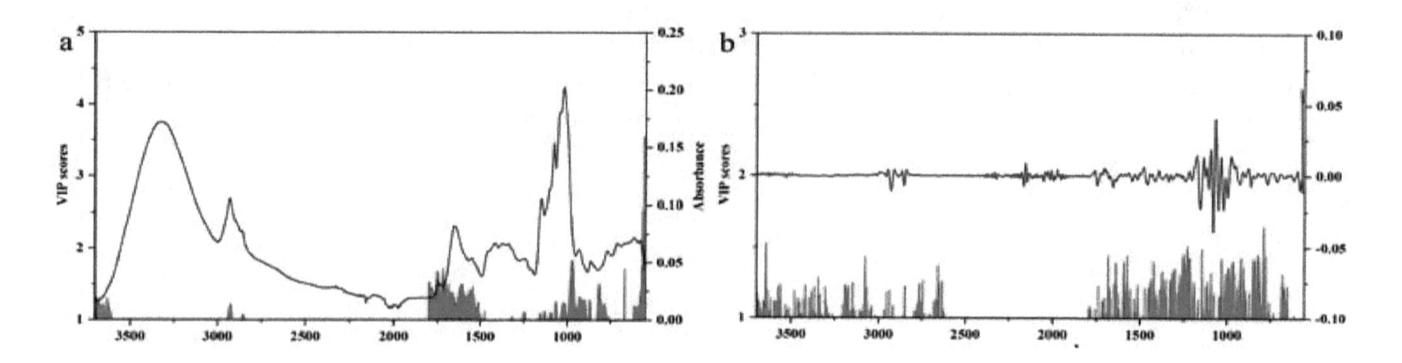

Figure 3. Variable importance for the projection (VIP) scores of the FT-MIR data of rhizomes for regional differences: (**a**) raw dataset, (**b**) standard normal variate–second derivative (SNV-SD) dataset.

RF models were established on raw and SNV-SD rhizome FT-MIR spectra data matrixes. The 1207 and 1202 variables were contained in raw and SNV-SD rhizome FT-MIR spectra datasets, respectively. For the two RF models of raw and SNV-SD rhizomes FT-MIR spectra, the initial number of trees (n_{tree}) were set as 2000 trees. The suitable value of n_{tree} was selected, based on the lowest total value, and the need to be assured of the lower values of the most classes. The 1328–1392 trees and 650–740 trees are the lowest ranges for n_{tree} of raw and SNV-SD rhizomes datasets, respectively, which are shown in Figure 4a,b. Besides, the optimal values 1383 and 951 trees were obtained for further selection of the suitable number of variable (m_{try}) values of the RF models, based on raw and SNV-SD rhizomes FT-MIR datasets, respectively. As shown in Figure 4c,d, the optimal m_{try} were calculated to be 33 and 36, according to the lowest out-of-bag (OOB) values for the raw and SNV-SD datasets, respectively. The suitable n_{tree}, combined with the optimal m_{try}, were used to select the most important variables.

To start with, all variables of the raw and SNV-SD datasets were sorted from the least important variables, to the most important variables, respectively. The 10-fold cross validation error rates of the RF model, based on raw and SNV-SD FT-MIR datasets of rhizomes *P. yunnanensis* samples are shown in Figure 5a,b. It was reduced sequentially by five variables for each step for the initial variables of 1207 and 1202, for raw and SNV-SD datasets, respectively. In both the range of 1–1207 and 1–1202 variables numbers, all important variables were divided into three regions. Among these regions, the 10-fold cross validation error rate values showed a reduced or incremental trend. When the 10-fold cross validation error rate shows a drop trend and then an upward trend, that number of variables at the turning point is likely to be the optimal number for the most importance variables. Hence, variable numbers of 207 and 292 with a lower than 10-fold cross validation error rate for 0.34202 and 0.08143 were selected, to establish the RF models of raw and SNV-SD rhizome FT-MIR spectra, respectively.

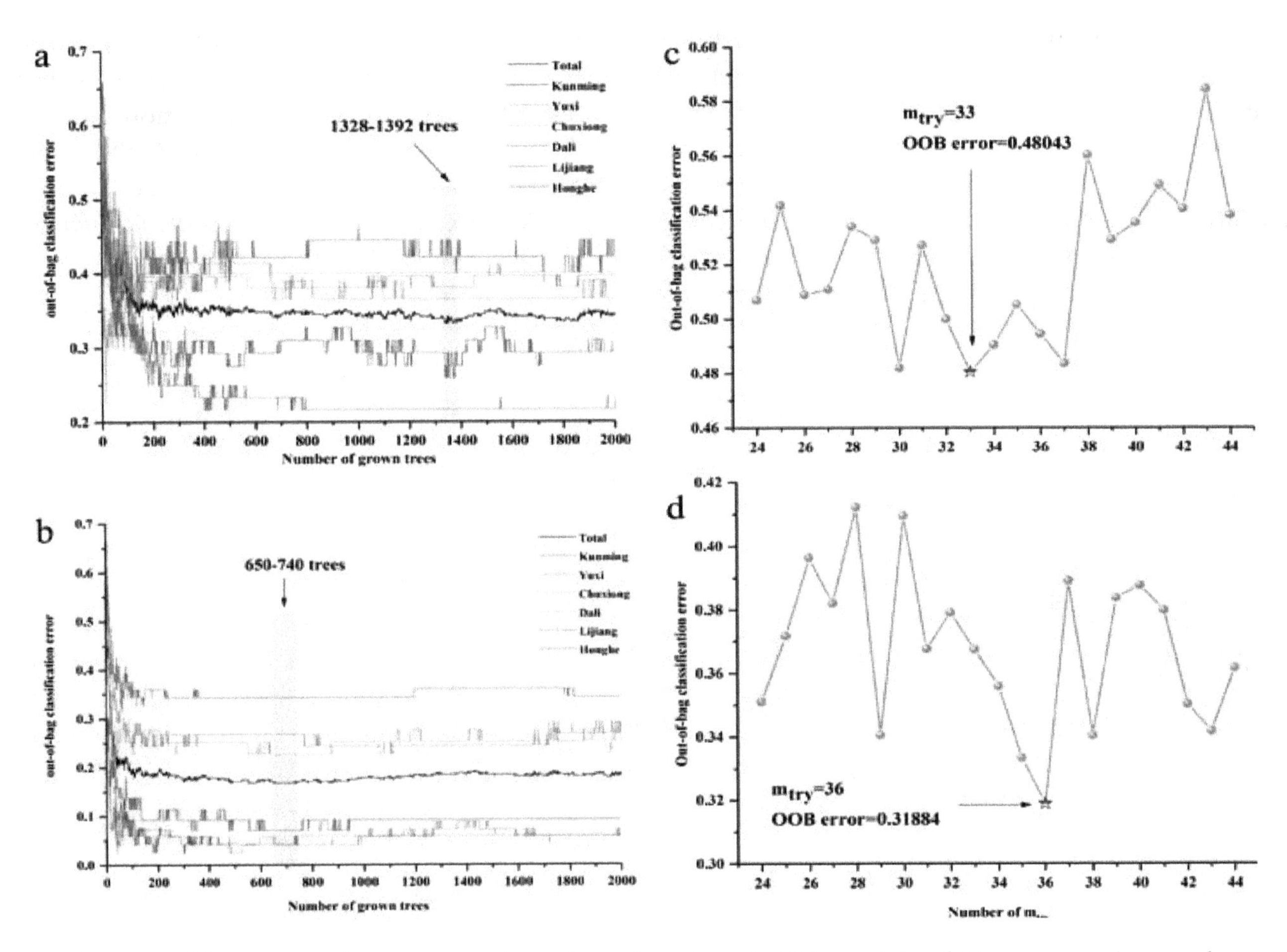

Figure 4. The n_{tree} and m_{try} screening of random forest (RF) models of *P. yunnanensis* samples before variables ranked by permutation accuracy importance: (**a**) n_{tree} of the raw rhizomes dataset, (**b**) n_{tree} of the SNV-SD rhizomes dataset, (**c**) m_{try} of the raw rhizomes dataset, (**d**) m_{try} of the SNV-SD rhizomes dataset.

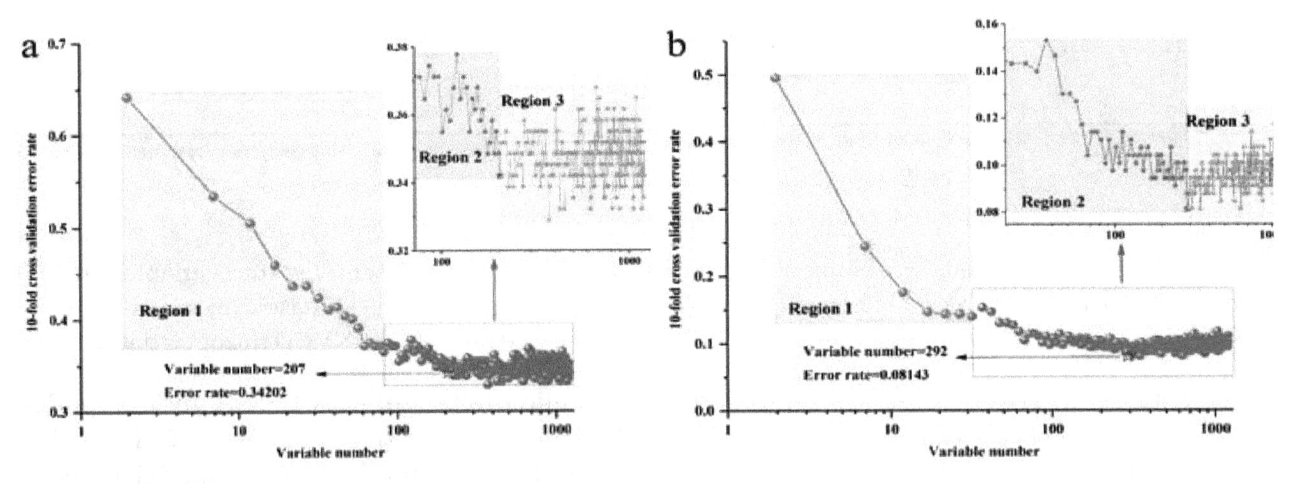

Figure 5. The 10-fold cross validation error rates of the RF model (sequentially reduce each five variables) based on *P. yunnanensis* samples: (**a**) raw rhizomes dataset, (**b**) SNV-SD rhizomes dataset.

When the most important variables were re-selected, forming the new data matrix, it was necessary for the reconstruction of optimal n_{tree} and m_{try} values for raw and SNV-SD FT-MIR spectra. The selecting process was the same as above. As shown in Figure 6, the 1011–1201 trees and 788–880 trees are the lowest ranges for n_{tree} of raw and SNV-SD rhizomes dataset, respectively. Finally, 1110 and 820 trees are selected for the optimal n_{tree}, as well as 19 and 26, are selected for the best m_{try} of raw and SNV-SD FT-MIR rhizome spectra of *P. yunnanensis* samples, respectively. These optimal n_{tree} and m_{try} were used to establish the RF model, and they obtained the accuracy of the calibration set and the validation set, respectively. It is undeniable that the variable selection process is important. The error rate for calibration set of raw datasets was reducing from 36.16% to 33.88%, and it was decreasing from 10.42% to 8.79% for the SNV-SD dataset. In addition, the geographical origin classification ability of the RF model, based on SNV-SD FT-MIR spectra of rhizome *P. yunnanensis* samples, was significantly better than that of the raw spectra.

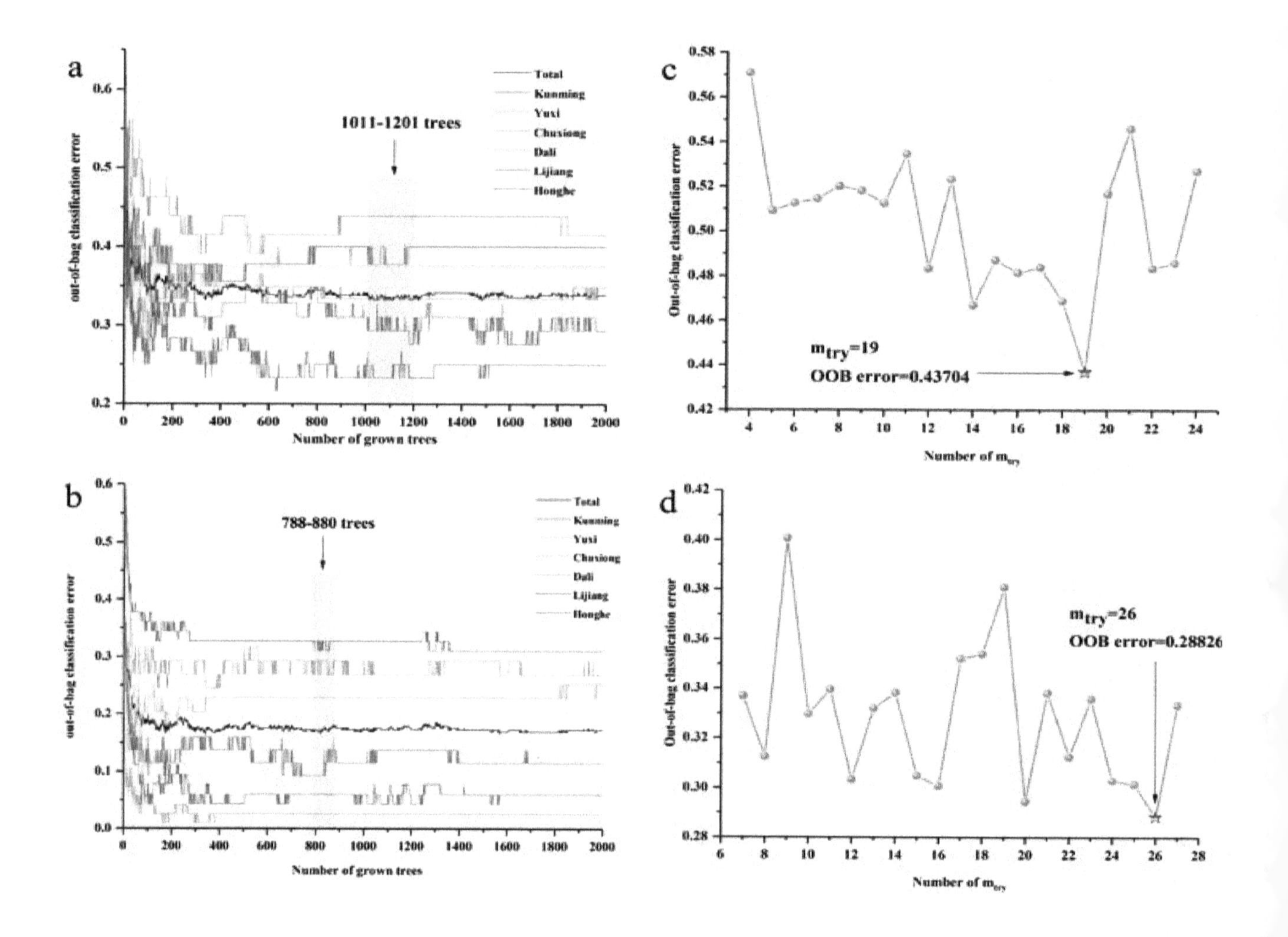

Figure 6. The n_{tree} and m_{try} screening of RF models of the *P. yunnanensis* samples after variables are ranked by permutation accuracy importance: (**a**) n_{tree} of the raw rhizomes dataset, (**b**) n_{tree} of the SNV-SD rhizomes dataset, (**c**) m_{try} of the raw rhizomes dataset, (**d**) m_{try} of the SNV-SD rhizomes dataset.

The parameters for each class of calibration set and validation set of the PLS-DA and RF models, based on raw and SNV-SD rhizomes FT-MIR spectra data matrixes, are shown in Table S2. The values for all parameters of each class of calibration set and the validation set for the PLS-DA model, based on raw FT-MIR data matrixes, were higher than that of the RF model, and they differ greatly. Additionally, all parameters for the RF model based on SNV-SD FT-MIR data matrixes were greatly enhanced and close to that of the PLS-DA model. Obviously, the parameters of two models for the SNV-SD data matrixes based on FT-MIR spectra were higher than those of raw data matrixes. However, the identification abilities and accuracy for two models based on rhizome FT-MIR spectra were needed for improvement.

2.2.2. Using Leaf FT-MIR Spectra Datasets

Raw leaf FT-MIR spectra dataset was preprocessed by SNV, SNV-FD, SNV-SD, and SD preprocessing methods to select the best pretreatment method. All parameters for these four kinds of preprocessing methods are displayed in Table S3. Similarity to rhizomes, all parameters for the preprocessed model of FT-MIR spectra for leaves are better than those of the raw data matrix. Besides, the SNV-SD pretreatment among all preprocessing methods was the best one for classifying the different origins of *P. yunnanensis* leaf samples, which possessed values of R^2, Q^2, RMSEE, RMSECV, accuracy and LVs that were more satisfactory than other pretreatment methods. For the following study of leaves, models established by raw data, and the best pretreatment (SNV-SD) FT-MIR spectra data were selected to study.

The VIP scores for values greater than 1 of the raw leaf FT-MIR data are shown in Figure S1a. The region of 1800–1700 cm^{-1} is the most important variable region for differentiating six geographical origins of *P. yunnanensis* by leaf FT-MIR spectra. The regions of 1700–1300 cm^{-1}, 1250–1100 cm^{-1}, and 1200–750 cm^{-1} almost possessed equally important degrees for differentiating various geographical origins of *P. yunnanensis* by leaf FT-MIR spectra. The bonds at these regions are also mainly assigned to oils, saccharides, steroid saponins, and flavonoids, saccharides, and glycoside compounds. What's more, the number of important variables of leaf VIP scores were more than those of rhizome VIP scores, which reflected the difference in chemical information in classifying *P. yunnanensis* samples from different regions. Based on the SNV-SD leaf FT-MIR data, the VIP scores for values greater than 1 are displayed in Figure S1b. Compared with the other three regions, variables important for the region of 1800–1700 cm^{-1} show greater importance. Similar, it was also demonstrated that each peak of leaf FT-MIR spectra was important to distinguish *P. yunnanensis* samples from a variety of geographical origins. However, a number of peaks were non-identified chemical compounds in the leaf FT-MIR spectra.

RF models were established on raw and SNV-SD leaf FT-MIR spectra datasets. To start with, the 1207 and 1201 variables were contained in the raw and SNV-SD leaf FT-MIR spectra matrixes, respectively. Similar to the rhizomes, the initial n_{tree} were set as 2000 trees for the RF models of raw and SNV-SD leaf FT-MIR spectra. As shown in Figure S2a,b, 947–961 trees and 980–1008 trees were selected to be the lowest ranges for n_{tree} of raw and SNV-SD leaf datasets, respectively. Additionally, the optimal values of 951 and 982 trees were selected for further selection of the suitable m_{try} values of RF models, based on raw and SNV-SD leaf FT-MIR datasets, respectively. As shown in Figure S2c,d, the optimal m_{try} were calculated to be 42 and 31, respectively.

Like rhizomes, all variables of the raw and SNV-SD matrixes of leaves were ranked from to the least important variables to the most important variables, respectively. The 10-fold cross-validation error rates of the RF model, based on the raw and SNV-SD FT-MIR datasets of leaf *P. yunnanensis* samples are shown in Figure S3a,b. In addition, in both the range of 1–1207 and 1–1201 variables numbers, all important variables, were also divided into three regions. Moreover, variable numbers of 157 and 441 with lower than 10-fold cross validation error rates for 0.36808 and 0.02280 were selected to establish the RF models of the raw and SNV-SD FT-MIR spectra, respectively. The 10-fold cross validation error rate of the SNV-SD matrix was far below that of the raw dataset.

Similar to rhizomes, the most important variables were as the new data matrixes, and meanwhile, the optimal n_{tree} and m_{try} values for raw and SNV-SD datasets were re-selected, respectively. The selection process was the same as above. As shown in Figure S4, the 1527–1607 trees and 898–966 trees were the lowest ranges for n_{tree} of raw and SNV-SD leaf datasets, respectively. Then, 1570 and 900 trees were selected for the optimal n_{tree}, as well as 18, and 18 were selected for the best m_{try} of the raw and SNV-SD datasets, respectively. Furthermore, these optimal n_{tree} and m_{try} were used to establish high-performance RF models. The error rate for the calibration set of raw datasets was reduced from 40.07% to 38.11%, and it decreased from 3.26% to 2.93% for the SNV-SD dataset. In addition, not only was the geographical origin classification ability of the RF model based on SNV-SD FT-MIR leaves spectra significantly better than that of the raw spectra, but higher performances were also obtained by the RF models of leaves than those of rhizomes.

Parameters of sensitivity (SENS), specificity (SPEC), accuracy (ACC), and the Matthews correlation coefficient (MCC) for each class of calibration set and validation set of PLS-DA and RF model, based on raw and SNV-SD leaf FT-MIR spectra data matrices are displayed in Table S4. Similar to the performance of parameters for the models of rhizomes, the values for all parameters of each class of calibration set and validation set for the PLS-DA model, based on raw leaf FT-MIR data matrices, were higher than that of the RF model. The identification ability of the SNV-SD PLS-DA model of the leaf data matrix almost reached the best ratings, and only samples collected from Yuxi and Dali were misclassified. Additionally, all parameters of validation set for the RF model based on the SNV-SD FT-MIR data matrixes were close, to the best, and only samples collected from Dali and Lijiang were misclassified. Additionally, parameters of two models for the SNV-SD data matrices based on FT-MIR spectra were higher than those of raw data matrixes. However, the classification performance for the PLS-DA and RF models on the basis of the leaf FT-MIR spectra were required for enhancement.

2.3. Regional Differences between VIP and Important Variables

The VIP and important variables of the RF and PLS-DA models of *P. yunnanensis* samples are displayed in Figure 7. In detail, Figure 7a,b are based on the raw FT-MIR spectra of rhizomes and leaves, respectively. Figure 7c,d are based on the SNV-SD FT-MIR spectra of rhizomes and leaves, respectively. The important variable numbers of the RF model of raw datasets for rhizomes and leaves were far more than those of the SNV-SD RF models. The variables with VIP values greater than 1 showed greater concentrations for several regions in the VIP scores based on raw rhizome and leaf matrixes, than those of the VIP scores of the SNV-SD datasets. From a comparison of the scatter of the most important variables between rhizomes and leaves, the number and distribution of important variables are different. It was demonstrated that various and different chemical profiles were contained between the rhizomes and leaves of *P. yunnanensis*. From the higher accuracy rate and the more uniform distribution of important variables of rhizomes or leaves in the PLS-DA model than those of rhizomes or leaves in the RF model, it was found that the PLS-DA was more suitable for the identification of geographical origins for *P. yunnanensis*.

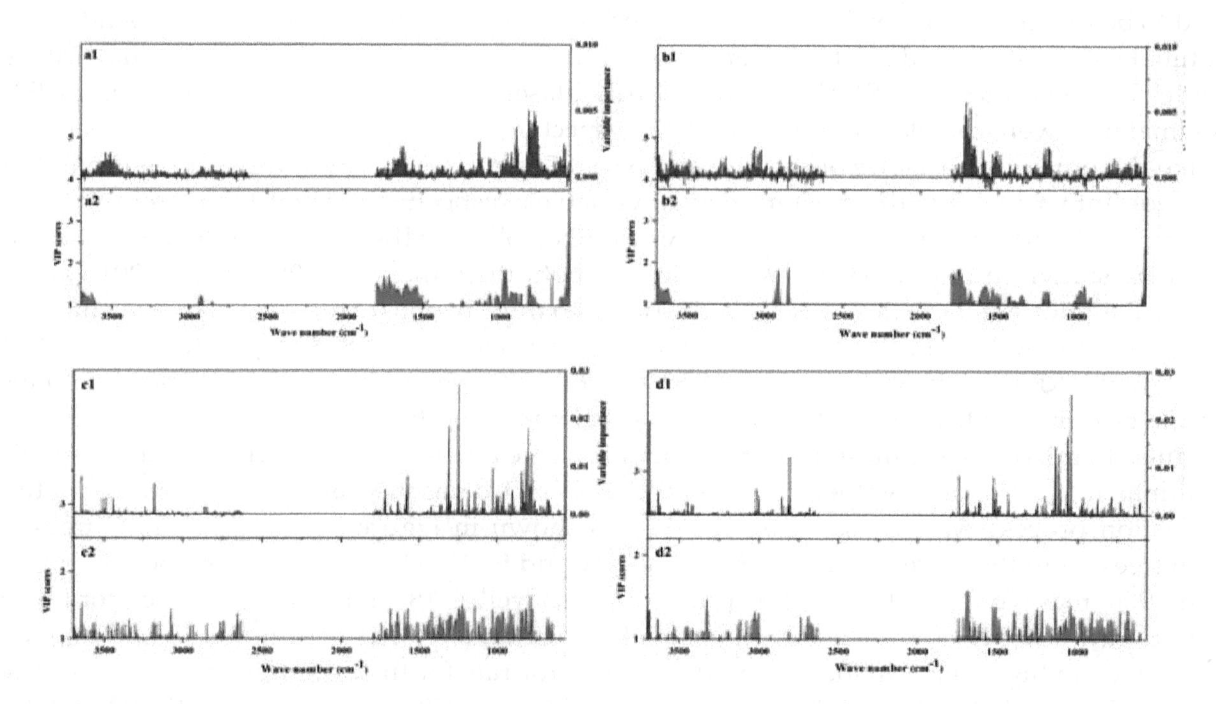

Figure 7. The importance variables (1) of RF models and the VIP values (2) of partial least squares discriminant analysis (PLS-DA) models of the *P. yunnanensis* samples: (**a**) the raw rhizomes dataset, (**b**) the raw leaves dataset, (**c**) the SNV-SD rhizomes dataset, (**d**) the SNV-SD leaves dataset.

2.4. Data Fusion Strategy

Despite the high performance obtained by PLS-DA, and the RF classification models of leaves of the FT-MIR spectra of *P. yunnanensis* samples, the 100% identification accuracy of the calibration set and the validation set were not acquired, and models' abilities needed further enhancement. Hence, the data fusion strategy was used to further improve the prediction abilities of PLS-DA and RF models. Data fusion were concatenated variables of FT-MIR spectra from different parts, forming a single matrix where row numbers were the analyzed sample quantities, and columns consisted of variables. In other words, the rhizome and leaf datasets were combined to establish the classification models.

The process for establishing the data fusion RF model was similar to the individual dataset. RF models were established based on raw and SNV-SD data fusion FT-MIR spectra datasets. A total of 2414 and 2403 variables were contained in the two matrices, respectively. As shown in Figure S5a,b, 356–399 trees and 375–404 trees were the lowest ranges for n_{tree} of raw and SNV-SD matrices, respectively. Additionally, the optimal values of 377 and 393 trees were selected for further selection of the suitable m_{try} values for raw and SNV-SD datasets, respectively. As shown in Figure S5c,d, the optimal m_{try} were 51 and 18, respectively. Similar to the individual dataset, all variables were in ascending order with importance. The 10-fold cross-validation error rates of the RF model for raw and SNV-SD data fusion datasets are shown in Figure S6a,b, respectively. Additionally, variable numbers of 69 and 288 with the lower 10-fold cross-validation error rates of 0.34853 and 0.02606 were selected to establish the data fusion RF models. Besides, the most important variables were the new data matrices, while re-selecting the optimal n_{tree} and m_{try} values for raw and SNV-SD data fusion datasets, respectively. As shown in Figure S7, the 1609–1660 trees and 98–125 trees were the lowest ranges for n_{tree} of the two datasets, respectively. Besides, 1652 and 104 trees, as well as 10 and 18, are selected for the best n_{tree} and m_{try}, respectively. Compared to the accuracy of the RF models between the raw and SNV-SD data fusion matrixes, the error rate for the calibration set of the raw dataset decreased from 37.46% to 37.13%, and decreased from 2.61% to 1.63% for the SNV-SD dataset. The classification abilities in the rhizome and lead data fusion RF model were better than in the individual dataset RF model.

From a comparison of parameters for SENS, SPEC, ACC, and MCC between the PLS-DA and RF models, based on data fusion strategy, the PLS-DA model had a better classification ability than that of the RF model. As shown in Table 1, the geographical origins identification abilities reached the best of each class calibration set and validation set for the PLS-DA model of the SNV-SD FT-MIR spectra. However, the parameter values were close to 100% for most classes of RF model. Hence, it could be demonstrated that the PLS-DA model was more suitable for tracing the different geographical origins of cultivated *P. yunnanensis*.

Table 1. The major parameters of PLS-DA and RF models of each class, based on the data fusion SNV-SD FT-MIR spectra datasets of *P. yunnanensis* samples.

Preprocessing	Set	Classes [a]	PLS-DA				RF			
			SENS	SPEC	ACC	MCC	SENS	SPEC	ACC	MCC
SNV-SD	Calibration set	1	1	1	1	1	1	0.996	0.997	0.987
		2	1	1	1	1	0.984	0.996	0.993	0.98
		3	1	1	1	1	0.975	1	0.997	0.986
		4	1	1	1	1	0.951	0.992	0.987	0.944
		5	1	1	1	1	0.9831	1	0.997	0.989
		6	1	1	1	1	1	0.996	0.997	0.99
	Validation set	1	1	1	1	1	1	1	1	1
		2	1	1	1	1	1	1	1	1
		3	1	1	1	1	1	1	1	1
		4	1	1	1	1	0.95	1	0.994	0.971
		5	1	1	1	1	1	0.992	0.994	0.979
		6	1	1	1	1	1	1	1	1

[a] 1: Kunming, 2: Yuxi, 3: Chuxiong, 4: Dali, 5: Lijiang, 6: Honghe. Sensitivity (SENS), specificity (SPEC), accuracy (ACC) and the Matthews correlation coefficient (MCC).

2.5. Hierarchical Clustering Analysis

HCA dendrograms based on average SNV-SD FT-MIR spectra datasets of rhizomes and leaves of *P. yunnanensis* from different geographical origins are presented in Figure 8a,b, respectively. It is obviously that all the six classes are grouping into two main clusters, both in the two HCA dendrograms. However, the clustering results among Kunming, Yuxi, Chuxiong, Dali, Lijiang and Honghe were obtained based on rhizomes and leaves FT-MIR spectral matrixes were different. As shown in Figure 9, the altitude is decreasing gradually from Northwest Yunnan to Southeast Yunnan. In addition, the two main clusters are influenced to some extent by the topography including altitude. Nevertheless, Kunming was cluster with Honghe and Yuxi in HCA dendrograms based on rhizomes dataset but cluster with Lijiang, Dali and Chuxiong of HCA plot based on leaves. It is demonstrated that the different chemical information between rhizomes and leaves of *P. yunnanensis* were influenced on the results of clustering.

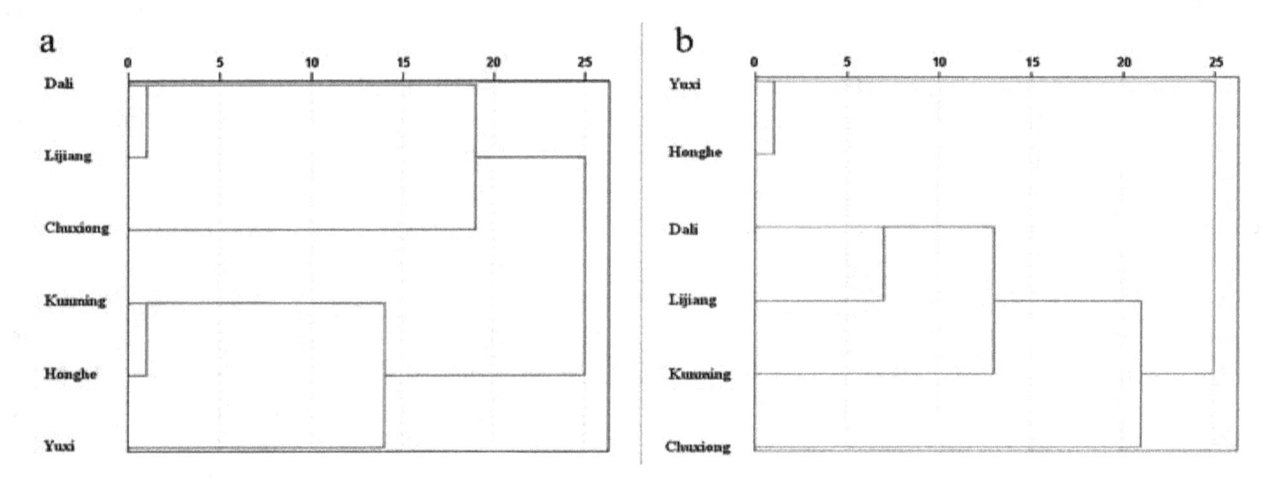

Figure 8. Dendrograms resulting of hierarchical cluster analysis (HCA) based on six geographical origins of *P. yunnanensis* samples: (**a**) the rhizomes dataset, (**b**) the leaves dataset.

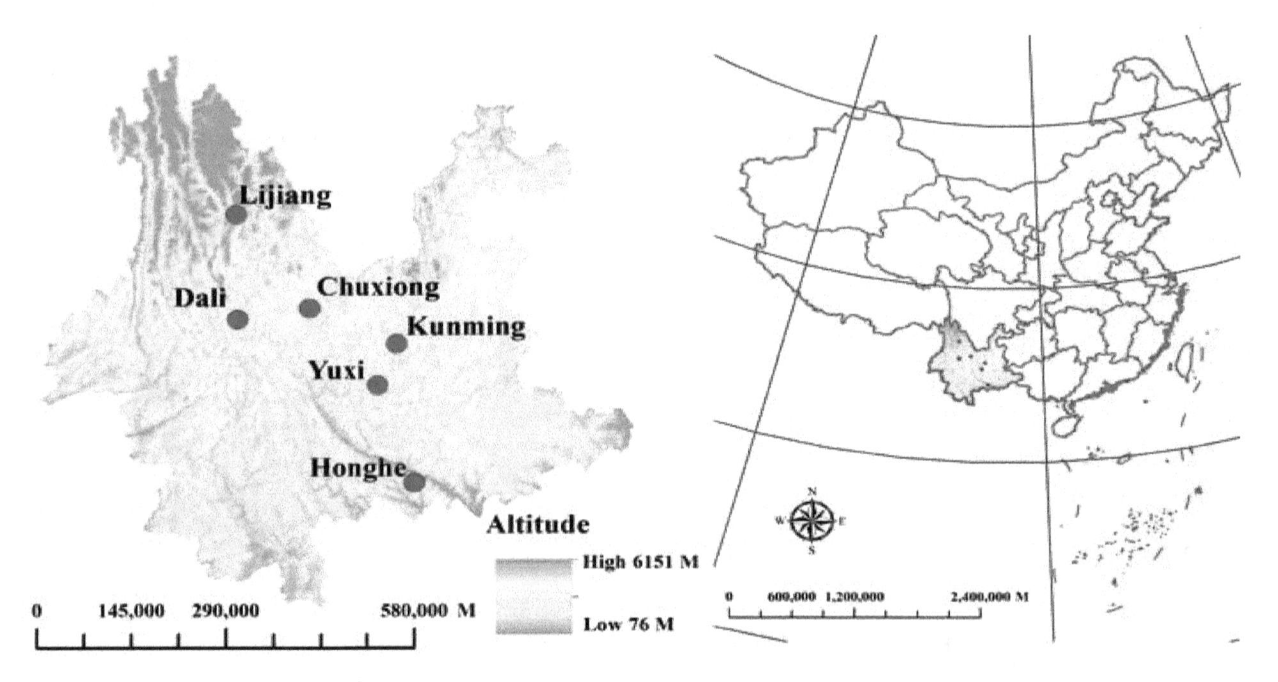

Figure 9. Location distribution of cultivated *P. yunnanensis* samples in Kunming, Yuxi, Chuxiong, Dali, Lijiang and Honghe, Yunnan Province.

3. Materials and Methods

3.1. Plant Material Preparation

In our experiment, rhizomes and leaves of 462 cultivated *P. yunnanensis* samples were collected from Kunming, Yuxi, Chuxiong, Dali, Lijiang, and Honghe cities in Yunnan Province; the collection locations and detailed information are shown in Figure 9 and Table S5. All samples were identified as *P. polyphylla* Smith var. *yunnanensis* (Franch.) Hand.-Mazz. by Professor Hang Jin (Institute of Medicinal Plants, Yunnan Academy of Agricultural Sciences, China). To start with, the different parts for each *P. yunnanensis* samples were separated and washed, then dried at 50 degrees Celsius. In addition, both rhizome and leaf samples were sifted through 100 mesh sieves, and stored in a relatively dry environment.

3.2. FT-MIR Spectral Acquisition

FT-MIR analysis uses a FTIR spectrometer with a DTGS detector equipped, combined with a ZnSe attenuated total reflectance accessory (Perkin Elmer, Norwalk, CT, USA). The FT-MIR spectra collection parameters and methods are referenced in our previous experiment [14]. The FT-MIR spectra recorded ranges of 4000–550 cm^{-1} with 4 cm^{-1} resolution and 16 scans, both for rhizomes and leaves of each of the *P. yunnanensis* samples. Three scans were repeated for all rhizomes and leaves samples. Moreover, it was required that a relatively constant temperature and humidity was provided during the assessment of the FT-MIR spectra.

3.3. Chemometrics Methods

3.3.1. Principal Component Analysis

PCA is an exploratory data analysis method and an unsupervised pattern recognition technique, which seeks for the optimum data distribution in a multivariate space [25–27]. The fundamental of PCA is that all the raw data are projected onto a two-dimensional sub-space, to ensure that information loss is minimized. The higher the front PCs, the higher the proportion of important variables represented. Generally, the first few PCs represent the most information. The first two or three PCs of all samples can be shown in two- or three-dimensional scores plots, and they further show the regularities of distribution for all the samples. Moreover, the relationship between the first two PCs and wave numbers can be shown by the loading plot.

3.3.2. Partial Least Squares Discriminant Analysis

PLS-DA, a binary classification algorithm from 0 to 1, is based on the PLS algorithm, to add category labels to achieve the effect of classification prediction, and it shows the relationship by multivariate projection between independent and dependent variables, which are expressed by X and Y, respectively [28,29]. Besides, LVs were one feature variable that were produced by an intermediate process in the PLS-DA method [30]. LVs are useful for us to analyze the important variables and information. The X matrix and target and important values in Y are more closely correlated than the noise or unimportant values in Y. Additionally, the VIP plot summarizes the importance of the variables, both to explain X, and to correlate to Y, meaning that variables with a VIP value greater of than 1 are important; as well, those that are greater than 0.5 and less than 1 may be important, depending on the circumstances. Hence, classifying samples by PLS-DA requires that variables possess numbers that are greater than the classification sample numbers, and there should be some correlation among the identified samples.

3.3.3. Random Forest

RF model, developed by Breiman in 2001, has been widely used to resolve classification problems in the field of food, and so on [31,32]. The RF model is based on the assembly classification or regression trees algorithm, and it shows a higher ability to resolve binary classification or regression issues [31]. The operational steps of the RF model can be roughly divided into the following five steps. Firstly, a spectra dataset was separated into two parts according to the ratio of 2 to 1, by the Kennard-stone (KS) algorithm by MATLAB 2017a (MathWorks, Natick, MA, USA) [33,34]. Two-thirds of the dataset was assigned as the calibration set (bootstrap samples), and one-third as the validation set (out-of-bag samples). The calibration set was used to obtain the optimal classification trees, and the validation set was applied to evaluate the ability of the FR model. Besides, the initial values of n_{tree} and m_{try} were defined as 2000, and the square root of the number of all variables, respectively. The optimal n_{tree} and m_{try} were both selected according to the lowest OOB classification error values. Thirdly, the most important variables were selected by a lower 10-fold cross-validation error rate, and as a new data matrix reimport. Fourth, the optimal n_{tree} and m_{try} were reselected according to the fundamental of step 2. Finally, the establishment of the final RF discrimination model was performed by using the optimized n_{tree} and m_{try} parameters. Step two to five were completed by R package (version 4.6–14).

3.3.4. Hierarchical Cluster Analysis

HCA clusters different categories at a certain distance, according to the degree of similarity of each class, which means that it could preliminarily identify a classification trend for each category [35]. Besides, the Person correlation coefficient was applied to measure the linear relationship between the distance variables. These analyses were completed by SPSS 20.0 software (IBM Corp., Armonk, NY, USA).

3.4. Data Analysis

The purpose of data analysis involves the reduction of the influence by noise and other factors from experiments and instruments on the raw FT-MIR spectra data. Firstly, the raw FT-MIR spectra were pretreated by advancing ATR (attenuated total reflection) correction, and absorbance was transformed from transmittance by OMNIC 9.7.7 (Thermo Fisher Scientific, Madison, WI, USA). Secondly, the best preprocessing method was selected among a combination of various pretreatment methods, including SNV, FD and SD, which can enhance the accuracy and feasibility for identification study [36,37]. SNV and its derivatives could decrease a part of the irrelevant interferences, such as high frequency random noise, the interference of light scattering, baseline drift, and unequal concentration, and so on, to improve the classification ability of the models. All these preprocessing methods were completed by SIMCA-P^+ 13.0 (Umetrics, Umea, Sweden). The datasets were separated into a calibration set and a validation set, with a rate of 2 to 1 by the KS algorithm, using MATLAB 2017a (The MathWorks), which was also used to establish the PLS-DA and RF models. In other words, the FTIR spectra of samples were divided into a calibration set (307 samples) and validation set (155 samples), as shown in Table S6.

Generally, parameters including RMSEE, RMSECV, and the accuracy of calibration sets Q^2 and R^2 were used to estimate the identification ability of the calibration model [38,39]. The optimal preprocessing model required lower values of RMSEE and RMSECV, as well as higher values of accuracy for the calibration sets R^2 and Q^2. Besides, the model may have poor robustness and over-fitting when the values of the root mean square error of prediction (RMSEP) are greater than that of RMSECV [12]. In addition, due to both the PLS-DA and RF models being able to obtain the vote matrices, the two models could calculate the values of true negative (TN), true positive (TP), false negative (FN), and false positive (FP), respectively. SENS (Equation (1)), SPEC (Equation (2)), ACC (Equation (3)), and MCC (Equation (4)) were the four parameters for each class, resulting in identification effects for different geographical origins of *P. yunnanensis* samples of PLS-DA and RF models. Obviously, this led to the higher values of these four parameters and a better identification ability for each class.

$$\text{SENS} = \frac{\text{TP}}{(\text{TP} + \text{FN})} \tag{1}$$

$$\text{SPEC} = \frac{\text{TN}}{(\text{TN} + \text{FP})} \tag{2}$$

$$\text{ACC} = \frac{(\text{TN} + \text{TP})}{(\text{TP} + \text{TN} + \text{FP} + \text{FN})} \tag{3}$$

$$\text{MCC} = \frac{(\text{TP} \times \text{TN} - \text{FP} \times \text{FN})}{\sqrt{(\text{TP} + \text{FP})(\text{TP} + \text{FN})(\text{TN} + \text{FP})(\text{TN} + \text{FN})}} \tag{4}$$

4. Conclusions

In our article, the geographical origin identification of 462 *P. yunnanensis* samples from Kunming, Yuxi, Chuxiong, Lijiang, Dali, and Honghe were analyzed by rhizome and leaf FT-MIR spectra, combined with PLS-DA, RF, and HCA methods. The chemical information differences between rhizomes and leaves were directly displayed on the FT-MIR spectra and the results of models. PLS-DA was more suitable for use in classifying the different geographical origins of *P. yunnanensis* than the RF model, in that it had the best identification ability and more uniformly distributed important variables. Besides, the order of classification ability from strong to weak is the data fusion dataset > leaves dataset > rhizomes dataset, which means that leaves can be used quickly and accurately to identify the geographical origin of *P. yunnanensis*, and more comprehensive information can be showed by multiple sources of chemical information.

Supplementary Materials: The following are available online. Figure S1: VIP scores of FT-MIR data of leaves for regional differences, Figure S2: The ntree and mtry screening of RF models of *P. yunnanensis* samples before variables are ranked by permutation accuracy importance, Figure S3: The 10-fold cross validation error rates of the RF model (sequentially reduced each five variables), based on *P. yunnanensis* samples, Figure S4: The ntree and mtry screening of RF models of *P. yunnanensis* samples after variables are ranked by permutation accuracy importance, Figure S5: The n_{tree} and m_{try} screening of RF models of *P. yunnanensis* samples before variables are ranked by permutation accuracy importance, Figure S6: The 10-fold cross validation error rates of the RF model (sequentially reduced each five variables) based on *P. yunnanensis* samples, Figure S7: The n_{tree} and m_{try} screening of RF models of *P. yunnanensis* samples after variables are ranked by permutation accuracy importance.

Author Contributions: Y.-F.P. and Y.-Z.W. developed the concept of the manuscript, Y.-F.P. performed the experiments, analyzed the data, and discussed the results, Q.-Z.Z. and Z.-T.Z. performed final corrections for this manuscript.

References

1. Cunningham, A.B.; Brinckmann, J.A.; Bi, Y.F.; Pei, S.J.; Schippmann, U.; Luo, P. *Paris* in the spring: A review of the trade, conservation and opportunities in the shift from wild harvest to cultivation of *Paris polyphylla* (Trilliaceae). *J. Ethnopharmacol.* **2018**, *222*, 208–216. [CrossRef] [PubMed]
2. Lu, H.; Xu, J.H.; Chen, R.P.; Yang, H.; Liu, Y.P. Status of the genus *Paris* L. re-sources of Yunnan and countermeasures for protection. *J. Yunnan Univ.* **2006**, *28*, 307–310.
3. State Pharmacopoeia Commission. *Chinese Pharmacopoeia*; Chemistry and Industry Press: Beijing, China, 2015.
4. Li, H. *The Genus Paris (Trilliaceae)*; Science Press: Beijing, China, 1998; pp. 12–16.
5. Yang, J.; Wang, Y.H.; Li, H. *Paris* qiliangiana (Melanthiaceae), a new species from Hubei, China. *Phytotaxa* **2017**, *329*, 193–196. [CrossRef]
6. Qin, X.J.; Yu, M.Y.; Ni, W.; Yan, H.; Chen, C.X.; Cheng, Y.C.; Li, H.; Liu, H.Y. Steroidal saponins from stems and leaves of *Paris polyphylla* var. *yunnanensis*. *Phytochemistry* **2016**, *121*, 20–29. [CrossRef] [PubMed]
7. Dai, X.W.; Feng, L.L.; Li, H.F. Analysis of differences and correlation of steroidal saponins in rhizomes and leaves of *Paris polyphylla* var. *yunnanensis* from different planting base. *Chin. J. Exp. Tradit. Med. Form.* **2018**, *24*, 41–48.
8. Yang, Y.G.; Zhang, J.; Zhao, Y.L.; Zhang, J.Y.; Wang, Y.Z. Quantitative determination and evaluation of *Paris polyphylla* var. *yunnanensis* with different harvesting times using UPLC-UV-MS and FT-IR spectroscopy in combination with partial least squares discriminant analysis. *Biomed. Chromatogr.* **2017**, *31*. [CrossRef] [PubMed]
9. Yang, Y.G.; Jin, H.; Zhang, J.; Zhang, J.Y.; Wang, Y.Z. Quantitative evaluation and discrimination of wild *Paris polyphylla* var. *yunnanensis* (Franch.) Hand.-Mazz from three regions of Yunnan Province using UHPLC-UV-MS and UV spectroscopy couple with partial least squares discriminant analysis. *J. Nat. Med.*

2017, *71*, 148–157. [CrossRef] [PubMed]
10. Chen, T.Z.; Wen, F.Y.; Zhang, T.; Yang, Y.X.; Fang, Q.M.; Zhang, H.; Xue, D. Evaluation of saponins in *Paris Polyphylla* var. *chinensis* from twenty-one growing areas. *Chin. Tradit. Patent Med.* **2017**, *39*, 2345–2350.
11. Wu, X.M.; Zhang, Q.Z.; Wang, Y.Z. Traceability of wild *Paris polyphylla* Smith var. *yunnanensis* based on data fusion strategy of FT-MIR and UV-Vis combined with SVM and random forest. *Spectrochim. Acta A* **2018**, *205*, 479–488. [CrossRef] [PubMed]
12. Pei, Y.F.; Wu, L.H.; Zhang, Q.Z.; Wang, Y.Z. Geographical traceability of cultivated *Paris polyphylla* var. *yunnanensis* using ATR-FTMIR spectroscopy with three mathematical algorithms. *Anal. Methods* **2018**. [CrossRef]
13. Gad, H.A.; El-Ahmady, S.H.; Abou-Shoer, M.I.; Al-Azizi, M.M. Application of chemometrics in authentication of herbal medicines: A review. *Phytochem. Anal.* **2012**, *24*, 1–24. [CrossRef] [PubMed]
14. Biancolillo, A.; Marini, F. Chemometrics applied to plant spectral analysis. In *Vibrational Spectroscopy for Plant Varieties and Cultivars Characterization*; Elsevier: Amsterdam, The Netherlands, 2018; pp. 69–104.
15. Li, J.; Zhang, J.; Zhao, Y.L.; Huang, H.Y.; Wang, Y.Z. Comprehensive quality assessment based specific chemical profiles for geographic and tissue variation in *Gentiana rigescens* using HPLC and FTIR method combined with principal component analysis. *Front. Chem.* **2017**, *5*. [CrossRef] [PubMed]
16. Qi, L.M.; Liu, H.G.; Li, J.Q.; Li, T.; Wang, Y.Z. Feature fusion of ICP-AES, UV-Vis and FT-MIR for origin traceability of *Boletus Edulis* mushrooms in combination with chemometrics. *Sensors* **2018**, *18*, 241. [CrossRef] [PubMed]
17. Li, Y.; Zhang, J.Y.; Wang, Y.Z. FT-MIR and NIR spectral data fusion: A synergetic strategy for the geographical traceability of *Panax notoginseng*. *Anal. Bioanal. Chem.* **2018**, *410*, 91–103. [CrossRef] [PubMed]
18. Wang, Y.; Huang, H.Y.; Zuo, Z.T.; Wang, Y.Z. Comprehensive quality assessment of *Dendrubium officinale* using ATR-FTIR spectroscopy combined with random forest and support vector machine regression. *Spectrochim. Acta A* **2018**, *205*, 637–648. [CrossRef] [PubMed]
19. Yang, Y.G.; Wang, Y.Z. Characterization of *Paris polyphylla* var. *yunnanensis* by infrared and ultraviolet spectroscopies with chemometric data fusion. *Anal. Lett.* **2018**, *51*, 1730–1742. [CrossRef]
20. Xie, J.D.; Sun, L. An overall quality evaluation of Paridis Rhizoma by multiple components determination based on the chemometrics. *Chin. J. Pharm. Anal.* **2015**, *35*, 1585–1590.
21. Zhang, S.S.; Liu, X.; Wang, J.F.; Yu, M.J.; Huang, Z.J.; Liu, Y.; Zhang, H. Determination of seven steroidal saponins in Paridis Rhizoma and polygerm varieties from different regions in Yunnan Province by UPLC and establishment of fingerprint. *Chin. Tradit. Herb. Drugs* **2016**, *47*, 4257–4263.
22. Zhang, J.Y.; Wang, Y.Z.; Zhao, Y.L.; Yang, S.B.; Zhang, J.; Yuan, T.J.; Wang, J.J.; Jin, H. Ultraviolet absorption spectrum analysis and identification of medicinal plants of *Paris*. *Spectrosc. Spectr. Anal.* **2012**, *32*, 2176–2180.
23. Sun, S.Q.; Zhou, Q.; Chen, J.B. *Analysis of Traditional Chinese Medicine by Infrared Spectroscopy*; Chemical Industry Press: Beijing, China, 2010.
24. Yang, L.F.; Ma, F.; Zhou, Q.; Sun, S.Q. Analysis and identification of wild and cultivated Paridis Rhizoma by infrared spectroscopy. *J. Mol. Struct.* **2018**, *1165*, 37–41. [CrossRef]
25. Wold, S.; Esbensen, K.; Geladi, P. Principal component analysis. *Chemometr. Intell. Lab. Syst.* **1987**, *2*, 37–52. [CrossRef]
26. Ringnér, M. What is principal component analysis? *Nat. Biotechnol.* **2008**, *26*, 303–304. [CrossRef] [PubMed]
27. Jolliffe, I.T. *Principal Component Analysis*, 2nd ed.; Springer: New York, NY, USA, 2002.
28. Ståle, L.; Wold, S. Partial least squares analysis with cross-validation for the two-class problem: A Monte Carlo study. *J. Chemometr.* **1987**, *1*, 185–196.
29. Indahl, U.G.; Martens, H.; Naes, T. From dummy regression to prior probabilities in PLS-DA. *J. Chemometr.* **2007**, *21*, 529–536. [CrossRef]
30. Nocairi, H.; Qannari, E.M.; Vigneau, E.; Bertrand, D. Discrimination on latent components with respect to patterns. Application to multicollinear data. *Comput. Stat. Data Anal.* **2005**, *48*, 139–147. [CrossRef]
31. Breiman, L. Random forests. *Mach Learn.* **2001**, *45*, 5–32. [CrossRef]
32. Amjad, A.; Ullah, R.; Khan, S.; Bilal, M.; Khan, A. Raman spectroscopy based analysis of milk using random forest classification. *Vib. Spectrosc.* **2018**, *99*, 124–129. [CrossRef]
33. Saptoro, A.; Tadé, M.O.; Vuthaluru, H. A modified Kennard-Stone algorithm for optimal division of data for developing artificial neural network models. *Chem. Prod. Process Model.* **2012**, *7*, 1–14. [CrossRef]

34. Rajer-Kanduč, K.; Zupan, J.; Majcen, N. Separation of data on the training and test set for modelling: A case study for modelling of five colour properties of a white pigment. *Chemometr. Intell. Lab. Syst.* **2003**, *65*, 221–229. [CrossRef]
35. Jain, A.K.; Dubes, R.C. Algorithms for clustering data. In *Technometrics*; Prentice-Hall, Inc.: Englewood Cliffs, NJ, USA, 1988.
36. Barnes, R.J.; Dhanoa, M.S.; Lister, S.J. Standard normal variate transformation and de-trending of near-infrared diffuse reflectance spectra. *Appl. Spectrosc.* **1989**, *43*, 772–777. [CrossRef]
37. Savitzky, A.; Golay, M.J. Smoothing and differentiation of data by simplified least squares procedures. *Anal. Chem.* **1964**, *36*, 1627–1639. [CrossRef]
38. Xie, L.J.; Ye, X.Q.; Liu, D.H.; Ying, Y.B. Quantification of glucose, fructose and sucrose in bayberry juice by NIR and PLS. *Food Chem.* **2009**, *114*, 1135–1140. [CrossRef]
39. Qi, L.M.; Zhang, J.; Liu, H.G.; Li, T.; Wang, Y.Z. Fourier transform mid-infrared spectroscopy and chemometrics to identify and discriminate *Boletus edulis* and *Boletus tomentipes* mushrooms. *Int. J. Food Prop.* **2017**, *20*, S56–S68. [CrossRef]

5

Approaching Authenticity Issues in Fish and Seafood Products by Qualitative Spectroscopy and Chemometrics

Sergio Ghidini, Maria Olga Varrà * and Emanuela Zanardi

Department of Food and Drug, University of Parma, Strada del Taglio 10, 43126 Parma, Italy; sergio.ghidini@unipr.it (S.G.); emanuela.zanardi@unipr.it (E.Z.)

* Correspondence: mariaolga.varra@studenti.unipr.it

Abstract: The intrinsically complex nature of fish and seafood, as well as the complicated organisation of the international fish supply and market, make struggle against counterfeiting and falsification of fish and seafood products very difficult. The development of fast and reliable omics strategies based on spectroscopy in conjunction with multivariate data analysis has been attracting great interest from food scientists, so that the studies linked to fish and seafood authenticity have increased considerably in recent years. The present work has been designed to review the most promising studies dealing with the use of qualitative spectroscopy and chemometrics for the resolution of the key authenticity issues of fish and seafood products, with a focus on species substitution, geographical origin falsification, production method or farming system misrepresentation, and fresh for frozen/thawed product substitution. Within this framework, the potential of fluorescence, vibrational, nuclear magnetic resonance, and hyperspectral imaging spectroscopies, combined with both unsupervised and supervised chemometric techniques, has been highlighted, each time pointing out the trends in using one or another analytical approach and the performances achieved.

Keywords: fish and seafood; food authentication; chemometrics; fingerprinting; wild and farmed; geographical origin; vibrational spectroscopy; absorption/fluorescence spectroscopy; nuclear magnetic resonance; hyperspectral imaging

---

1. Introduction

The demand for fish and seafood products has increased notably during the last years, mostly as a consequence of the new special attention paid by consumers towards healthier food. The technological development that has invested the whole fisheries sector has additionally contributed to overcome the well-known obstacles to export fish and seafood worldwide, deriving from the high vulnerability of the products, to the point that today more than 35% of all caught and cultured fish is traded across national boundaries [1]. The growing competitiveness of the sector and diversification in fish supply chain have, in turn, led to the presence of a huge variety of look-alike products on the international market, whose global quality features are, however, quite different. More than 700 different species of fish, 100 of molluscan, and 100 of crustacean are, in fact, used as food for humans [2].

In this scenario, what is remarkable is that consumers demand not only for more fish, but for even safer and higher-quality fish, whilst the deliberate or accidental lack of transparency about the identity of products and fraudulent or negligent activities continue to grow. Based on what has been recently reported by the Food and Agriculture Organization, fish and related products have become among the most vulnerable to fraud category of food. Nevertheless, the effective monitoring of illicit practices in the fisheries sector is hampered by the increasing spread of highly processed fish products, in which the presence of different types of fraud can be hidden with ease [3].

The voluntary substitution of commercially valuable fish species with lower quality ones, represents the most recurrent form of fish fraud, although substitution can also take place accidentally when species look so similar that they are mistaken for each other. The geographical provenance and the production process are other current authenticity topics concerning fish and seafood products, whose falsification which is hard to bring to light, has a negative economic impact. Despite being economically motivated, mislabelling concerning these issues may occasionally represent a risk to public health. The illegal commercialisation of poisonous fish species (*Tetraodontidae, Molidae, Diodontidae,* and *Canthigasteridae* families) or the replacement of certain kinds of raw fish fillets with gastro-intestinal toxic fish (i.e., those belonging to the *Gempylidae* family) are just some of many examples. Likewise, occurrence of some harmful marine biotoxins may be linked to the geographical distribution of the producing organisms [4], while the presence of higher levels of heavy metals or residues of antibiotic and pesticides are more likely to be found in farmed products than in wild ones [5–7].

Ensuring a clear discrimination of the authenticity of fish and seafood is of special concern today not only for consumers, but also for producers, traders, and industries. Traceability throughout the whole production chain and at all stages of the market, covered by Regulations 178/2002/EC [8], 1005/2008/EC [9], and 1224/2009/EC [10], is considered to be the starting point for the assurance of a high level of safety and quality of food and ingredients, as it represents the basic instrument not only for preventing illegal activities, but also for protecting consumers through the opportunity to access information about the exact nature and characteristics of fish. Specific regulations for the provision of information to consumers [11], and the requirement to uniquely identify fish and seafood on the label [12], play also an essential role in providing more transparency regarding the nature of the products, as they allow consumers to make informed choices and further contribute to the implementation of seafood traceability. As a matter of fact, labels of all unprocessed and some processed fishery and aquaculture products must include information on both the commercial and scientific names of the species, whether the fish has been caught or farmed, the catch or the production area, the fishing gear used, whether the product has been defrosted, and the date of minimum durability (where appropriate). Many other voluntary claims can also be reported on the label, including the date of catch/harvest for wild/aquaculture products, information about the production techniques and practices, and environmental and ethical information [12].

All the claimed declarations appearing on the label must always be checked to verify whether they are truthful. Therefore, in spite of the utility of the traceability system, the fisheries sector needs effective methods to address the problem of fish authenticity and ensure product quality. Innovative analytical approaches based on the evaluation of total spectral properties, are rapidly gaining ground at all levels of current food authenticity research, thanks to their ability to simultaneously provide lots of information related to physical and chemical characteristics of the food matrix. Recent advances in chemometrics, moreover, have represented a major turning point in the dissemination of 'fingerprinting strategies', as they allow for the study of all the genetic, environmental, and other external factors influencing food identity, and to bypass many obstacles related to the application of conventional techniques [13]. This way, chemometrics can be now considered an essential tool for differentiation of similar samples according to the authentication issues of interest.

Until now, several spectroscopic techniques in conjunction with chemometrics have been used as rapid, simple, and cheap tools for fish quality and authenticity testing. Among these, vibrational (near-infrared (NIR), mid-infrared (MIR), Raman), fluorescence or absorption ultraviolet-visible (UV–Vis), and nuclear magnetic resonance (NMR) spectroscopies, together with hyperspectral imaging (HSI) spectroscopy, represent the most used techniques, even if they are still being developed.

Based on this background, the present review article has been designed to highlight the uses and developments of fast and reliable omics strategies based on UV–Vis, NIR, MIR, Raman, NMR, and HSI spectroscopies, with the attempt to address the key authenticity challenges within the fish and seafood sector. To this end, a brief discussion concerning basilar concepts underlying these techniques has been provided, and has been accompanied by a short overview about the implementation of several chemometric tools, in order to highlight the potential benefits in extracting relevant information from spectral data.

The main body of this review focuses specifically on the application, over the years, of spectroscopy and chemometrics to distinguish products in accordance with the species, production method (wild or farmed), farming system (conventional or organic; intensive, semi-intensive, or extensive), geographical provenance (different FAO areas and countries of origin), and the processing technique (fresh or fresh/thawed) that at present, correspond to the key authenticity concerns for which there must be ongoing and effective monitoring.

2. A Conceptual Framework of Spectroscopy and Chemometrics

Spectroscopy is the study of electromagnetic radiation interacting with matter, which can be absorbed, transmitted, or scattered on the basis of both the specific frequency of the radiation and the physical/chemical nature of the matter. When absorbed, radiation leads to a change in the energy states of atoms, nuclei, molecules, or crystals that make up matter, inducing an electronic, vibrational, or rotational transition, depending on the energy of the incident radiation [14]. When the radiation, at a specific frequency, is scattered by molecules (as in Raman spectroscopy), some changes can occur in the energy of the incident photon, which transfers parts of its energy to the matter. In any case, the result of these interactions is a spectrum enclosing many features of the matter analysed, which, when properly interpreted with the help of chemometrics, can be used in a great number of different applications. In choosing the most appropriate spectroscopic method to be used, consideration should be given to some factors, which go beyond the purely analytical purposes: the physical state and chemical composition of the sample, sensitivity, specificity, and overall accuracy of the technique, scale of operation, time of analysis, and cost/availability of the instrumentation [15].

For the sake of conciseness, the main features related to spectroscopic techniques used mostly in the food authentication field are summarised in Table 1.

Table 1. Comparison of different spectroscopic techniques used for food authentication purposes: summary of the main characteristics.

Spectroscopic Technique		Wavelength Range (nm)	Interaction Light-matter	Basic Principle	Sensitive Compounds	Information Obtained	Applications	Possible Limitations
UV–Vis	UV Vis	2×10^2–4×10^2 4×10^2–7.5×10^2	Absorption/emission	Electronic transitions	Double-conjugated bonds; isolated double, triple, peptide bonds; aromatic and carbonyl groups	Molecular structure	Qualitative/ quantitative	Need of sample preparation pH and temperature interferences
IR[1]:	NIR MIR	7.5×10^2–2.5×10^3 2.5×10^3–2.5×10^4	Absorption	Vibrations/rotations of molecular bonds (changes in dipole moments)	Polar bonds (N–H, C–H, O–H, S–H, C–O)	Chemical bonds and physical structure	Qualitative/ quantitative	Water interferences Overlapping of spectral peaks
Raman		2.5×10^3–1.0×10^6	Scattering	Vibrations of molecular bonds (changes in polarizability)	Non-polar double or triple bonds (C = C, C ≡ C)	Chemical bonds and physical structure	Qualitative/ quantitative	Fluorescence and photodecomposition interferences Low-intensity Peaks
HSI		Varying by spectroscopic modules	Absorption/emission/scattering	Varying by vibrational spectroscopic modules	Varying by vibrational spectroscopic modules	Varying by vibrational spectroscopic modules	Qualitative/ quantitative/ spatial	Varying by vibrational spectroscopic modules
NMR		5.0×10^8–7.5×10^9	Absorption	Nuclear spin changes	Nuclei having a proper magnetic field (spin quantum number ≠ 0 [2]	Regio/stereo chemistry of molecules	Qualitative/ quantitative/ structural	Cost of the equipment

[1] Infrared (IR) electromagnetic regions taken into consideration do not include far-infrared (FIR) range (2.5×10^4–1.0×10^5 nm) since it is not commonly used in food authentication studies. [2] H-1, C-13, and P-31 are the most frequently investigated nuclei in food science-related nuclear magnetic resonance (NMR) applications.

2.1. *UV–Vis Absorption and Fluorescence Emission Spectroscopy*

UV–Vis spectroscopy involves the electronic excitation of molecules containing specific chromophore groups, which results from the absorption of photons at two wavelength regions of the electromagnetic spectrum. In the absorption mode, the amount of light retained by the sample is measured, while in the fluorescence mode the amount of light emitted after absorption is taken into consideration [15]. Typically, the UV–Vis spectrum is characterised by broad absorption or emission peaks which reflect the molecular composition of the matrix: by exploiting the unicity absorption or emission patterns of the entire spectrum, or by measuring the absorbance or fluorescence intensity of the analyte at one wavelength, this spectrum can be used for many food analytical qualitative and quantitative applications, respectively [16,17].

2.2. *IR Spectroscopy*

Infrared spectroscopy involves three different sub-regions of the electromagnetic spectrum, namely NIR, MIR, and FIR, whose absorption by samples results in vibrations of atoms in molecular bonds [18]. These vibrations give out a great amount of information related not only to chemical bonding, but also to the general molecular conformation, structure, and intermolecular interactions within the sample [19]. This way, IR spectra enclose the total sample composition, whose pattern of peaks distribution represents a unique signature profile and whose intensity of bands is linked to the concentration of specific compounds [20,21].

The NIR spectrum of food samples results from absorption by molecular bonds containing prevalently light atoms and it is characterised by the presence of broad and overlapping overtone and combination bands [22,23]. By contrast, spectral signature in the MIR region is characterised by the presence of more intense and delineated bands, whose position and intensity are more informative of molecule's concentration in the sample [24,25]. Here too, the spectral profile is complex and data mining is very difficult without the use of multivariate data analysis. Finally, with reference to FIR spectroscopy, it is noted that no applications to food authentication are currently available since it relates to molecules containing halogen atoms, organometallic compounds, and inorganic compounds, whose interest is more limited within the context of food research [26].

2.3. *Raman Spectroscopy*

Raman spectroscopy is a molecular vibration technique based on the inelastic Raman scattering, a physical effect that comes with molecular vibrations and triggers a change in the polarizability of the molecule [27]. In particular, this kind of spectroscopy focuses on the measurement of those small fractions of the radiation which is scattered by specific categories of compounds at higher or lower frequencies than incident photons. The typical Raman spectrum, showing intensities of the scattered light versus the wavelengths of the Raman shift, is characterised by sharp and well-resolved bands, which provide information about molecular structure and composition of the matter analysed.

For a long time after its discovery, Raman spectroscopy has been poorly exploited in food applications, by reason of several analytical disadvantages and interference (see Table 1). These drawbacks have now been overcome thanks to the overall technological improvement of Raman equipment: by way of example, surface-enhanced Raman spectroscopy (SERS) has recently made it possible to surmount hurdles related to faint scattering signals [28].

2.4. *Hyperspectral Imaging*

HSI is a technique cobbling together spectroscopy and computer vision to give useful information concerning the physicochemical characteristics of samples in relation to their specific spatial distribution. Briefly, HSI systems provide several hyperspectral images of the tested sample, corresponding to three-dimensional data containers, of which each sub-image is a map showing spatial distribution of the sample constituents in relation to each single wavelength [29,30].

Over the recent years, the steady usage growth of HIS technology in the field of food research has been mainly driven by the availability of different instrumental configurations that exploit fluorescence, absorbance, or light scattering phenomena. On the other side, application of spectral imaging technologies is not at all widespread in the food industry, due to a variety of factors ranging from high costs and low availability of instrumentations, to the computation speed and necessity of expertise by users [31].

2.5. NMR Spectroscopy

NMR spectroscopy is a very versatile technique for food analysis and its untargeted applications have become very popular. The first reason for NMR popularity is that the composition of the matter under study can be perfectly mapped out by the overall NMR spectral profiles, thus giving a comprehensive view for the identification of all major and minor food components [32]. At the same time, the area of the NMR spectral bands is directly proportional to the number of nuclei that produce the signal, so the technique is also well-suited for quantitative purposes. Additionally, despite relatively high NMR equipment costs and spectra interpretation difficulties, NMR spectroscopy is one of the only techniques available that can provide information about the regio/stereo chemistry of molecules [33].

On the basis of the physical state of the matter and on the intended aim of NMR application, different methodologies involving the use of NMR have been optimized. Among these, high-resolution NMR, low-field NMR, solid-state NMR, liquid-state NMR, and NMR imaging are the most used ones, any of which requires specific instrumentation and different approaches to sample preparation, data acquisition, and processing [34].

2.6. Qualitative Chemometric Methods

Raw spectra resulting from spectroscopic analyses are usually characterised by broad and unresolved bands containing too much information, some of which are certainly useful and need to be retained, but some of which hamper the correct data interpretation and need to be removed. Recent advances in chemometrics have marked an important milestone in spectra analysis, since they have simplified the identification of hidden interrelations between variables providing the key for discrimination and classification of samples [20,35]. In other words, qualitative chemometrics methods help to recognise similarities and dissimilarities within spectral data, which can be used to confirm the authenticity or detect adulteration of food samples [36].

Based on the explorative or predictive nature of the methodology, qualitative chemometric techniques are usually classified into unsupervised and supervised techniques. While unsupervised techniques are independent of prior knowledge of class membership of samples to perform classification, supervised techniques call for such knowledge. Brief descriptions of the principles behind the chemometric techniques which are being used to a greater extent are provided below.

2.6.1. Spectral Pre-Treatments

Pre-treatment of spectral data is recognized as being fully integrated into the chemometric set-up itself. Prior to the development of chemometric models, raw spectroscopic data are suggested to be pre-processed by applying some corrections, aimed to enhance spectral properties and minimize the fraction of systematic variation which does not contain relevant information to the discrimination of samples. One such systematic variation is the sum of different physical effects which arise during instrumental acquisition of spectra (e.g., light scattering or background fluorescence phenomena), which are responsible for the appearance, especially in solids samples, of multiplicative, additive, and non-linearity effects (e.g., overlapping bands, baseline shifts/drifts, random noise) [37].

Thus, pre-processing algorithms are usually classified into signal correction methods (e.g., multiplicative scatter correction, MSC; standard normal variate, SNV), differentiation methods (first, second, or third order derivation), and filtering-based methods (e.g., orthogonal signal correction, OSC;

orthogonal wavelet correction, OWAVEC) [38]. While signal correction and filtering-based methods are conceived to retain only the spectral information mainly by suppressing the light-scattering effects, derivative-based methods also help to reduce the spectral complexity through the separation of the broad overlapping bands.

A more detailed description of spectral pre-processing techniques can be widely found in the literature [37,39,40]. Either way, it is essential to point out that spectral filters are most often concatenated to exploit the effects of each one, but this concatenation might increase model complexity and background noise, resulting in an inaccurate chemometric modelling of data and, thus, wrong predictions. For this reason, it is recommended to customize the selection of the pre-treatments prior to performing chemometric analysis according to the spectroscopic technique used and the sample characteristics, trying to restrict, whenever possible, their number.

2.6.2. Unsupervised Methods

Unsupervised methods look at the study of variability among samples for the purpose of identifying their natural characteristics and possible similarities among them, without the need to provide any information about the class to which samples belong.

Between the various available techniques, principal component analysis (PCA) is the most used one. PCA is a quite basic projection method able to reduce the original correlated variables into a smaller number of new uncorrelated latent variables (known as principal components), containing as much systematic variation as possible of the original data [41]. Score plot outputs deriving from PCA applications show in a simple and intuitive graphical way the hidden structures among samples, the interrelations among variables and between samples and variables, the probable presence of any outliers, and possible groupings or dispersion of sample according to specific class membership.

Hierarchical cluster analysis (HCA) is another frequently employed unsupervised method, based on the splitting of samples into different clusters. This splitting is based on the degree of analogy among samples and it is generally performed by evaluating the Mahalanobis or Euclidean distance between the same samples. The hierarchical approach followed is thus aimed at constructing a ladder, in which the most closely related samples are first classified into small groups, and then progressively assembled into bigger groups including less similar samples [35]. Results of HCA are graphically expressed by tree diagrams (dendrograms) showing relationships among clusters; nevertheless, despite being easily computable, dendrograms are often misunderstood, since the number of clusters to be considered is arbitrary, making the interpretation of results more subjective than objective.

2.6.3. Supervised Methods

Supervised techniques require the previous knowledge of the class membership of the samples tested, which can be used to develop predictive models able to discriminate and classify future unidentified samples. There are several different chemometric techniques belonging to the category of the supervised methods, most of which require a training set (to find classification rules for the sample), and a test set (to assess the predictability of the model developed) [42].

Linear discriminant analysis (LDA) and quadratic discriminant analysis (QDA) are variance-based methods which use Euclidean distance to find those combinations of the original variables determining maximum separation among the different groups of samples [20]. Both techniques presume that the measurements within each class are normally distributed, but while LDA supposes that dispersion (covariance) is identical for all the classes, QDA, on the contrary, allows the possibility of different dispersion to be present within different classes [35]. Although QDA is considered an extension of LDA, there are some common limitations, for instance the risks of overfitting and failing in classification, especially when the samples size for each class in unbalanced.

K-nearest neighbors (k-NN) clustering is one of the simplest method to discriminate samples on the basis of the distance among them. After choosing the adequate number of k-neighbor samples, the algorithm identifies the k-nearest samples of known class membership to select the classification of unknown samples. This method, unlike LDA and QDA, does not require any prior assumption and its success is independent of the homogeneity of sample numbers in each tested class [43].

Among supervised machine learning approaches, support vector machines (SVM) are particularly advantageous when samples classification is complicated by non-linearity and high dimensional space. The core of the method is the use of specific functions for pattern analysis (kernel algorithms), through which the margin of separation between classes is maximised and complex classification problems that are not linear in the initial dimension (but may be at high dimensional spaces) are resolved [20].

Similarly, artificial neural networks (ANN) is a machine learning method characterised by the ability to adapt to the data, providing classification also in the presence of non-linearity input–output relationships. Structured and organized in a less complex way than SVM, ANN usually generate a more rapid response at a lower computational cost; these efforts, however, are counterbalanced by a reduction in accuracy [20,44]. Nevertheless, ANN suffers from poor data generalisation and, by consequence, it is inclined to return model's overfitting errors. This tendency to overfitting is the main reason why accurate ANN computation analyses call for a very high number of samples to be considered, and at the same time, require strict internal and external validations to be performed, where the training set and the test set should enclose as much similar variability as possible [45].

Soft independent modelling of class analogy (SIMCA) is an alternative pattern recognition method which first performs individual PCA on the samples for each class they must be assigned to, in order to compress original variables into a smaller number of new principal components. Principal components and critical distances computed are then used to delineate a confidence limit for each class. Unknown samples are then assigned to the class to which they get close by projection into the resulting multidimensional space [36]. SIMCA is particularly useful when samples belong to several different classes; since maximum class-separation is not covered by the method, the interpretation of the outcomes may be difficult, if not impossible [20].

Regression-based supervised discriminant analyses exploit specific classification algorithms to model the interrelations existing between measured variables (i.e., spectra) and qualitative parameters (i.e., class membership), such that maximum separation between the different groups of samples is achieved. Partial least square-discriminant analysis (PLS-DA) and orthogonal partial least square-discriminant analysis (OPLS-DA) belong to this category of techniques. PLS-DA involves a standard PLS regression to find interrelations between the X-matrix (containing measured variables) and Y-matrix (containing categorical variables) by building new variables (latent variables). These interrelations allow not only to classify new samples into one of the Y-groups based on measured spectrum, but also to identify variables that mostly contribute to the classification. Although PLS-DA has the advantage of modelling noisy and highly collinear data efficiently, the technique is often unsuccessful when the non-related (orthogonal) variability in the X-matrix is substantial, since it hinders the correct interpretation of the results [20]. This drawback can be overcome by the application of OPLS-DA, through which the orthogonal variability within the X-matrix is separated from the related (predicted) variability and then modelled apart. Consequently, if samples cannot be discriminated along the predictive direction, the orthogonal variability may be handled to increase the effectiveness of discrimination among classes [46].

3. Authenticating Fish and Seafood through the Application of Qualitative Spectroscopy and Chemometrics

Spectroscopic and chemometric analyses have been used over the years for many applications in fishery research, those in the authentication field being among the most promising ones. Some of the works concerning the flexibility of spectroscopy in fish and seafood analysis have already been reviewed by different authors [24,25,47–49], but they have mainly centred on illustration of the advances of the available techniques for quality attributes assessment, as well as on the advantages and limitations of the single type of technique over traditional methods.

Therefore, in the following section, more attention has been paid to the resolution, on a case-by-case basis, of the weightiest authentication issues in the fish and seafood sector, namely species substitution, geographical origin falsification, production method or farming system misrepresentation, and fresh for frozen/thawed product substitution, each time pointing out the trends in using one or another method as well as the discrimination performances achieved, which are considered to be the most intuitive parameters used for chemometric models diagnostics. An overview of the most frequently investigated authentication issues in the fishery sector and the trend of using each spectroscopic technique over the years by the scientific community are plotted in Figures 1 and 2, respectively.

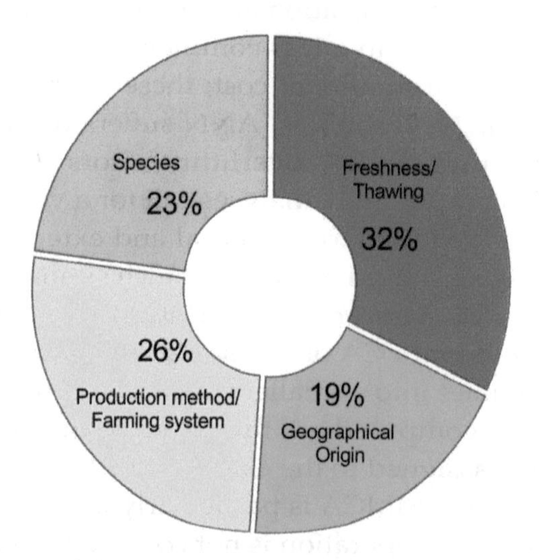

Figure 1. Percentage distribution of the authenticity issues covered by the scientific literature reviewed in the present work. Data were collected in February 2019 from the web search engine Google Scholar (search criteria: time period: "any time", and keywords: "fish and/or seafood"; "authenticity"; "spectroscopy"; "chemometrics".

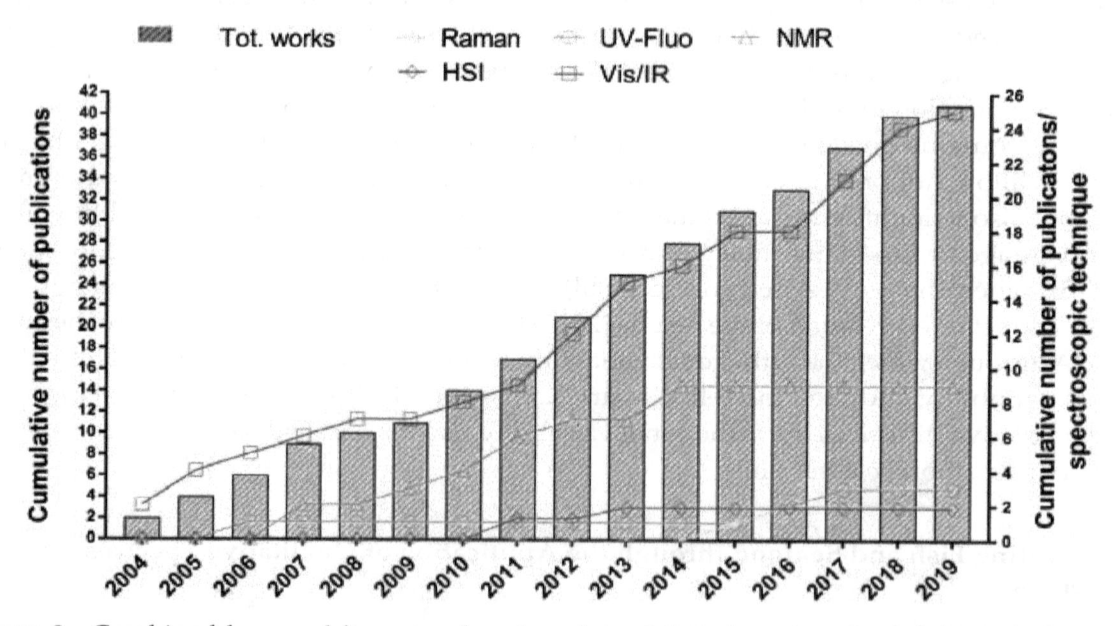

Figure 2. Combined bars and lines graph, where bars (plotted against the left Y-axis) show the cumulative number of scientific works concerning the use of spectroscopy and chemometrics for fish authentication purposes, and lines (plotted against the right Y-axis) show the cumulative number of works using each spectroscopic technique. Data were collected in February 2019 from the web search engine Google Scholar (search criteria: time period: "any time", and keywords: "fish and/or seafood"; "authenticity"; "spectroscopy"; "chemometrics".

3.1. *Species Substitution*

Substitution or counterfeit of high-value fish species with low-value ones has many quality and safety implications. Therefore, the confirmation of scientific and commercial names declared on the label through the use of rapid and low-cost methods is increasingly popular in food research.

3.1.1. Application of Vibrational Spectroscopy

An early study explored Vis-NIR spectroscopy as a tool to detect the counterfeit of Atlantic blue crabmeat (*Callinectes sapidus*) with blue swimmer crabmeat (*Portunus pelagicus*) in 10% increments, taking into consideration their different commercial values [50]. Qualitative chemometric analysis was performed on 400–2498 nm Vis-NIR spectra (previously subjected to different pretreatments to evaluate the effects on model performance), by means of a full-spectrum PCA and a sequential-spectrum PCA. As a result, both the first derivative-pretreated full spectra and second derivative-pretreated sequential spectra, highlighted a trend of samples towards moving from the left part to the right part of the PCA score plot with increased adulteration levels, but authors identified the sequential approach, using 400–1700 nm second derivative spectra, as being the most informative and, thus, the most suitable approach [50].

Based on the fact that the past several years have seen a sharp rise in the interest towards the portability of instruments, which may provide greater flexibility especially in on-line, in-line, and at-line routine quality control, a study performed by O'Brien et al. (2013), explored the ability of a hand-held NIR spectrometer to give positive results of discrimination between high-value and low-value whole fish and fish fillet species [51]. In particular, the objective was to discriminate between two different species of mullet (red mullet from mullet), cod (winter cod from cod), and trout (samlet from salmon trout). NIR spectra (906–1648 nm) obtained from skin (whole fish) and meat (fish fillets), were first pre-processed and then elaborated by PCA and SIMCA analysis. Successful PCA results were achieved only in separating the whole mullet samples, but the discrimination performances improved significantly also for mullet fillets after the application of the SIMCA analysis. PCA failed to discriminate both whole cod and cod fillets, but here too, SIMCA predictions provided a correct assignment of the tested fish samples. Similar outcomes for samlet from salmon trout were achieved [51]. Thus, although PCA investigation failed, SIMCA supervised analysis clearly outlined the possibility to authenticate high quality fish species which are potentially substitutable with lower-quality alternatives. Still in the context of the use of hand-held and compact NIR devices, a broader attempt to distinguish fillets and patties of Atlantic cod (*Gadus morhua*) from those of haddock (*Melanogrammus aeglefinus*) was recently made [52]. Raw fillets and patties of the two fish species were scanned at 950–1650 nm (by the portable instrument) or at 800–2222 nm (by a benchtop instrument) and after being pre-treated with SNV, MSC, or Savitzky–Golay smoothing (SG) coupled with first or second derivative, they were elaborated by means of supervised LDA and SIMCA analysis. Regardless of instrumentation used, the best LDA models were computed on the MSC spectra of both fillets and patties, since the correct classification rate in the external validation step reached 100% [52]. SIMCA class-modelling strategy obtained 100% correctly classified SNV, SG-first derivative, or SG-second derivative fillets spectra acquired by benchtop NIR, and 100% correctly classified MSC fillets spectra acquired with a portable NIR [52]. As for patties, samples acquired by benchtop NIR and portable NIR were 100% correctly classified when spectra were subjected to SG-first derivative or SG-second derivative, and SNV or MSC, respectively. The worst SIMCA outcomes in prediction for patties and fillets were obtained for SG-second derivative spectra acquired with the portable instrument. Despite these results, no significant differences in the performances of the two instruments tested were found, thus confirming equivalent discrimination powers also in processed product.

Different species of freshwater fish of the Cyprinidae family, namely black carp (*Mylopharyngodon piceus*), grass carp (*Ctenopharyngodon idellus*), silver carp (*Hypophthalmichthys molitrix*), bighead carp (*Aristichthys nobilis*), common carp (*Cyprinus carpio*), crucian (*Carassius auratus*), and bream (*Parabramis pekinensis*), were also investigated by NIR spectroscopy [53]. Fish samples were scanned in the

1000–1799 nm region, MSC pre-treated, and pre-reduced in dimensionality by different methods, including PCA, PLS, and fast Fourier transform (FFT). In this case, LDA models were built by using only nine pre-selected spectra wavelengths from the entire spectrum and results obtained showed a good prediction ability of the adopted strategy: PCA-LDA and FFT-LDA models, in fact, showed 100% accuracy, specificity, sensitivity, and precision, even if most of the information was not taken into account by calculation [53].

Zhang et al. (2017) attempted to classify marine fish surimi by 1100–2500 nm NIR spectroscopy, according to the species by which products were composed, namely white croaker (*Argyrosomus argentatus*), hairtail (*Trichiurus haumela*), and red coat (*Nemipterus virgatus*) [54]. According to results obtained from PCA of the pre-processed spectra, the presence of a well-defined and separated cluster associated with red coat surimi species was observed, but the separation of the other two species of surimi samples was not clear [54]. However, as regards LDA results, 100% correct classification rate for external validation datasets after MSC pre-treatment was achieved, demonstrating once again the greater effectiveness of supervised analyses compared to unsupervised ones.

Species authenticity was also studied by comparing FT-NIR and FT-MIR spectra of red mullet and plaice fillets (higher-value species) to those of Atlantic mullet and flounder fillets (lower-value species) [55]. LDA and SIMCA analysis applied to differently pre-treated NIR and MIR spectra (800–2500 nm and 2500–14,300 nm spectral ranges, respectively), clearly discriminated Atlantic mullet fillets from those of the more valuable red mullet. While LDA gave a 100% correct classification percentage in prediction (irrespective of the spectroscopic technique considered), sensitivity and specificity higher than 70% and 100%, respectively, were calculated for FT-NIR spectra subjected to SIMCA analysis [55]. Poorer, but acceptable, results were obtained for flounder and plaice fillets discrimination: in this case, FT-IR spectroscopy showed the best discrimination power, with a prediction ability higher than 83% and a specificity of 100%.

The usefulness of NIR spectroscopy was explored to identify different fish species used to make fishmeal under industrial conditions. The 1100–2500 nm raw or second derivative NIR spectra of samples containing salmon, blue whiting, and other (i.e., mackerel or herring) fish species were elaborated by PCA, LDA, and DPLS (PLS-DA). Models developed correctly classify, on average, more than 80% of the fish meal samples into the three groups assigned according to the fish species [56].

In contrast to the multiple applications of NIR spectroscopy, only one study explored the discrimination abilities of MIR spectroscopy [57]. This study coupled SG- and SNV-pre-treated MIR spectra (2500–20,000 nm) with chemometrics (PCA) to specifically detect adulteration of Atlantic salmon (*Salmo salar*) mini-burgers with different percentage (from 0 to 100%, in steps of 10%) of Rainbow trout (*Onconrhynchus mykiss*). The resulting 11 formulations of salmon burgers were grouped into 11 distinct clusters, even when the samples were stored for different periods of time before acquisition [57].

Only two applications of Raman spectroscopy concerning fish species authentication are available. The aim of the first study was to discriminate 12 different fish fillets of different species by using pre-treated Raman spectra in the range 300–3400 cm^{-1} (about 3940–33,333 nm) recorded by a Raman spectrometer equipped with a 532 nm laser exciting source [58]. HCA analysis applied to the Raman spectra revealed the presence of three major clusters, one corresponding to fish from the Salmonidae family (rainbow trout and Chum salmon), one corresponding to various freshwater fish (zander, Nile perch, pangasius, and European seabass), and one corresponding to various saltwater fish (Atlantic herring, Atlantic pollock, Alaska pollock, Atlantic cod, blue grenadier, and yellowfin tuna). Within these large clusters, spectra were also grouped according to their species in sub-clusters, with a high degree of accuracy of the spectral classification on species level (95.8%) [58]. Similarly, PCA analysis performed on 5000–50,000 nm Raman spectra (acquired by using a 785 nm laser exciting source) discriminated among horse mackerel (*Trachurus trachurus*), European anchovy (*Engraulis encrasicolus*), Bluefish (*Pomatamus saltatrix*), Atlantic salmon (*Salmo salar*), and flying gurnard (*Trigla lucerna*) samples. In this case, however, the study was less rapid and more elaborate since the spectral acquisition was performed on the previously extracted lipid fraction of fish [59].

3.1.2. Application of NMR Spectroscopy

Muscle lipids of four different species of fish belonging to the Gadoid family, namely cod (*Gadus morhua*), haddock (*Melanogrammus aeglifinus*), saithe (*Pollachius virens*), and pollack (*P. pollachius*), were subjected to ^{13}C-NMR spectroscopic analysis of phospholipid profiles, in order to authenticate samples according to the species [60]. As a result, supervised LDA and Bayesian belief network (BBN) performed on the resulting ^{13}C-NMR spectral peaks provided 78% and 100% of the correctly classified samples, respectively [60]. Other applications of NMR and chemometrics concerning fish species discrimination were not reported in literature until now. In our opinion, the method should be further explored in view of the several potentials and benefits provided, despite disadvantages deriving from the need of sample preparation prior to analysis.

3.2. *Production Method and Farming System Misrepresentation*

The differentiation of the production method of fish and seafood is another relevant aspect in certifying authenticity and traceability. During the last few years, the wild fish catches have been decreasing compared to the aquaculture production, thus supply of the market in farmed products has been growing very fast. From a compositional and organoleptic point of view, a wild fish is quite different from an aquaculture one, and this diversity is inevitably reflected on the different economic value of the two types of products [61–63]. By way of example, wild fish is usually characterised by higher levels of muscle protein, saturated, and polyunsaturated fatty acids, while farmed fish by a higher content of total lipid and monounsaturated fatty acids [64,65]. Consequently, the illegal substitution of higher-value wild fish with lower-value farmed fish is not an uncommon occurrence. Additionally, aquaculture fish consist of a number of high-variable products (i.e., extensively, semi-intensively, or intensively farmed fish, as well as organic or conventional farmed fish), whose final characteristics, since influenced by the husbandry environment and, above all, by the diet, are slight and very difficult to identify. This the reason is why the authentication of the production method (wild or farmed, organic or conventional), but also of the farming system of the aquaculture products is of extreme importance from the standpoint of fraud prevention and transparency towards consumers.

3.2.1. Application of Vibrational Spectroscopy

Among various vibrational spectroscopic methods applied to differentiate production processes and farming systems of fish, NIR is once again the most widely used. No application of UV or Raman spectroscopy, to the best of our knowledge, are currently available.

Ottavian et al. (2012) proposed a comparison between the classification performances of wild and farmed European sea bass obtained by three different NIR spectroscopic/chemometric approaches, and the classification performances obtained using only chemical and morphometric features [66]. The use of 1100–2500 nm raw spectra, WPTER-pre-treated spectra (wavelet packet transform for efficient pattern recognition), or of some parameters predicted by building a regression-based model, were found to be equivalent in terms of predictability assessed by PLS-DA and no differences between classification obtained by these models and classification obtained by using only chemical and morphometric data was observed. Moreover, authors identified (by using the variable influence of projection indexes, VIP) the wavelengths related to the absorbance of fat, fatty acids, and water as most influential in differentiating the production process of the fish tested.

More recently, the systems behind the production of European sea bass, was also investigated by applying unsupervised PCA and supervised OPLS-DA to 1100–2500 nm NIR spectra [67]. PCA built to SNV-SG-second derivative spectral data did not return a clear separation of groups, mainly as a consequence of the fact that the intraclass variability among samples was higher than the among-class variability between samples. A correct classification rate of 100% for both wild and farmed sea bass was instead achieved by OPLS-DA, and, in this case, authors found VIP indexes related to proteins exerting a greater contribution to the variance between the two types of fish. A deeper insight into the different farming systems of aquaculture samples, moreover, showed the ability of NIR and OPLS-DA to authenticate 67%, 80%, 100% of extensively, semi-intensively, and intensively-reared subjects, respectively, thanks above all to the spectral bands associated with protein absorption [67]. Concrete tank-cultured sea bass were also successfully discriminated from sea cage-cultured sea bass during storage, by means of Vis-NIR spectroscopy coupled with PLS-DA [68]. The best performances (87% of correct classification), were observed for spectral measurements performed at 48 h post mortem [68]. However, the greater contributions of the wavelengths to the PLS discrimination of samples analysed at 48 h post mortem were different from those of samples analysed at 96 h post, thus classification by farming system may have been affected also by other unrelated factors, such as the well-known compositional changes occurring during shelf life.

Authentication by NIR and SIMCA analysis of European sea bass raised in extensive ponds, semi-intensive ponds, intensive tanks, and intensive sea-cages, was also performed both on fresh fillets and freeze-dried fillets [69]. Authors found that freeze-drying the samples gave the best classification outcomes. The same results were obtained when classifying fresh minced fillets and freeze-dried fillets of farmed European sea bass according to the semi-intensive conventional or the organic production system [70]. SIMCA classification based on second-derivative spectra (1100–2500 nm) of samples, in fact, generated good results when fitted on the freeze-dried fillets (65–75% of correct classification), and worse results when performed on fresh fillets (20–25% of correct classification) [70]. All these results are particularly informative about problems posed by water when analysing high-moisture foods like fish. One of the main drawbacks of NIR spectroscopy is, in fact, the difficulty in separating relevant from useless information from spectra, in which peaks of water are predominant. These peaks, when included in chemometric calculations may hinder reliable features related to functional groups of molecules of interest and, thus, produce misleading results, especially when samples only slightly differ, such as in the case of fish reared under different conditions.

Following these principles, NIR spectroscopy was also used to directly authenticate freeze-dried rainbow trout fillets by rearing farm and, at the same time, to check whether NIR discriminating capability changed between raw and cooked freeze-dried fillets [71]. Rainbow trout samples came from three different aquaculture systems, varying in average well water temperatures, of which one consisted in indoor rearing at 11–14 °C, one in outdoor rearing at 9–11 °C, and one in outdoor rearing at 3–14 °C. Results for classification by farm (using SNV and second derivative 1100–2500 nm spectra of raw samples) showed approximately 97–100% of accuracy, with k-NN analysis giving the best overall statistical performances and PLS-DA the worst ones. As for cooked freeze-dried samples discrimination, the accuracy was approximately the same as those obtained for raw samples (90–100% for LDA, QDA, k-NN and 80% for PLS-DA), highlighting that the cooking process did not alter the capabilities of the technique to discriminate the sample by rearing farm [71].

3.2.2. Application of NMR Spectroscopy

Several applications of NMR spectroscopy aimed at authenticating the production process or the farming system were found in literature. In particular, proton (^{1}H) NMR spectroscopy can be used to analyse lipid mixtures such as fish oil, requiring simple preparation of samples and short time of spectra acquisition and providing a great deal of useful information [72]. Thus, considering that fish flesh lipids are the main compounds changing on the basis of the feeding regime, many attempts to use ^{1}H-NMR to identify the production process or the farming system were made. One of the earliest studies used SVM to elaborate ^{1}H-NMR spectra, and it was highly effective in predicting the wild or the farmed origin of salmon from different European countries [72]. Similarly, encouraging results were achieved through the combination of ^{1}H-NMR fingerprinting of lipids from gilthead sea bream with more complex chemometric data analyses [73]. The only unsupervised PCA applied on raw or processed ^{1}H-NMR spectral profiles returned, in fact, a clear separation between wild and farmed samples, which was found to be linked to methyl and methylene protons, together with methylene and methyne protons in unsaturated fatty acids [73]. Moreover, LDA variables selection allowed classification of 100% of the tested wild and farmed samples, and results from probabilistic neural network (PNN) analyses further reinforced the findings that such class discriminations were readily feasible.

If the previous studies were performed on fresh raw fish, other studies were intended to evaluate any differences in classification outcomes deriving from various degrees of fish processing. Lipids extracted from different types of processed Atlantic salmon products (frozen, smoked, and canned) were subjected to ^{1}H-NMR fingerprinting to develop models for determining labelling authenticity (wild/farmed) of these products [74]. SIMCA analysis applied to 138 pre-selected spectral peaks of NMR data, correctly classified as 100% of wild and 100% farmed samples, thanks mostly to the influence of a higher content of linoleic and oleic acid in farmed salmon compared to wild salmon [74]. A higher content of unsaturated fatty acids (and especially $n{-}3$ polyunsaturated fatty acids) was also found to play a special role in the discrimination between wild and farmed specimens of gilthead sea breams [75]. The influence exercised by these compounds was studied though the application of a supervised OPLS-DA to the whole lipid fingerprinting data obtained by ^{1}H-NMR spectroscopy. Just like SIMCA classification did in the previous study, OPLS-DA also led to a perfect separation of samples, but with the great advantage of being able to highlight the most effective variables in discrimination in the simplest of ways.

The ^{1}H-NMR molecular profiles of gilthead sea bream fish specimens produced according to different farming systems, have also been investigated, to seek out differences among three different kinds of aquaculture practices (cage, tank, and lagoon), but also any variations in the molecular patterns after a 16-day storage time under ice [76]. PCA-score plot of the pre-treated spectra showed a clear separation of fresh samples from ice-stored samples. At the same time, three distinct sub-clusters for each of the storage times, corresponding to the three farming systems investigated, highlighted the ability of the proposed methods to detect those molecular changes taking place during fish storage and exploited them for authentication purposes.

Another different NMR approach retrieved from the published literature concerned the use of carbon-13 (^{13}C) NMR instead of ^{1}H-NMR. Authors combined ^{13}C-NMR spectra of muscle lipids of Atlantic salmon with PNN and SVM chemometric elaborations, to discriminate between farmed and wild samples and obtained excellent discrimination performances (98.5% and 100.0% of correctly classified samples, respectively) [77]. Despite ^{13}C-NMR signals being generally much weaker than those provided by ^{1}H-NMR (as well as time of analysis is often longer), useful and complementary information can be obtained by this technique.

3.3. Geographical Origin Falsification

Proving the geographical origin authenticity of fish and seafood often involves the use of multi-disciplinary and cross-disciplinary approaches which take account of the environmental and genetic backgrounds affecting fish final characteristics [78]. Several published scientific researches concerning the use of spectroscopic methods pointed out the usefulness in classification of fish and seafood according to country or FAO area of origin.

3.3.1. Application of Vibrational Spectroscopy

Unlike the other authentication issues discussed above, NIR spectroscopy has been less explored for fish geographical origin identification. The reason, probably, is the great difficulty experienced in modelling total variability of NIR spectra and uniquely steering it to provenance, since provenance is the sum of a huge amount of different intrinsic or extrinsic factors (genetic, growth pattern, feeding regime, muscular activity, water temperature and salinity, etc.).

A traceability model able to predict the geographical origin of Chinese tilapia fillets coming from four different Chinese provinces, was developed by NIR spectroscopy [79]. SIMCA analysis, performed on 1000–2500 nm spectra of the minced samples, allowed more than 80% of fillets from Guangdong, Hainan, and Fujian provinces and 75% of fillets from the Fujian province to be correctly and exclusively assigned to the corresponding area of origin. Several locations in the Northern China Sea and East China Sea, from which sea cucumber (*Apostichopus japonicus*) come from, were also identified by using NIR spectroscopy [80]. In this case, authors found pre-treated (SNV or MSC, and second derivative) 1000–1800 nm spectra to give the best performance in PCA, since 100% correct classification rate was obtained both in the internal calibration model and in the external validation model. Similarly, 100% of sea cucumber analysed by means of diffuse reflectance MIR spectroscopy (fingerprint 5800–16,600 nm region) combined with SIMCA, were discriminated by the Chinese geographical region of provenance [81].

The last available application of NIR spectroscopy concerned the authentication of European sea bass according to Western, Central, or Eastern Mediterranean Sea provenances, by using OPLS-DA as a classification technique [67]. Results showed an overall discrimination performance of 89% according to these geographical origins, with 100% of Eastern, 88% of Central, and 85% of Western Mediterranean Sea samples being correctly classified. The VIP index analysis, moreover, identified lipid-associated bands as the most influential variables on the samples geographic discrimination.

3.3.2. Application of NMR Spectroscopy

Masoum et al. (2007) proposed a method for the origin authentication of Atlantic salmon based on ^{1}H-NMR and SVM of spectra extracted from samples coming from Canada, Alaska, Faroes, Ireland, Iceland, Norway, Scotland, and Tasmania. SVM returned a low degree of misclassification (4.6%) and, thus, an excellent correct classification rate for all the salmon samples [72]. Likewise, Aursand et al. (2009), used NMR combined with pattern recognition techniques to assess the geographical origin of Atlantic salmon and to verify the origin of market samples [77]. Here too, muscle lipids were extracted from tissues of fish coming from the same origins as those previously listed, but on the contrary, lipid composition was studied by ^{13}C-NMR coupled with PNN or SVM. This time, although the PNN- and SVM-based approaches used returned different correct classification rates (93.8% and 99.3%, respectively), a comparable classification accuracy between the two methodologies approaches was observed [77]. The ^{1}H-NMR lipid fingerprint, elaborated by LDA or PNN, allowed also to differentiate 76.2–100% of wild and farmed gilthead sea bream samples coming from Italy, Greece, Croatia, Turkey, and the Mediterranean Sea (for wild specimens), with better classification rates when PNN was applied [73]. Farmed gilthead sea bream specimens coming from five geographically distinct sites of Sardinia (Italy) and Greece were also discriminated by means of ^{1}H-NMR lipid fingerprint [75]. In this case, the fraction of unwanted variability related to the different production system of samples (off-shore sea cages and lagoon) was successfully overlooked thanks to the application of the OPLS-DA and, although authors did not provide statistical outcomes from internal or external classification, the significance of the clusters observed in the score plot was confirmed by bootstrap statistical analysis. The highest bootstrap values (indicating a well-defined class separation) were obtained for discrimination between Greek and Sardinian fish (100%), while lower but meaningful bootstrap values were obtained for discrimination among samples coming from different Sardinian offshore sea cage farms (68–57%) [75].

One last interesting application of ^{1}H-NMR dealt with the geographical authentication of bottarga, a fish-derived product consisting of salted and dried mullet (*Mugil Cephalus*) roe [82]. Low-molecular weight metabolites of aqueous extracts of samples, were analysed by PCA in order to identify clusters corresponding to one of the specific geographical provenances studied, namely FAO 37.1.3, FAO 34, FAO 41, FAO 31, and one unknown provenance. Results from PCA confirmed the possibility to characterise bottarga samples having different geographical origins, since samples with the same known geographical origin were closely clustered in the same region of the PCA scores plot, and those of different origin were far away from each other.

3.4. *Discrimination between Fresh and Frozen/Thawed Fish and Seafood*

Fish is commonly processed by freezing in order to be preserved from deterioration. Frozen fish, however, is usually characterised by much lower quality and commercial value compared to fresh fish. Therefore, fraudulent practices consisting in the substitution of fresh with frozen/thawed products are not uncommon events [83]. Considering that labelling of fish must state if the fish is fresh, frozen, or previously frozen (or refreshed), discriminating fresh from frozen/thawed products is one of the most important authenticity issues. The differentiation between fresh and frozen/thawed products is hampered by difficulties in detecting those tiny physical and chemical variations occurring during freeze storage, which, moreover, do not cause any perceptible organoleptic change [83,84]. Therefore, the rapid confirmation of fish freshness by spectroscopy has been widely studied during the last few years and several published researches are currently available.

3.4.1. Application of Fluorescence and Vibrational Spectroscopy

Front-face fluorescence spectroscopy is one of the earliest spectroscopic techniques historically applied to differentiate fresh from frozen/thawed fish. It has been demonstrated that typical changes in fluorescence spectra of aromatic amino acids, nucleic acids, and nicotinamide adenine dinucleotide (NADH) occur during storage, as a consequence of several reactions involving free amino acids and carbonyl compounds of reducing sugars, formaldehyde (produced from trimethylamine oxide), and malondialdehyde (produced from oxidation of fish lipids during storage). Therefore, changes in fluorescence of fish samples may be considered as fingerprints for fresh and aged fish fillet identification [85]. The fluorescence emission spectra of tryptophan (305–400 nm) recorded directly on whiting fillets and elaborated by factorial discriminant analysis (FDA) led to correct classification rates of 62.5% and 70.8% in the calibration and validation set, respectively. NADH fluorescence spectra (360–570 nm), indeed, were found to have a higher potential to differentiate fresh from frozen/thawed products as they allowed to achieve 100% of correct discrimination for both calibration and validation set [85]. More recently, the same authors confirmed the success of a similar methodology in authenticating freshness of sea bass samples. Fluorescence emission spectra at 340 and 380 nm, elaborated by FDA, led to 94.87% of total correct classification rate [86]. Additionally, the elaboration of NADH fluorescence spectra by Fisher's linear discriminant analysis, was stated as a reliable method to rapidly discriminate fresh and frozen/thawed large yellow croaker fillets, since 100% of total correct classification rate was achieved [87].

More applications of IR spectroscopy are reported in the published literature. Uddin and Okazaki (2004) used NIR reflectance spectroscopy on dry extract of horse mackerel specimens to evaluate freshness [88]. Both PCA (using 1100–2500 nm spectra) and SIMCA analysis (using only three selected wavelengths which were strongly related to protein content) successfully discriminated 100% of fresh and frozen/thawed samples. Thereafter, the same authors performed further investigations on fresh and frozen/thawed red sea bream by using Vis-NIR spectroscopy in the 400–1100 nm region [89]. In this case, raw spectra were used to build an LDA model, by which 100% classification accuracy in prediction was reached. PLS-DA of SG-smoothed spectra (670–1100 nm) of shrimps subjected to different treatments (including ice, water, and brine at various salt concentrations), also led to 100% of fresh and frozen/thawed samples to be authenticated [90].

Another study was directed to compare classification ability of Vis-NIR (380–1080 nm) and NIR (1100–2500) spectroscopy in authenticating fresh and frozen/thawed swordfish and, through the application of PLS-DA, it was found that in this case, Vis-NIR spectra gave better results in the external validation (≥96.7% of correctly classified samples) [91]. Although worse outcomes were obtained by only using the NIR region, the technique, combined with SVMs, also authenticated 93% of fresh and 83% of frozen/thawed sole (*Solea vulgaris*) samples [92]. Again, high accuracy (90%) and sensitivity (80%) in prediction were observed for the discrimination of fresh and frozen/thawed tuna sample by Vis-NIR spectral analysis (350–2500 nm) combined with PLS-DA [93], while better and more homogenous SIMCA prediction results were obtained when using MIR (2500–14,300 nm) instead of

NIR (800–2500 nm) regions for the discrimination between fresh and previously frozen Atlantic mullet fillets [94].

Ottavian et al. (2013) proposed an interesting three-step approach based only on NIR spectra and latent variable modelling techniques to develop a species-independent classifier able to simultaneously discriminate between fresh and frozen/thawed fish and, remarkably, overall classification accuracy of the method ranged between 80% and 91%, based on the strategy adopted and the instrument used [94]. By contrast, the only MIR region was found to be useful for determining whether whiting fish fillets have been frozen/thawed: when FDA was applied to the 3300–3570 nm MIR subregion (usually related to fatty acids absorption), 87.5% of sample spectra in the validation set was correctly identified [95].

Finally, one single application of Raman spectroscopy to the authentication of fresh fish is now available [59]. Lipid fraction of fish from several species (horse mackerel, European anchovy, bluefish, Atlantic salmon, red mullet, and flying gurnard) was extracted from three samples batches (fresh samples, once frozen/towed samples, and twice frozen/thawed samples), and then collected by a Raman spectrometer along the 5000–50,000 nm spectral range and using a 785 nm laser exciting source. Chemometric analysis, performed by PCA, identified three different clusters in the score plot, each corresponding to one of the three batches of fish investigated [59].

3.4.2. Application of Hyperspectral Imaging Spectroscopy

Discrimination between fresh and frozen/thawed cod fillet was studied by Vis-NIR/HSI, using both a handheld interactance probe and an imaging spectrometer (for automatic online analysis at typical industrial speeds) [96]. Spectra resulting from the two instruments were pre-treated (SNV and second derivative) and statistically analysed by applying the Rosenblatt's perceptron linear classifier to the first and third principal component of the imaging data. Results showed that fresh cod fillets can be completely separated from fresh/thawed cod fillets using only a few wavelengths in the Vis region, mainly related to the oxidation of haemoglobin and myoglobin which occur during freezing/thawing [97]. Similarly, hyperspectral data from Vis-NIR/HSI (380–1030 nm) combined with least square-SVMs, returned an average correct classification rate of 91.67% for fresh and frozen/thawed halibut fillets [97].

3.4.3. Application of NMR Spectroscopy

NMR spectroscopy is considered to be a useful and suitable tool for the discrimination of fresh from frozen/thawed fish, since NMR signals are sensitive enough to changes in water mobility and its interaction with other molecules [98]. NMR spectroscopy has been already widely exploited to identify the various modifications in fish tissues occurring during freezing and thawing of fish [99–102]; however, as far as we know, no application of this technique for fish freshness authentication is currently available.

4. Critical Aspects and Limitations to Overcome

The food scientists' interest towards the development of reliable methods for the resolution of several food authenticity issues is well documented by the increasing number of scientific works which, albeit through different methodologies, have attempted to address the same problems. It is clear from the analysis of the latest literature that spectroscopy combined with chemometrics is just one of the many untargeted strategies adopted: chromatographic, MS-based, as well as bio-molecular and sensory techniques have been already widely exploited and have demonstrated their exceptional multipurpose qualities for fish authenticity testing [78,103–108].

These techniques are known to share certain common disadvantages, such as the long time needed for analysis, high costs of the equipment, the need of sample preparation prior to analysis, destructiveness, and the demand for qualified personnel. On the other hand, as they become more consolidated within the research community, these techniques excel by their higher accuracy, specificity, and sensitivity compared to spectroscopic ones, to the point that many of them are used in food official controls. Despite this, the attractiveness of spectroscopy and chemometrics is evidenced by not only by the large literature provided in the present review, but also by several other applications covering a wide range of food and foodstuffs: fruits and vegetables, honey, wine, edible oils and fats, cereal and cereal-based products, milk, and dairy products [109–114] have been successfully investigated and authenticated by means of spectroscopy.

Having said that, some critical reflections should be made about the problems related to the use of spectroscopy and chemometrics, which still have not been overcome. In accordance to what has been already reported and to our opinion, the research papers analysed were found to be highly variable to each other in terms of analytical set-up (e.g., sample pre-processing, spectral ranges, spectra pre-treatments, resolutions, number of samples tested, and statistical elaboration). This variability, as easily understood from Section 3, is further worsened by the fact that only a few of the works analysed reported in-depth statistical outputs and, where present, they were not comparable to each other.

A critical and objective evaluation of these works is also severely hampered by a lack, in certain cases, of comprehensive data with regards to the validation of the results. Alongside the internal cross-validation, the external validation of the qualitative chemometric model is, in our opinion, a crucial point in assessing the overall goodness of the classifiers and avoiding misleading interpretations. The last aspect which should be emphasised is that a detailed description of the characteristics of the sample dataset was not often reported and the lack of standardisation of external factors (e.g., storage times and conditions), may have interfered with spectral analysis, possibly affecting the robustness of the model. In this scenario, a recommendation for future works is to consider the intrinsically natural variability of the fish products (as well as those of all other foodstuffs), and to organise the sampling in such a way that as much of the expected variability of samples is collected during the calibration stage. That way, the robustness of the models can make their way to the spread of applications also in the industrial sector.

As a final remark, no technique should be universally regarded as the optimal solution. However, the possibility of using UV, IR, Raman, and NMR spectroscopies with no distinction for food authentication purposes is still an obstacle to overcome, and therefore, in accordance to our experience, untargeted NIR spectroscopy represents the most versatile option thanks to its high sensitivity to organic molecules of food, cost-effectiveness, and ease of use. Additionally, the use of NIR spectroscopy with supervised chemometric method, able to separate relevant from non-relevant spectral variation like OPLS-DA, should be encouraged since the interpretability of results is enhanced.

5. Conclusions and Prospects for the Future

Recent increases in the complexity and competitiveness of the fishery and seafood sectors, have resulted in the presence, on the international market, of a huge variety of fresh and processed products, but at the same time, have meant that the risk of fraud deriving from substitution among look-alike products is now exponentially higher than it was even a few years ago. Thus, ensuring the truthfulness of fish and seafood claims concerning their quality and origin, has become an exceptionally important topic, firstly with a view to enable consumers to make informed decisions.

The overview presented in this review clearly highlights the effective support provided by analytical approaches based on spectroscopy and multivariate data analysis for the evaluation and monitoring of fish and seafood products authenticity. Fluorescence, vibrational, NMR, and HSI spectroscopic applications have been discussed, with an accent on the trends toward their use for several authentication purposes. In this connection, IR spectroscopy has been the most exploited technique, especially in studies concerning species and fresh for frozen/thawed products substitutions. NMR, instead, has shown many applications in the field on the production method, farming system, and geographical origin identification. By contrast, Raman and HIS have provided very encouraging results in some fish authentication fields, but their overall potential has so far been largely ignored.

Rapidity, non-destructive nature, ease of use, and high-throughput measurements make the spectroscopic non-targeted approach an ideal tool for quality control operations, especially in the context of daily routine and screening analysis in the food industry, and as a possible substitute of traditional analytical techniques. Thanks to the technological development of the spectroscopic instrumentation, the availability of miniaturized and portable devices on the market is rapidly growing, and this will contribute to an additional growth of applications in the food sector. On the other hand, these analytical strategies in the official control of foodstuffs are still far from being effectively applied, largely due to the need of a strict validation to assure further reliability and robustness of results before implementation as standalone tools. For these reasons, standardisation of the working conditions, optimisation of the chemometric software, and creation of large databases for data-sharing and for encouraging greater cooperation between food scientists, represent important current research fields and future challenges to be faced.

Author Contributions: All authors contributed equally to this work.

Abbreviations

ANN	artificial neural networks;
BBN	Bayesian belief network;
FIR	far-infrared;
FDA	factorial discriminant analysis;
FFT	fast Fourier transform;
FT	Fourier transform;
HCA	hierarchical cluster analysis;
HSI	hyperspectral imaging;
IR	infrared;
k-NN	k-nearest neighbors;
LDA	linear discriminant analysis;
LW-NIR	long-wave near infrared;
MIR	mid-infrared;
NMR	nuclear magnetic resonance;
MSC	multiplicative scatter correction;
NIR	near-infrared;
OPLS-DA	orthogonal partial least square-discriminant analysis;
PCA	principal component analysis;
PLS-DA	partial least square-discriminant analysis;
PNN	probabilistic neural network;
QDA	quadratic factorial analysis;
SERS	surface-enhanced Raman spectroscopy;
SG	Savitzky–Golay smoothing;
SIMCA	soft independent modelling of class analogy;
SNV	standard normal variate;
SVM	support vector machine;
SW-NIR	short-wave near infrared;
UV	ultraviolet;
Vis	visible.

References

1. FAO. *The State of World Fisheries and Aquaculture 2018–Meeting the Sustainable Development Goals*; FAO: Rome, Italy, 2018; Volume 35, pp. 52–62. ISBN 978-92-5106-029-2.
2. Rehbein, H.; Oehlenschläger, J. Basic Facts and Figures. In *Fishery products: Quality, Safety and Authenticity*; Rehbein, H., Oehlenschläger, J., Eds.; John Wiley & Sons: Chichester, West Sussex, UK, 2009; Volume 1, pp. 1–18. ISBN 978-1-4051-4162-8.
3. FAO. *Overview of Food Fraud in the Fisheries Sector, by Alan Reilly; Fisheries and Aquaculture Circular No. 1165*; FAO: Rome, Italy, 2018; pp. 4–6. ISBN 978-92-5-130402-0.
4. Van Dolah, F.M. Marine Algal Toxins: Origins, Health Effects, and Their Increased Occurrence. *Environ. Health Perspect.* **2000**, *108*, 133–141. [CrossRef]
5. Fallah, A.A.; Saei-Dehkordi, S.S.; Nematollahi, A.; Jafari, T. Comparative study of heavy metal and trace element accumulation in edible tissues of farmed and wild rainbow trout (Oncorhynchus mykiss) using ICP-OES technique. *Microchem. J.* **2011**, *98*, 275–279. [CrossRef]
6. Okocha, R.C.; Olatoye, I.O.; Adedeji, O.B. Food safety impacts of antimicrobial use and their residues in aquaculture. *Public Health Rev.* **2018**, *39*, 1–22. [CrossRef]
7. Kelly, B.C.; Ikonomou, M.G.; Higgs, D.A.; Oakes, J.; Dubetz, C. Flesh residue concentrations of organochlorine pesticides in farmed and wild salmon from British Columbia, Canada. *Environ. Toxicol. Chem.* **2011**, *30*, 2456–2464. [CrossRef]
8. Council regulation (EC) No 178/2002 of the European Parliament and of the Council of 28 January2002 laying down the general principles and requirements of food law, establishing the European Food Safety Authority and laying down procedures in matters of food safety. *Off. J. Eur. Commun.* **2002**, *31*, 1–24.
9. Council regulation (EC) No 1005/2008 of 29 September 2008 establishing a Community system to prevent, deter and eliminate illegal, unreported and unregulated fishing, amending Regulations (EEC) No 2847/93, (EC) No 1936/2001 and (EC) No 601/2004 and repealing Regulations (EC) No 1093/94 and (EC) No 1447/1999. *Off. J. Eur. Union* **2008**, *286*, 1–32.
10. Council regulation (EC) No 1224/2009 of 20 November 2009 establishing a Community control system for ensuring compliance with the rules of the common fisheries policy, amending Regulations (EC) No 847/96, (EC) No 2371/2002, (EC) No 811/2004, (EC) No 768/2005, (EC) No 2115/2005, (EC) No 2166/2005, (EC) No 388/2006, (EC) No 509/2007, (EC) No 676/2007, (EC) No 1098/2007, (EC) No 1300/2008, (EC) No 1342/2008 and repealing Regulations (EEC) No 2847/93, (EC) No 1627/94 and (EC) No 1966/2006. *Off. J. Eur. Union* **2009**, *343*, 1–50.
11. Regulation (EU) No 1169/2011 of the European Parliament and of the Council of 25 October 2011 on the provision of food information to consumers, amending Regulations (EC) No 1924/2006 and (EC) No 1925/2006 of the European Parliament and of the Council, and repealing Commission Directive 87/250/EEC, Council Directive 90/496/EEC, Commission Directive 1999/10/EC, Directive 2000/13/EC of the European Parliament and of the Council, Commission Directives 2002/67/EC and 2008/5/EC. *Off. J. Eur. Union* **2011**, *304*, 18–63.
12. Regulation (EU) No 1379/2013 of the European Parliament and of the Council of 11 December 2013 on the common organisation of th emarkets in fishery and aquaculture products, amending Council Regulations (EC) No 1184/2006 and (EC) No 1224/2009 and repealing Council Regulation (EC) No 104/2000. *Off. J. Eur. Union* **2013**, *354*, 12–14.
13. Esslinger, S.; Riedl, J.; Fauhl-Hassek, C. Potential and limitations of non-targeted fingerprinting for authentication of food in official control. *Food Res. Int.* **2014**, *60*, 189–204. [CrossRef]
14. Picò, Y. Near-Infrared, Mid-Infrared, and Raman Spectroscopy. In *Chemical Analysis of Food: Techniques and Applications*; Picò, Y., Ed.; Academic Press: San Francisco, CA, USA, 2012; Volume 1, pp. 59–91. ISBN 978-0-1238-4862-8.
15. Schrieber, A. Introduction to Food Authentication. In *Modern Techniques for Food Authentication*; Sun, D.W., Ed.; Academic Press: San Francisco, CA, USA, 2008; pp. 1–21.
16. Penner, M.H. Basic Principles of Spectroscopy. In *Food Analysis, Food Science Text Series*, 2nd ed.; Nielsen, S.S., Ed.; Springer: Cham, Switzerland; New York, NY, USA, 2017; pp. 79–88. ISBN 978-3-319-45776-5.
17. Strasburg, G.M.; Ludescher, R.D. Theory and applications of fluorescence spectroscopy in food research. *Trends Food Sci. Technol.* **1995**, *6*, 69–75. [CrossRef]

18. Xu, J.L.; Riccioli, C.; Sun, D.W. An Overview on Nondestructive Spectroscopic Techniques for Lipid and Lipid Oxidation Analysis in Fish and Fish Products. *Compr. Rev. Food Sci. Food Saf.* **2015**, *4*, 466–477. [CrossRef]
19. Rodriguez-Saona, L.E.; Allendorf, M.E. Use of FTIR for rapid authentication and detection of adulteration of food. *Ann. Rev. Food Sci. Technol.* **2011**, *2*, 467–483. [CrossRef] [PubMed]
20. Rodriguez-Saona, L.E.; Giusti, M.M.; Shotts, M. *Advances in Infrared Spectroscopy for Food Authenticity Testing*; Downey, G., Ed.; Woodhead Publishing: Duxford, UK, 2016; ISBN 978-0-08-100220-9.
21. Lohumi, S.; Lee, S.; Lee, H.; Cho, B.K. A review of vibrational spectroscopic techniques for the detection of food authenticity and adulteration. *Trends Food Sci. Technol.* **2015**, *1*, 85–98. [CrossRef]
22. Blanco, M.; Villarroya, I.N.I.R. NIR spectroscopy: A rapid-response analytical tool. *TrAC Trends Anal. Chem.* **2002**, *21*, 240–250. [CrossRef]
23. Cen, H.; Yong, H. Theory and application of near infrared reflectance spectroscopy in determination of food quality. *Trends Food Sci. Technol.* **2007**, *18*, 72–83. [CrossRef]
24. Cheng, J.H.; Dai, Q.; Sun, D.W.; Zeng, X.A.; Liu, D.; Pu, H.B. Applications of non-destructive spectroscopic techniques for fish quality and safety evaluation and inspection. *Trends Food Sci. Technol.* **2013**, *34*, 18–31. [CrossRef]
25. Cozzolino, D.; Murray, I. A review on the application of infrared technologies to determine and monitor composition and other quality characteristics in raw fish, fish products, and seafood. *Appl. Spectrosc. Rev.* **2012**, *47*, 207–218. [CrossRef]
26. Stuart, B.H. Spectral Analysis. In *Infrared Spectroscopy: Fundamentals and Applications*; Stuart, B.H., Ed.; Jhon Wiley & Sons Ltd.: Chichester, WS, UK, 2004; pp. 47–48. ISBN 0-470-85427-8.
27. Boyaci, I.H.; Temiz, H.T.; Geniş, H.E.; Soykut, E.A.; Yazgan, N.N.; Güven, B.; Uysal, R.S.; Bozkurt, A.G.; Ilaslan, K.; Torun, O.; et al. Dispersive and FT-Raman spectroscopic methods in food analysis. *RSC Adv.* **2015**, *5*, 56606–56624. [CrossRef]
28. Zheng, J.; He, L. Surface-Enhanced Raman Spectroscopy for the Chemical Analysis of Food. *Compr. Rev. Food Sci. Food Saf.* **2014**, *13*, 317–328. [CrossRef]
29. Feng, Y.Z.; Sun, D.W. Application of hyperspectral imaging in food safety inspection and control: A review. *Crit. Rev. Food Sci. Nutr.* **2012**, *52*, 1039–1058. [CrossRef]
30. Wu, D.; Sun, D.W. Advanced applications of hyperspectral imaging technology for food quality and safety analysis and assessment: A review—Part I: Fundamentals. *Innov. Food Sci. Emerg. Technol.* **2013**, *19*, 1–14. [CrossRef]
31. Roberts, J.; Power, A.; Chapman, J.; Chandra, S.; Cozzolino, D. A short update on the advantages, applications and limitations of hyperspectral and chemical imaging in food authentication. *Appl. Sci.* **2018**, *8*, 505. [CrossRef]
32. Sacchi, R.; Paolillo, L. NMR for Food Quality and Traceability. In *Advances in Food Diagnostics*; Nollet, L.M.L., Toldrà, F., Eds.; Blackwell Publishing: Oxford, UK, 2007; Volume 1, pp. 101–107. ISBN 978-0-4702-7780-5.
33. Hatzakis, E. Nuclear Magnetic Resonance (NMR) Spectroscopy in Food Science: A Comprehensive Review. *Compre. Rev. Food Sci. Food Saf.* **2019**, *18*, 189–220. [CrossRef]
34. Sobolev, A.P.; Circi, S.; Mannina, L. Advances in Nuclear Magnetic Resonance Spectroscopy for Food Authenticity Testing. In *Advances in Food Authenticity Testing*; Downey, G., Ed.; Woodhead Publishing: Duxford, UK, 2016; Volume 1, pp. 147–170. [CrossRef]
35. Oliveri, P.; Simonetti, R. Chemometrics for food authenticity applications. In *Advances in Food Authenticity Testing*; Downey, G., Ed.; Woodhead Publishing: Duxford, UK, 2016; pp. 701–728. [CrossRef]
36. Manley, M.; Baeten, V. Spectroscopic technique: Near infrared (NIR) spectroscopy. In *Modern Techniques for Food Authentication*; Sun, D.W., Ed.; Academic Press: San Francisco, CA, USA, 2008; Volume 1, pp. 51–102.
37. Rinnan, Å.; Van Den Berg, F.; Engelsen, S.B. Review of the most common pre-processing techniques for near-infrared spectra. *Trends Anal. Chem.* **2009**, *28*, 1201–1222. [CrossRef]
38. Pereira, A.C.; Reis, M.S.; Saraiva, P.M.; Marques, J.C. Madeira wine ageing prediction based on different analytical techniques: UV–vis, GC-MS, HPLC-DAD. *Chemom. Intell. Lab. Syst.* **2011**, *105*, 43–55. [CrossRef]
39. Rinnan, Å. Pre-processing in vibrational spectroscopy–when, why and how. *Anal. Methods* **2014**, *6*, 7124–7129. [CrossRef]
40. Engel, J.; Gerretzen, J.; Szymańska, E.; Jansen, J.J.; Downey, G.; Blanchet, L.; Buydens, L.M.C. Breaking with trends in pre-processing? *TrAC Trends Anal. Chem.* **2013**, *50*, 96–106. [CrossRef]

41. Ziegel, E.R. A User-Friendly Guide to Multivariate Calibration and Classification. *Technometrics* **2004**, *46*, 108–111. [CrossRef]
42. Voncina, D.B. Chemometrics in analytical chemistry. *Nova Biotechnol.* **2009**, *9*, 211–216. [CrossRef]
43. Berrueta, L.A.; Alonso-Salces, L.M.; Heberger, K. Supervised pattern recognition in food analysis. *J. Chromatog. A* **2007**, *1158*, 196–214. [CrossRef] [PubMed]
44. Ahmad, A.R.; Khalid, M.; Yusof, R. Machine Learning Using Support Vector Machines. In Proceedings of the International Conference on Artificial Intelligence in Science and Technology, Hobart, Australia, 19–21 September 2002.
45. Tetko, I.V.; Livingstone, D.J.; Luik, A.I. Neural network studies. 1. Comparison of overfitting and overtraining. *J. Chem. Inf. Comput. Sci.* **1995**, *35*, 826–833. [CrossRef]
46. Bylesjö, M.; Rantalainen, M.; Cloarec, O.; Nicholson, J.K.; Holmes, E.; Trygg, J. OPLS Discriminant Analysis, Combining the strengths of PLS-DA and SIMCA classification. *J. Chemom.* **2007**, *20*, 341–351. [CrossRef]
47. He, H.J.; Wu, D.; Sun, D.W. Nondestructive Spectroscopic and Imaging Techniques for Quality Evaluation and Assessment of Fish and Fish Products. *Crit. Rev. Food Sci. Nutr.* **2015**, *55*, 864–886. [CrossRef] [PubMed]
48. Uddin, M.; Okazaki, E. Applications of Vibrational Spectroscopy to the Analysis of Fish and Other Aquatic Food Products. *Handb. Vib. Spectro.* **2006**, 439–459. [CrossRef]
49. Fiorino, G.M.; Garino, C.; Arlorio, M.; Logrieco, A.F.; Losito, I.; Monaci, L. Overview on Untargeted Methods to Combat Food Frauds: A Focus on Fishery Products. *J. Food Qual.* **2018**, *3*, 1–13. [CrossRef]
50. Gayo, J.; Hale, S.A.; Blanchard, S.M. Quantitative analysis and detection of adulteration in crab meat using visible and near-infrared spectroscopy. *J. Agric. Food Chem.* **2006**, *54*, 1130–1136. [CrossRef]
51. O'Brien, N.; Hulse, C.A.; Pfeifer, F.; Siesler, H.W. Near infrared spectroscopic authentication of seafood. *J. Near Infrared Spectrosc.* **2013**, *21*, 299–305. [CrossRef]
52. Grassi, S.; Casiraghi, E.; Alamprese, C. Handheld NIR device: A non-targeted approach to assess authenticity of fish fillets and patties. *Food Chem.* **2018**, *243*, 382–388. [CrossRef]
53. Lv, H.; Xu, W.; You, J.; Xiong, S. Classification of freshwater fish species by linear discriminant analysis based on near infrared reflectance spectroscopy. *J. Near Infrared Spectrosc.* **2017**, *25*, 54–62. [CrossRef]
54. Zhang, X.Y.; Hu, W.; Teng, J.; Peng, H.H.; Gan, J.H.; Wang, X.C.; Sun, S.Q.; Xu, C.H.; Liu, Y. Rapid recognition of marine fish surimi by one-step discriminant analysis based on near-infrared diffuse reflectance spectroscopy. *Int. J. Food Prop.* **2017**, *20*, 2932–2943. [CrossRef]
55. Alamprese, C.; Casiraghi, E. Application of FT-NIR and FT-IR spectroscopy to fish fillet authentication. *LWT–Food Sci. Technol.* **2015**, *63*, 720–725. [CrossRef]
56. Cozzolino, D.; Chree, A.; Scaife, J.R.; Murray, I. Usefulness of Near-infrared reflectance (NIR) spectroscopy and chemometrics to discriminate fishmeal batches made with different fish species. *J. Agric. Food Chem.* **2005**, *53*, 4459–4463. [CrossRef] [PubMed]
57. Sousa, N.; Moreira, M.; Saraiva, C.; de Almeida, J. Applying Fourier Transform Mid Infrared Spectroscopy to Detect the Adulteration of Salmo salar with Oncorhynchus mykiss. *Foods* **2018**, *7*, 55. [CrossRef]
58. Rašković, B.; Heinke, R.; Rösch, P.; Popp, J. The Potential of Raman Spectroscopy for the Classification of Fish Fillets. *Food Anal. Methods* **2016**, *9*, 1301–1306. [CrossRef]
59. Velioğlu, H.M.; Temiz, H.T.; Boyaci, I.H. Differentiation of fresh and frozen-thawed fish samples using Raman spectroscopy coupled with chemometric analysis. *Food Chem.* **2015**, *173*, 283–290. [CrossRef]
60. Standal, I.B.; Axelson, D.E.; Aursand, M. 13C NMR as a tool for authentication of different gadoid fish species with emphasis on phospholipid profiles. *Food Chem.* **2010**, *121*, 608–615. [CrossRef]
61. Gabr, H.R.; Gab-Alla, A.A.F.A. Comparison of biochemical composition and organoleptic properties between wild and cultured finfish. *J. Fish. Aquat. Sci.* **2007**, *2*, 77–81. [CrossRef]
62. Grigorakis, K.; Taylor, K.D.A.; Alexis, M.N. Organoleptic and volatile aroma compounds comparison of wild and cultured gilthead sea bream (Sparus aurata): Sensory differences and possible chemical basis. *Aquaculture* **2003**, *225*, 109–119. [CrossRef]
63. Grigorakis, K. Compositional and organoleptic quality of farmed and wild gilthead sea bream (Sparus aurata) and sea bass (Dicentrarchus labrax) and factors affecting it: A review. *Aquaculture* **2007**, *272*, 55–75. [CrossRef]
64. Lenas, D.; Chatziantoniou, S.; Nathanailides, C.; Triantafillou, D. Comparison of wild and farmed sea bass (*Dicentrarchus labrax* L) lipid quality. *Procedia Food Sci.* **2011**, *1*, 1139–1145. [CrossRef]

65. Fuentes, A.; Fernández-Segovia, I.; Serra, J.A.; Barat, J.M. Comparison of wild and cultured sea bass (Dicentrarchus labrax) quality. *Food Chem.* **2010**, *119*, 1514–1518. [CrossRef]
66. Ottavian, M.; Facco, P.; Fasolato, L.; Novelli, E.; Mirisola, M.; Perini, M.; Barolo, M. Use of near-infrared spectroscopy for fast fraud detection in seafood: Application to the authentication of wild European sea bass (Dicentrarchus labrax). *J. Agric. Food Chem.* **2012**, *60*, 639–648. [CrossRef]
67. Ghidini, S.; Varrà, M.O.; Dall'Asta, C.; Badiani, A.; Ianieri, A.; Zanardi, E. Rapid authentication of European sea bass (*Dicentrarchus labrax* L.) according to production method, farming system, and geographical origin by near infrared spectroscopy coupled with chemometrics. *Food Chem.* **2019**, *280*, 321–327. [CrossRef]
68. Costa, C.; D'Andrea, S.; Russo, R.; Antonucci, F.; Pallottino, F.; Menesatti, P. Application of non-invasive techniques to differentiate sea bass (Dicentrarchus labrax, L. 1758) quality cultured under different conditions. *Aquac. Int.* **2011**, *19*, 765–778. [CrossRef]
69. Xiccato, G.; Trocino, A.; Tulli, F.; Tibaldi, E. Prediction of chemical composition and origin identification of european sea bass (*Dicentrarchus labrax* L.) by near infrared reflectance spectroscopy (NIRS). *Food Chem.* **2004**, *86*, 275–281. [CrossRef]
70. Trocino, A.; Xiccato, G.; Majolini, D.; Tazzoli, M.; Bertotto, D.; Pascoli, F.; Palazzi, R. Assessing the quality of organic and conventionally-farmed European sea bass (*Dicentrarchus labrax*). *Food Chem.* **2012**, *131*, 427–433. [CrossRef]
71. Dalle Zotte, A.; Ottavian, M.; Concollato, A.; Serva, L.; Martelli, R.; Parisi, G. Authentication of raw and cooked freeze-dried rainbow trout (Oncorhynchus mykiss) by means of near infrared spectroscopy and data fusion. *Food Res. Int.* **2014**, *60*, 180–188. [CrossRef]
72. Masoum, S.; Malabat, C.; Jalali-Heravi, M.; Guillou, C.; Rezzi, S.; Rutledge, D.N. Application of support vector machines to 1H NMR data of fish oils: Methodology for the confirmation of wild and farmed salmon and their origins. *Anal. Bioanal. Chem.* **2007**, *387*, 1499–1510. [CrossRef]
73. Rezzi, S.; Giani, I.; Héberger, K.; Axelson, D.E.; Moretti, V.M.; Reniero, F.; Guillou, C. Classification of gilthead sea bream (Sparus aurata) from1H NMR lipid profiling combined with principal component and linear discriminant analysis. *J. Agric. Food Chem.* **2007**, *55*, 9963–9968. [CrossRef]
74. Capuano, E.; Lommen, A.; Heenan, S.; de la Dura, A.; Rozijn, M.; van Ruth, S. Wild salmon authenticity can be predicted by 1H-NMR spectroscopy. *Lipid Technol.* **2012**, *24*, 251–253. [CrossRef]
75. Melis, R.; Cappuccinelli, R.; Roggio, T.; Anedda, R. Addressing marketplace gilthead sea bream (*Sparus aurata* L.) differentiation by 1H NMR-based lipid fingerprinting. *Food Res. Int.* **2014**, *63*, 258–264. [CrossRef]
76. Picone, G.; Balling, S.; Engelsen, F.; Savorani, S.; Testi, S.; Badiani, A.; Capozzi, F. Metabolomics as a powerful tool for molecular quality assessment of the fish Sparus aurata. *Nutrients* **2011**, *3*, 212–227. [CrossRef] [PubMed]
77. Aursand, M.; Standal, I.B.; Praél, A.; Mcevoy, L.; Irvine, J.; Axelson, D.E. 13C NMR pattern recognition techniques for the classification of atlantic salmon (*salmo salar* L.) according to their wild, farmed, and geographical origin. *J. Agric. Food Chem.* **2009**, *57*, 3444–3451. [CrossRef] [PubMed]
78. Abbas, O.; Zadravec, M.; Baeten, V.; Mikuš, T.; Lešić, T.; Vulić, A.; Prpić, J.; Jemeršić, L.; Pleadin, J. Analytical methods used for the authentication of food of animal origin. *Food Chem.* **2018**, *246*, 6–17. [CrossRef]
79. Liu, Y.; Ma, D.H.; Wang, X.C.; Liu, L.P.; Fan, Y.X.; Cao, J.X. Prediction of chemical composition and geographical origin traceability of Chinese export tilapia fillets products by near infrared reflectance spectroscopy. *LWT–Food Sci. Technol.* **2015**, *60*, 1214–1218. [CrossRef]
80. Guo, X.; Cai, R.; Wang, S.; Tang, B.; Li, Y.; Zhao, W. Non-destructive geographical traceability of sea cucumber (Apostichopus japonicus) using near infrared spectroscopy combined with chemometric methods. *R. Soc. Open Sci.* **2018**, *5*. [CrossRef] [PubMed]
81. Wu, Z.; Tao, L.; Zhang, P.; Li, P.; Zhu, Q.; Tian, Y.; Yang, T. Diffuse reflectance mid-infrared Fourier transform spectroscopy (DRIFTS) for rapid identification of dried sea cucumber products from different geographical areas. *Vib. Spectro.* **2010**, *53*, 222–226. [CrossRef]
82. Locci, E.; Piras, C.; Mereu, S.; Cesare Marincola, F.; Scano, P. 1H NMR metabolite fingerprint and pattern recognition of mullet (Mugil cephalus) bottarga. *J. Agric. Food Chem.* **2011**, *59*, 9497–9505. [CrossRef]
83. Verrez-Bagnis, V.; Sotelo, C.G.; Mendes, R.; Silva, H.; Kappel, K.; Schröder, U. Methods for Seafood Authenticity Testing in Europe. In *Bioactive Molecules in Food*; Springer: Cham, Switzerland; New York, NY, USA, 2018; Volume 1, pp. 1–55.

84. Tokur, B.; Ozkütük, S.; Atici, E.; Ozyurt, G.; Ozyurt, C.E. Chemical and sensory quality changes of fish fingers, made from mirror carp (*Cyprinus carpio* L., 1758), during frozen storage (−18 °C). *Food Chem.* **2006**, *99*, 335–341. [CrossRef]
85. Karoui, R.; Thomas, E.; Dufour, E. Utilisation of a rapid technique based on front-face fluorescence spectroscopy for differentiating between fresh and frozen–thawed fish fillets. *Food Res. Int.* **2006**, *39*, 349–355. [CrossRef]
86. Karoui, R.; Hassoun, A.; Ethuin, P. Front face fluorescence spectroscopy enables rapid differentiation of fresh and frozen-thawed sea bass (Dicentrarchus labrax) fillets. *J. Food Eng.* **2017**, *202*, 89–98. [CrossRef]
87. Gao, Y.; Tang, H.; Ou, C.; Li, Y.; Wu, C.; Cao, J. Differentiation between fresh and frozen-thawed large yellow croaker based on front-face fluorescence spectroscopy technique. *Trans. Chin. Soc. Agric. Eng.* **2016**, *32*, 279–285. [CrossRef]
88. Uddin, M.; Okazaki, E. Classification of fresh and frozen-thawed fish by near-infrared spectroscopy. *J. Food Sci.* **2004**, *69*, C665–C668. [CrossRef]
89. Uddin, M.; Okazaki, E.; Turza, S.; Yumiko, Y.; Tanaka, M.; Fukuda, Y. Non-destructive visible/NIR spectroscopy for differentiation of fresh and frozen-thawed fish. *J. Food Sci.* **2005**, *70*, C506–C510. [CrossRef]
90. Zhang, A.; Cheng, F. Identification of fresh shrimp and frozen-thawed shrimp by Vis/NIR spectroscopy. In Proceedings of the 2nd International Conference on Nutrition and Food Sciences IPCBEE, Singapore, 27–28 July 2013. [CrossRef]
91. Fasolato, L.; Balzan, S.; Riovanto, R.; Berzaghi, P.; Mirisola, M.; Ferlito, J.C.; Serva, L.; Benozzo, F.; Passera, R.; Tepedino, V.; Novelli, E. Comparison of visible and near-infrared reflectance spectroscopy to authenticate fresh and frozen-thawed swordfish (xiphias gladius L). *J. Aquat. Food Prod. Technol.* **2012**, *21*, 493–507. [CrossRef]
92. Fasolato, L.; Manfrin, A.; Corrain, C.; Perezzani, A.; Arcangeli, G.; Rosteghin, M.; Serva, L. Assessment of quality-parameters and authentication in sole (solea vulgaris) by NIRS (Near infrared reflectance spectroscopy). *Ind. Aliment.* **2008**, *47*, 355–361.
93. Reis, M.M.; Martínez, E.; Saitua, E.; Rodríguez, R.; Perez, I.; Olabarrieta, I. Non-invasive differentiation between fresh and frozen/thawed tuna fillets using near infrared spectroscopy (Vis-NIRS). *LWT– Food Sci. Technol. Int.* **2017**, *78*, 129–137. [CrossRef]
94. Ottavian, M.; Fasolato, L.; Facco, P.; Barolo, M. Foodstuff authentication from spectral data: Toward a species-independent discrimination between fresh and frozen-thawed fish samples. *J. Food Eng.* **2013**, *119*, 765–775. [CrossRef]
95. Karoui, R.; Lefur, B.; Grondin, C.; Thomas, E.; Demeulemester, C.; De Baerdemaeker, J.; Guillard, A.S. Mid-infrared spectroscopy as a new tool for the evaluation of fish freshness. *Int. J. Food Sci. Technol.* **2007**, *42*, 57–64. [CrossRef]
96. Sivertsen, A.H.; Kimiya, T.; Heia, K. Automatic freshness assessment of cod (Gadus morhua) fillets by Vis/Nir spectroscopy. *J. Food Eng.* **2011**, *103*, 317–323. [CrossRef]
97. Zhu, F.; Zhang, D.; He, Y.; Liu, F.; Sun, D.W. Application of Visible and Near Infrared Hyperspectral Imaging to Differentiate Between Fresh and Frozen-Thawed Fish Fillets. *Food Bioprocess Technol.* **2013**, *6*, 2931–2937. [CrossRef]
98. Aursand, M.; Veliyulin, E.; Standal, I.B.; Falch, E.; Aursand, I.G.; Erikson, U. Nuclear magnetic resonance. In *Fishery Products, Quality, Safety and Authenticity*; Rehbein, H., Oehlenschläger, J., Eds.; Wiley-Blackwell: Oxford, UK, 2009; Volume 1, pp. 252–266.
99. Nott, K.P.; Evans, S.D.; Hall, L.D. The effect of freeze-thawing on the magnetic resonance imaging parameters of cod and mackerel. *LWT- Food Sci. Technol. Int.* **1999**, *32*, 261–268. [CrossRef]
100. Howell, N.; Shavila, Y.; Grootveld, M.; Williams, S. High-resolution NMR and magnetic resonance imaging (MRI) studies on fresh and frozen cod (Gadus morhua) and haddock (Melanogrammus aeglefinus). *J. Sci. Food Agric.* **1996**, *72*, 49–56. [CrossRef]
101. Foucat, L.; Taylor, R.G.; Labas, R.; Renou, J.P. Characterization of frozen fish by NMR imaging and histology. *Am. Lab.* **2001**, *33*, 38–43.
102. Aursand, I.G.; Veliyulin, E.; Böcker, U.; Ofstad, R.; Rustad, T.; Erikson, U. Water and salt distribution in Atlantic salmon (Salmo salar) studied by low-field 1H NMR, 1H and 23Na MRI and light microscopy: Effects of raw material quality and brine salting. *J. Agric. Food Chem.* **2008**, *57*, 46–54. [CrossRef] [PubMed]

103. Mazzeo, M.F.; Siciliano, R.A. Proteomics for the authentication of fish species. *J. Proteomics* **2016**, *147*, 119–124. [CrossRef] [PubMed]
104. Li, L.; Boyd, C.E.; Sun, Z. Authentication of fishery and aquaculture products by multi-element and stable isotope analysis. *Food Chem.* **2016**, *194*, 1238–1244. [CrossRef] [PubMed]
105. Esteki, M.; Simal-Gandara, J.; Shahsavari, Z.; Zandbaaf, S.; Dashtaki, E.; Vander Heyden, Y. A review on the application of chromatographic methods, coupled to chemometrics, for food authentication (Chromatography-chemometrics in food authentication). *Food Control.* **2018**, *93*, 165–182. [CrossRef]
106. Primrose, S.; Woolfe, M.; Rollinson, S. Food forensics: Methods for determining the authenticity of foodstuffs. *Trends Food Sci. Technol.* **2010**, *21*, 582–590. [CrossRef]
107. Haynes, E.; Jimenez, E.; Pardo, M.A.; Helyar, S.J. The future of NGS (Next Generation Sequencing) analysis in testing food authenticity. *Food Control.* **2019**, *101*, 134–143. [CrossRef]
108. Asensio, L.; González, I.; García, T.; Martín, R. Determination of food authenticity by enzyme-linked immunosorbent assay (ELISA). *Food Control.* **2018**, *19*, 1–8. [CrossRef]
109. Sobolev, A.; Mannina, L.; Proietti, N.; Carradori, S.; Daglia, M.; Giusti, A.M.; Antiochia, R.; Capitani, D. Untargeted NMR-based methodology in the study of fruit metabolites. *Molecules* **2015**, *20*, 4088–4108. [CrossRef] [PubMed]
110. Yang, H.; Irudayaraj, J.; Paradkar, M.M. Discriminant analysis of edible oils and fats by FTIR, FT-NIR and FT-Raman spectroscopy. *Food Chem.* **2005**, *93*, 25–32. [CrossRef]
111. Dos Santos, C.A.T.; Pascoa, R.N.; Lopes, J.A. A review on the application of vibrational spectroscopy in the wine industry: From soil to bottle. *TrAC Trends Anal. Chem.* **2017**, *88*, 100–118. [CrossRef]
112. Maione, C.; Barbosa, F., Jr.; Barbosa, R.M. Predicting the botanical and geographical origin of honey with multivariate data analysis and machine learning techniques: A review. *Comput. Electron. Agric.* **2019**, *157*, 436–446. [CrossRef]
113. Cozzolino, D. An overview of the use of infrared spectroscopy and chemometrics in authenticity and traceability of cereals. *Food Res. Int.* **2014**, *60*, 262–265. [CrossRef]
114. Kamal, M.; Karoui, R. Analytical methods coupled with chemometric tools for determining the authenticity and detecting the adulteration of dairy products: A review. *Trends Food Sci. Technol.* **2015**, *46*, 27–48. [CrossRef]

6

Determination of Flavonoid Glycosides by UPLC-MS to Authenticate Commercial Lemonade

Ying Xue [1,2,3,†], Lin-Sen Qing [2,†], Li Yong [3], Xian-Shun Xu [3], Bin Hu [3], Ming-Qing Tang [1,2] and Jing Xie [1,*]

1 School of Pharmacy, Chengdu Medical College, Chengdu 610500, China
2 Chengdu Institute of Biology, Chinese Academy of Sciences, Chengdu 610041, China
3 Sichuan Provincial Center for Disease Control and Prevention, Chengdu 610041, China
* Correspondence: xiejing@cmc.edu.cn
† These authors contributed equally to this work.

Academic Editors: Marcello Locatelli, Angela Tartaglia, Dora Melucci, Abuzar Kabir, Halil Ibrahim Ulusoy and Victoria Samanidou

Abstract: So far, there is no report on the quality evaluation of lemonade available in the market. In this study, a sample preparation method was developed for the determination of flavonoid glycosides by ultra-performance liquid chromatography–mass spectrometry (UPLC-MS) based on vortex-assisted dispersive liquid-liquid microextraction. First, potential flavonoids in lemonade were scanned and identified by ultra-performance liquid chromatography–time of flight mass spectrometry (UPLC-TOF/MS). Five flavonoid glycosides were identified as eriocitrin, narirutin, hesperidin, rutin, and diosmin according to the molecular formula provided by TOF/MS and subsequent confirmation of the authentic standard. Then, an ultra-performance liquid chromatography–triple quadrupole mass spectrometry (UPLC-QqQ/MS) method was developed to determine these five flavonoid glycosides in lemonade. The results showed that the content of rutin in some lemonade was unreasonably high. We suspected that many illegal manufacturers achieved the goal of low-cost counterfeiting lemonade by adding rutin. This suggested that it was necessary for relevant departments of the state to make stricter regulations on the quality standards of lemonade beverages.

Keywords: vortex-assisted dispersive liquid-liquid microextraction; flavonoid glycoside; UPLC-MS; counterfeiting lemonade

---

1. Introduction

Lemon (*Citrus limon* L.) is considered the third most important citrus species in the world [1], with a large spectrum of biological activities that include antioxidant, antimicrobial, antiviral, antifungal, and antidiabetic activities [2,3], generating a large variety of healthy foods. Flavonoids are widely contained in lemon, conferring the typical taste and biological activities to lemon. According to the aglycone structures, flavonoids are divided into four classes: flavanones, flavones, flavonols, and flavans. Flavanones are the most abundant flavonoids, which are usually present in the 7-*O*-diglycoside form. Lemon flavanones are present in glycoside or aglycone forms. Among the phytochemicals, hesperetin and eriodictyol are the most abundant types of aglycones and rutinoside is the most abundant types of glycoside forms [4,5]. It has been reported that hesperidin and eriocitrin were the most abundant flavonoids in all the lemon juices studied and far exceed others [6–8].

Due to the high cost of fruit, counterfeiting of fruit juice has become a common problem in the industry. The three most common forms of counterfeiting are: (1) When a kind of cheaper fruit is used to replace all or part of it, (2) when a monomeric compound contained in the fruit with another cheaper source is added, and (3) when it is completely made up of additives such as artificial sweeteners, preservatives, and colors [9]. As the products produced with the first two counterfeiting methods contain some natural characteristic ingredients, they can generally meet the national testing standards [10]. However, such kinds of counterfeit juice not only seriously affect consumer confidence in the juice market, but may also cause a series of food safety problems. In addition to pure lemon juice, lemonade containing lemon ingredients occupies an increasing market share in the beverage market. Thus, it is of great scientific significance and commercial value to identify the authenticity of lemonade available in the market.

Some methods for analyzing lemon juice have been reported, such as nuclear magnetic resonance [11], $^{13}C/^{12}C$ isotope ratios [12], capillary electrochromatography (CEC) [13], and HPLC [6,7,14]. Among them, HPLC was considered as the most reliable method for determining flavonoids with high selectivity and sensitivity. Lemonade beverages currently available in the market contain a large number of additives besides a small amount of lemon juice. Therefore, a new sample preparation method is required to selectively separate and enrich low-content flavonoids from lemonade, so as to identify the authenticity of lemonade.

At present, sample preparation methods of flavonoids can be divided into liquid-liquid extraction (LLE) and solid phase extraction (SPE) [15–17]. However, they have some inherent disadvantages. For example, LLE needs a substantial amount of toxic solvents and is time-consuming. SPE materials are expensive and have poor reusability [18]. The dispersive liquid-liquid microextraction (DLLME) method developed in recent years can make up for these disadvantages [19–21]. DLLME can not only separate and enrich target analyte from aqueous solution, but also reduce or even eliminate the matrix interference of samples. Therefore, DLLME is considered to be an effective pretreatment method for food samples with the advantages of less solvent consumption, simple operation, high enrichment factor, etc. In order to improve the work efficiency by speeding up the mass transfer process and reducing the balance time, some assistant emulsification methods were also applied to improve the performance of DLLME, such as ultrasound-assisted [22], vortex-assisted [23], air-assisted [24], and microwave-assisted [25] DLLME. Currently, there are some studies on sample preparation of flavonoids by DLLME. However, as far as we know, there is no research on flavonoids in lemonade.

In this work, the sample preparation of flavonoids in lemonade was firstly performed by the vortex-assisted dispersive liquid-liquid microextraction (VA-DLLME) method. Then, the structure and content of flavonoids in lemonade available on the market from eight different manufacturers were identified and determined by ultra-performance liquid chromatography–time of flight mass spectrometry (UPLC-TOF/MS) and ultra-performance liquid chromatography–triple quadrupole mass spectrometry (UPLC-QqQ/MS), respectively. Finally, the counterfeiting phenomenon of lemonade was evaluated according to the determination results of flavonoids. As far as we know, this study was the first determination of flavonoid glycosides by UPLC-MS to authenticate commercial lemonade available in the market.

2. Results and Discussion

2.1. *Identification of Flavonoid Glycosides by UPLC-TOF/MS*

The time of flight mass spectrometer (TOF MS) was used to scan and identify potential flavonoids in lemonade for the first time in this work. As one of the most common high-resolution MS, TOF MS can determine the exact molecular formula of the target compound, thus identifying the structure in a complex matrix. After the target compound was located and identified, the triple quadrupole mass spectrometer (QqQ MS) was an excellent choice for subsequent quantitative analysis [26].

In this study, according to the calculation based on the molecular formula by TOF and the subsequent confirmation of the authentic standard under the same chromatographic conditions, 5 flavonoid glycosides in lemonade available in the market were located and identified (Figure 1), which were eriocitrin, narirutin, hesperidin, rutin, and diosmin, respectively. As shown in Table 1, the error of each compound in high-resolution MS is within ±5 ppm, which is the acceptable error limit for structure confirmation [27].

	R_1	R_2
Eriocitrin	OH	H
Nariutin	H	H
Hesperidin	OH	OCH_3

	R_3	R_4	R_5
Rutin	OH	O-rutinose	OH
Diosmin	O-rutinose	H	OCH_3

Figure 1. Chemical structures of eriocitrin, narirutin, hesperidin, rutin, and diosmin.

Table 1. UPLC-MS parameters of five analytes in the negative ion-scan mode.

Analyte	TOF/MS			QqQ/MS	
	Quasi-Molecular Ion (*m*/*z*)	**Error (ppm)**	**Product Ion (*m*/*z*)**	**Parent Ion (*m*/*z*)**	**Product Ion (*m*/*z*)**
eriocitrin	595.16788	1.7	287.0586, 151.0065	595	287 *, 151 #
narirutin	579.17238	0.8	271.0612	579	271 *, 151 #
hesperidin	609.18386	2.2	301.0737	609	301 *, 286 #
rutin	609.14689	1.3	301.0383, 300.0281	609	300 *, 271 #
diosmin	607.16784	1.0	299.0582, 284.0345	607	299 *, 284 #

Note: * quantitative ion, # qualitative ion.

2.2. The Selection of VA-DLLME Conditions

Since the extraction conditions have a crucial influence on the performance of VA-DLLME, single-factor experiments were carried out to select the extraction conditions of the amount of ethyl acetate and acetonitrile. In the present study, recoveries of 5 flavonoid glycosides were assessed by means of fixing one variable and changing the other two variables. The results are shown in Figure 2. Due to structural differences, the recoveries of the 5 flavonoid glycosides were different, but the overall trend was relatively consistent. Based on the investigation of single-factor experiments, the VA-DLLME condition was set as 1 mL of lemonade, 500 μL acetonitrile, and 1.5 mL ethyl acetate.

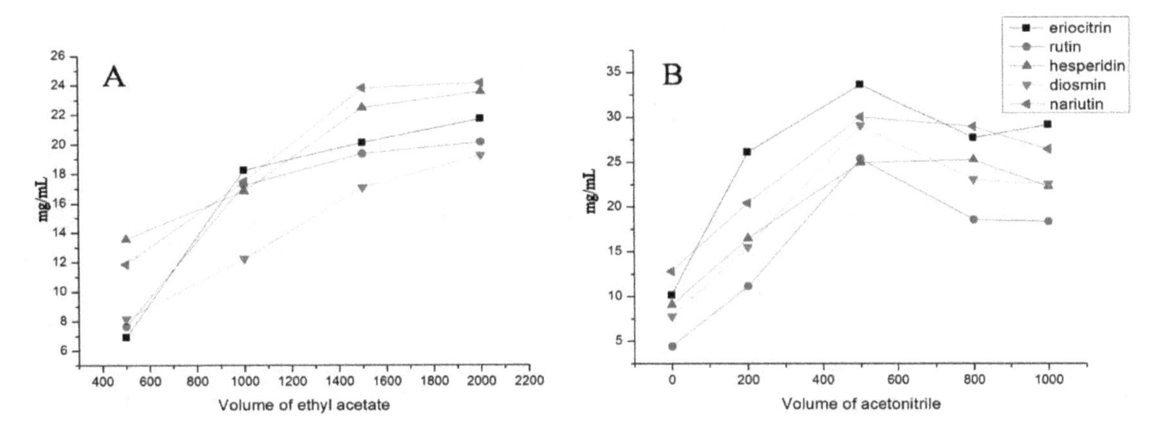

Figure 2. The evaluation of extraction conditions of the amount of ethyl acetate (**A**) and acetonitrile (**B**).

2.3. Determination of Flavonoid Glycosides by UPLC–QqQ/MS

All 5 flavonoid glycosides are acidic compounds. Therefore, acid mobile phase could increase the separating degree, symmetry factor, and the number of theoretical plates. Considering the ion suppression induced by a high concentration of acid, 0.2% formic acid was finally added into the mobile phase [28]. In order to optimize the MS condition of 5 flavonoid glycosides in the present study, all of these target analytes were tested in direct infusion mode using the full-scan MS method, respectively. It was found that the negative mode was more sensitive and selective than the positive mode. By optimizing mass spectrum variables, including the vaporizer temperature, sheath gas pressure, aux gas pressure, the parent/product ion pairs, collision energy, and S-Lens value, two stable product ions with high sensitivity were selected for MRM analysis (Table 1). The representative mass spectra of lemonade samples are shown in Figure 3.

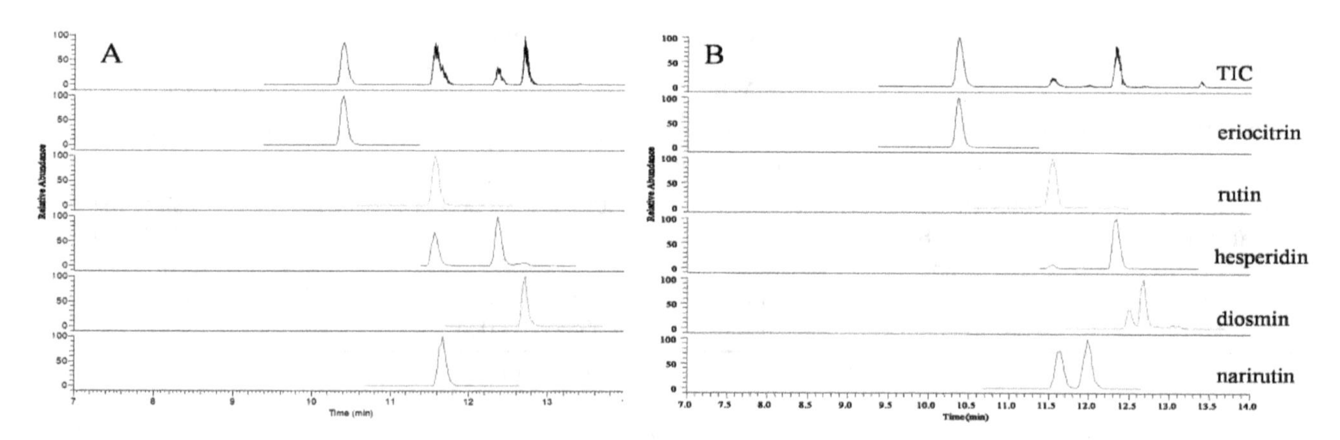

Figure 3. UPLC–QqQ/MS total ion count chromatograms of five flavonoid glycosides standards (**A**) and commercial lemonade sample (**B**).

2.4. Speculation on the Possible Counterfeiting Means of Lemonade

A total of 8 batches of lemonade samples purchased from local supermarkets was determined by the proposed UPLC-QqQ/MS method. The contents of 5 flavonoid glycosides are shown in Table 2. The content of total flavonoid glycosides in lemonade varies greatly. On the surface, it seems that the higher the content of total flavones, the higher the amount of lemon juice added in lemonade, which means the better the quality of the product. However, after further analysis of the content of monomeric compounds, it was found that the main ingredients in S1–S4 were flavanone glycosides (mainly hesperidin and eriocitrin) and the content of flavonol glycosides (mainly rutin) was relatively low. This result is consistent with the distribution characteristics of flavonoid glycosides in *Citrus* L. With regard to S5–S8, the content of rutin is extremely high and hesperidin as a characteristic ingredient of *Citrus* L. is not detected (nd). Hesperidin was the predominant flavonoid glycoside in lemon reported by the previous study. For example, Mannan et al. reported values of 67 ± 15 mg/L for hesperidin in 38 natural lemon juices, showing that the absence of this compound in lemonade shows it to be a possible counterfeit [29]. Under normal circumstances, the content of rutin in lemon should not exceed the content of hesperidin. Xi reported the contents of hesperidin and rutin in juice varied from 105.5 to 210.3 μg/g and nd to 3.82 μg/g, respectively [30]. Due to the abnormal phenomenon in our work, we have reason to suspect that S5–S8 were counterfeited as there was no or only a small trace of lemon juice and had instead a large amount of rutin added to meet the national testing standards (colorimetric assay by UV-Vis) of fruit juice products. Rutin is widely distributed in the plant kingdom.

It was reported that its content in *Sophora japonica* L. was up to 37.8% [31]. Therefore, only with a simple separation process the commercialized low-cost supply of rutin can be realized [32]. For example, the price of rutin reagent supplied by Aladdin is ¥368/100 g and if it is a crude extract of food-grade, the price will be even lower. According to the testing method of total flavonoids in fruit juice beverage specified by national standard, rutin also has an obvious response in the colorimetric assay by UV-Vis at a wavelength of 420 nm. Therefore, illegal businessmen achieved the goal of low-cost counterfeiting lemonade by adding rutin.

Table 2. The contents of five flavonoid glycosides in eight lemonade samples (μg/100 mL).

Sample No.	Eriocitrin	Rutin	Hesperidin	Diosmin	Narirutin	Total
S1	0.04	0.27	1.00	nd	nd	1.31
S2	0.92	0.34	3.82	0.66	nd	5.75
S3	1.73	0.29	16.33	2.95	nd	21.30
S4	28.96	6.01	28.30	0.28	0.74	49.07
S5	2.66	191.54	nd	0.64	nd	194.83
S6	0.04	243.71	nd	nd	nd	243.75
S7	0.05	264.24	nd	nd	nd	264.29
S8	0.35	470.00	nd	nd	nd	470.35

3. Material and Methods

3.1. Chemicals and Reagents

A total of eight lemonade samples were purchased from local supermarkets. A total of five authentic standards of eriocitrin, narirutin, hesperidin, rutin, and diosmin were obtained from Chengdu Push Bio-technology Co., Ltd. (Chengdu, China). The Milli-Q water purification system was used to prepare ultra-pure water for UPLC analysis (Millipore, Bedford, MA, USA). Formic acid and acetonitrile of LC/MS grade for UPLC-MS analysis were purchased from Sigma-Aldrich. Ethyl acetate, ether, dichloromethane, methanol, acetone, and acetonitrile of analytical grade were purchased from Sinopharm Chemical Reagent Co., Ltd. (Shanghai, China).

3.2. Preparation of Standard Solution

Stock solutions of five target analytes (eriocitrin, narirutin, hesperidin, rutin, and diosmin) were prepared by dissolving each 10 mg authentic standard in 10 mL of methanol. Then, 250 μL of each of the five stock solutions was transferred to a 50-mL volumetric flask and diluted with 20% methanol to obtain the mixed stock solution. Next, 500 μL of mixed stock solution was transferred to a 50-mL volumetric flask and diluted with 20% methanol to obtain the working solution I with a concentration of approximately 50 ng/mL. Finally, mixed working solutions II–V were obtained by diluting working solution I with respective concentrations of about 20.0 ng/mL, 10.0 ng/mL, 5.0 ng/mL, and 2 ng/mL. All the solutions were stored in a refrigerator at 4 °C before use.

3.3. Sample Preparation by the VA-DLLME Procedure

Accurately add 1 mL of lemonade to a 4 mL centrifuge tube, then add 500 μL of acetonitrile and 1.5 mL ethyl acetate, then vortex for 30 s. After centrifugation, the upper organic phase was transferred. The extraction was repeated once using another 1.5 mL of ethyl acetate and the combined solvent of the upper organic phase was removed by a Termovap Sample Concentrator. The resulting residue was re-dissolved in 1 mL of 20% methanol and filtered through a 0.22 μm filter for UPLC-MS analysis.

3.4. UPLC–MS Analysis

3.4.1. Identification of Flavonoid Glycosides by UPLC-TOF/MS

The Shimadzu UPLC ((Shimadzu, Kyoto, Japan) system consists of an online degasser (DGU-20A5R), an auto-sampler (SIL-30AC), two pumps (LC-30AD), and a column oven (CTO-30aHE). Chromatographic separation was performed on a Waters BEHC18 analytical column (2.1 × 100 mm, 1.7 μm, Waters, Milford, MA, USA) at 40 °C. The mobile phase consisted of 0.2% formic acid and acetonitrile. The linear gradient elution with a constant flow rate of 0.2 mL/min was 10%~10%~40%~95%~10% acetonitrile at 0~1~10~13~15 min. The sample solution and mixed working solutions of 5 μL were injected into the UPLC system by the auto-sampler.

TOF/MS measurements in negative ion mode were performed on a 4600 Q-TOF mass spectrometer (AB Sciex, Concord, CA, USA) equipped with an electrospray ionization (ESI) source with the following parameters: Ion source gas 1 (GS1) at 50 psi, ion source gas 1 (GS1) (N_2) at 50 psi, curtain gas at 35 psi, temperature at 500 °C, and ionspray voltage floating at −4500 V. The mass range was set to *m/z* 100–800. The system was operated under Analyst 1.6 and Peak 2.0 (AB Sciex, Concord, CA, USA) and used an APCI negative calibration solution to calibrate the instrument's mass accuracy in real-time.

3.4.2. Determination of Flavonoid Glycosides by UPLC-QqQ/MS

Chromatographic separation was the same as that used in UPLC-TOF/MS analysis described above. QqQ/MS measurements in negative ion mode were accomplished by a triple quadrupole mass spectrometer equipped with an ESI source (Thermo Fisher Scientific, San Jose, CA, USA). The determination of the target analytes was performed in a multi-reaction monitoring mode. The MS parameters were as follows: Vaporizer temperature and capillary temperature both 350 °C, aux gas pressure of 10 Arb, sheath gas pressure of 40 Arb, ion sweep gas pressure of 2 Arb, discharge current of 4.0 μA, and spray voltage of −2000 V. Data collection and processing were conducted with Thermo Xcalibur Workstation (Version 2.2, Thermo).

3.5. Analytical Figures of Merit

Method validation was performed according to the above UPLC–QqQ/MS conditions. After it was determined by the mixed working solutions I–V, the calibration curves of five analytes were obtained as shown in Table 3 by taking the concentration of each authentic standard as the abscissa (x) and the corresponding peak area as the ordinate (y), respectively. The limit of detection (LOD) and the limit of quantification (LOQ) were measured by a gradual dilution process of the standard stock solutions until the signal-to-noise ratio of 3:1 and 10:1, respectively. The precision was evaluated by standard working solution III, which was tested within one day to determine the intra-day precision and was tested within 3 days to determine the inter-day precision. The repeatability was evaluated by analyzing six independent portions of sample S4 with parallel running. The recovery was carried out by spiking an amount of about 1:1 of authentic standards to six independent portions of sample S4 with parallel running. The validation results are summarized in Table 3, which show that the present developed UPLC–QqQ/MS method meets the requirements of quantitative analysis and was appropriate for the determination of five flavonoid glycosides in lemonade. The analytical figures of merit were compared with those of several other quantitative methods reported for flavonoid glycosides in lemon as shown in Table 4.

Table 3. The results of method validation.

Analyte	Regression Equation ($y = ax + b, r^2$)	Linear Range (ng/mL)	LOD (ng/mL)	LOQ (ng/mL)	Precision (RSD, $n = 6$)		Repeatability ($n = 6$)		Recovery ($n = 6$)	
					Intra-Day	Inter-Day	Mean (μg/100 mL)	RSD	Mean	RSD
eriocitrin	$y = 523.81x - 82.61, 0.996$	2.01–50.3	0.70	2.01	1.36%	3.53%	28.9	3.22%	88.5%	3.93%
rutin	$y = 1034.77x - 1160.22, 0.996$	2.44–60.9	0.81	2.44	2.51%	3.79%	6.01	4.62%	89.9%	4.51%
hesperidin	$y = 513.03x + 252.28, 0.998$	2.01–50.4	0.70	2.10	1.97%	2.58%	28.3	3.47%	88.7%	5.30%
diosmin	$y = 769.76x + 84.74, 0.995$	2.14–53.4	0.71	2.14	2.02%	3.17%	0.28	5.47%	102%	2.61%
narirutin	$y = 556.25x + 148.69, 0.997$	2.02–51.2	0.70	2.02	1.48%	3.69%	0.74	4.86%	92.8%	4.47%

Table 4. Comparison of analytical methods reported for determination of flavonoid glycosides in lemon.

Method	Analyte	Linea Range	LOD	LOQ	Recovery
CEC [13]	eriocitrin, narirutin, hesperidin	5–200 μg/mL	2.5 μg/mL	5 μg/mL	71–112%
HPLC/UV [33]	narirutin, hesperidin, diosmin	0.25–20 μg/mL	-	0.1 μg/mL	-
HPLC/UV [34]	narirutin	2–50 mg/L	1.25 mg/L	2.5 mg/L	83%
	hesperidin	2–50 mg/L	1.0 mg/L	2.5 mg/L	74%
HPLC/UV [35]	eriocitrin	1.01–50.50 μg/mL	0.02 μg/mL	0.065 μg/mL	103.10%
	narirutin	0.505–10.10 μg/mL	0.024 μg/mL	0.18 μg/mL	99.14%
	hesperidin	5.00–100.00 μg/mL	0.04 μg/mL	0.132 μg/mL	99%
	rutin	0.101–10.100 μg/mL	0.079 μg/mL	0.263 μg/mL	98.37%
UPLC/UV [36]	eriocitrin	0.5–130 mg/L	6 μg/kg	-	90.50%
	narirutin	0.05–300 mg/L	5 μg/kg	-	87.40%
	hesperidin	0.05–500 mg/L	8 μg/kg	-	92.70%
	rutin	0.05–310 mg/L	5 μg/kg	-	88.40%
	diosmin	0.01–200 mg/L	8 μg/kg	-	100.80%

4. Conclusions

In this study, five flavonoid glycosides of eriocitrin, narirutin, hesperidin, rutin, and diosmin in lemonade were identified and determined by UPLC-TOF/MS and UPLC-QqQ/MS, respectively. By estimating the content characteristics of flavonoid glycosides in the samples, we highly suspected that some lemonade available in the market was counterfeited: Cheap rutin was added to increase the content of "total flavonoids of lemon". This indicates that besides using total flavonoids, the content of multiple flavonoid compounds should be included in the quality standard of lemonade in the future.

Author Contributions: Conceptualization, L.-S.Q. and J.X.; Data curation, Y.X.; Formal analysis, Y.X. and M.-Q.T.; Funding acquisition, L.-S.Q.; Methodology, Y.X.; Project administration, J.X.; Resources, L.Y., B.H., and X.-S.X.; Supervision, L.Y., B.H., X.-S.X., and J.X.; Validation, Y.X.; Writing—original draft, L.-S.Q.; Writing—review & editing, J.X.

References

1. Papoutsis, K.; Pristijono, P.; Golding, J.B.; Stathopoulos, C.E.; Scarlett, C.J.; Bowyer, M.C.; Vuong, Q.V. Impact of different solvents on the recovery of bioactive compounds and antioxidant properties from lemon (*Citrus limon* L.) pomace waste. *Food Sci. Biotechnol.* **2016**, *25*, 971–977. [CrossRef] [PubMed]
2. Kim, J.; Jayaprakasha, G.K.; Patil, B.S. Lemon (*Citrus lemon* L. Burm) as a source of unique bioactive compounds. *Acta Hortic.* **2014**, *1040*, 377–380. [CrossRef]
3. Proteggente, A.R.; Pannala, A.S.; Paganga, G.; Buren, L.v.; Wagner, E.; Wiseman, S.; Put, F.v.d.; Dacombe, C.; Rice-Evans, C.A. The antioxidant activity of regularly consumed fruit and vegetables reflects their phenolic and vitamin C composition. *Free Radic. Res.* **2002**, *36*, 217–233. [CrossRef] [PubMed]
4. Tripoli, E.; Guardia, M.L.; Giammanco, S.; Majo, D.D.; Giammanco, M. Citrus flavonoids: Molecular structure, biological activity and nutritional properties: A review. *Food Chem.* **2007**, *104*, 466–479. [CrossRef]
5. Del Río, J.A.; Fuster, M.D.; Gómez, P.; Porras, I.; García-Lidón, A.; Ortuño, A. Citrus limon: A source of flavonoids of pharmaceutical interest. *Food Chem.* **2004**, *84*, 457–461. [CrossRef]
6. Abad-García, B.; Garmón-Lobato, S.; Sánchez-Ilárduya, M.B.; Berrueta, L.A.; Gallo, B.; Vicente, F.; Alonso-Salces, R.M. Polyphenolic contents in Citrus fruit juices: Authenticity assessment. *Eur. Food Res. Technol.* **2014**, *238*, 803–818. [CrossRef]
7. Abad-García, B.; Berrueta, L.A.; Garmón-Lobato, S.; Urkaregi, A.; Gallo, B.; Vicente, F. Chemometric characterization of fruit juices from spanish cultivars according to their phenolic compound contents: I. Citrus fruits. *J. Agric. Food Chem.* **2012**, *60*, 3635–3644. [CrossRef]
8. Kawaii, S.; Tomono, Y.; Katase, E.; Ogawa, K.; Yano, M. Quantitation of flavonoid constituents in citrus fruits. *J. Agric. Food Chem.* **1999**, *47*, 3565–3571. [CrossRef]
9. Zhang, M.; Li, X.X.; Jia, H.F.; Wang, X. The research on detection techniques of adulteration about fruit juice in China. *Food Res. Dev.* **2016**, *37*, 205–208.
10. National Standards of China. *General Analytical Methods for Beverage*; National Standards of China: GB/T 12143-2008; Standards Press of China: Beijin, China, 2009.
11. Salazar, M.O.; Pisano, P.L.; González Sierra, M.; Furlan, R.L.E. NMR and multivariate data analysis to assess traceability of argentine citrus. *Microchem. J.* **2018**, *141*, 264–270. [CrossRef]
12. Guyon, F.; Auberger, P.; Gaillard, L.; Loublanches, C.; Viateau, M.; Sabathié, N.; Salagoïty, M.H.; Médina, B. $^{13}C/^{12}C$ isotope ratios of organic acids, glucose and fructose determined by HPLC-co-IRMS for lemon juices authenticity. *Food Chem.* **2014**, *146*, 36–40. [CrossRef] [PubMed]
13. Desiderio, C.; De Rossi, A.; Sinibaldi, M. Analysis of flavanone-7-O-glycosides in citrus juices by short-end capillary electrochromatography. *J. Chromatogr. A* **2005**, *1081*, 99–104. [CrossRef] [PubMed]
14. Vaclavik, L.; Schreiber, A.; Lacina, O.; Cajka, T.; Hajslova, J. Liquid chromatography–mass spectrometry-based metabolomics for authenticity assessment of fruit juices. *Metabolomics* **2012**, *8*, 793–803. [CrossRef]
15. Qing, L.S.; Xiong, J.; Xue, Y.; Liu, Y.M.; Guang, B.; Ding, L.S.; Liao, X. Using baicalin-functionalized magnetic nanoparticles for selectively extracting flavonoids from *Rosa chinensis*. *J. Sep. Sci.* **2011**, *34*, 3240–3245. [CrossRef] [PubMed]
16. Qing, L.S.; Xue, Y.; Liu, Y.M.; Liang, J.; Xie, J.; Liao, X. Rapid magnetic solid-phase extraction for the selective determination of isoflavones in soymilk using baicalin-functionalized magnetic nanoparticles. *J. Agric. Food Chem.* **2013**, *61*, 8072–8078. [CrossRef] [PubMed]

17. Qing, L.S.; Xue, Y.; Zhang, J.G.; Zhang, Z.F.; Liang, J.; Jiang, Y.; Liu, Y.M.; Liao, X. Identification of flavonoid glycosides in Rosa chinensis flowers by liquid chromatography–tandem mass spectrometry in combination with ^{13}C nuclear magnetic resonance. *J. Chromatogr. A* **2012**, *1249*, 130–137. [CrossRef] [PubMed]
18. Andrade-Eiroa, A.; Canle, M.; Leroy-Cancellieri, V.; Cerdà, V. Solid-phase extraction of organic compounds: A critical review. part II. *TrAC Trends Anal. Chem.* **2016**, *80*, 655–667. [CrossRef]
19. Mousavi, L.; Tamiji, Z.; Khoshayand, M.R. Applications and opportunities of experimental design for the dispersive liquid–liquid microextraction method—A review. *Talanta* **2018**, *190*, 335–356. [CrossRef] [PubMed]
20. Sajid, M.; Alhooshani, K. Dispersive liquid-liquid microextraction based binary extraction techniques prior to chromatographic analysis: A review. *TrAC Trends Anal. Chem.* **2018**, *108*, 167–182. [CrossRef]
21. Rykowska, I.; Ziemblińska, J.; Nowak, I. Modern approaches in dispersive liquid-liquid microextraction (DLLME) based on ionic liquids: A review. *J. Mol. Liq.* **2018**, *259*, 319–339. [CrossRef]
22. Homem, V.; Alves, A.; Alves, A.; Santos, L. Ultrasound-assisted dispersive liquid–liquid microextraction for the determination of synthetic musk fragrances in aqueous matrices by gas chromatography–mass spectrometry. *Talanta* **2016**, *148*, 84–93. [CrossRef] [PubMed]
23. Xue, Y.; Xu, X.S.; Yong, L.; Hu, B.; Li, X.D.; Zhong, S.H.; Li, Y.; Xie, J.; Qing, L.S. Optimization of vortex-assisted dispersive liquid-liquid microextraction for the simultaneous quantitation of eleven non-anthocyanin polyphenols in commercial blueberry using the multi-objective response surface methodology and desirability function approach. *Molecules* **2018**, *23*, 2921.
24. Li, G.; Row, K.H. Air assisted dispersive liquid–liquid microextraction (AA-DLLME) using hydrophilic–hydrophobic deep eutectic solvents for the isolation of monosaccharides and amino acids from kelp. *Anal. Lett.* **2019**, 1–15. [CrossRef]
25. Mahmoudpour, M.; Mohtadinia, J.; Mousavi, M.M.; Ansarin, M.; Nemati, M. Application of the microwave-assisted extraction and dispersive liquid–liquid microextraction for the analysis of PAHs in smoked rice. *Food Anal. Methods* **2017**, *10*, 277–286. [CrossRef]
26. Chen, C.; Xue, Y.; Li, Q.M.; Wu, Y.; Liang, J.; Qing, L.S. Neutral loss scan - based strategy for integrated identification of amorfrutin derivatives, new peroxisome proliferator-activated receptor gamma agonists, from *Amorpha Fruticosa* by UPLC-QqQ-MS/MS and UPLC-Q-TOF-MS. *J. Am. Soc. Mass Sp.* **2018**, *29*, 685–693. [CrossRef] [PubMed]
27. Gross, M.L. Accurate masses for structure confirmation. *J. Am. Soc. Mass Sp.* **1994**, *5*, 57. [CrossRef]
28. Xie, J.; Li, J.; Liang, J.; Luo, P.; Qing, L.S.; Ding, L.S. Determination of contents of catechins in oolong teas by quantitative analysis of multi-components via a single marker (QAMS) method. *Food Anal. Methods* **2017**, *10*, 363–368. [CrossRef]
29. Hajimahmoodi, M.; Moghaddam, G.; Mousavi, S.; Sadeghi, N.; Oveisi, M.; Jannat, B. Total antioxidant activity, and hesperidin, diosmin, eriocitrin and quercetin contents of various lemon juices. *Trop. J. Pharm. Res.* **2014**, *13*, 951–956. [CrossRef]
30. Xi, W.; Lu, J.; Qun, J.; Jiao, B. Characterization of phenolic profile and antioxidant capacity of different fruit part from lemon (*Citrus limon* Burm.) cultivars. *J. Food Sci. Technol.* **2017**, *54*, 1108–1118. [CrossRef] [PubMed]
31. Tan, J.; Li, L.Y.; Wang, J.R.; Ding, G.; Xu, J. Study on quality evaluation of *Flos Sophorae* Immaturus. *Nat. Prod. Res. Dev.* **2018**, *30*, 138–146+174.
32. Chua, L.S. A review on plant-based rutin extraction methods and its pharmacological activities. *J. Ethnopharmacol.* **2013**, *150*, 805–817. [CrossRef] [PubMed]
33. Kanaze, F.I.; Gabrieli, C.; Kokkalou, E.; Georgarakis, M.; Niopas, I. Simultaneous reversed-phase high-performance liquid chromatographic method for the determination of diosmin, hesperidin and naringin in different citrus fruit juices and pharmaceutical formulations. *J. Pharm. Biomed. Anal.* **2003**, *33*, 243–249. [CrossRef]
34. Belajová, E.; Suhaj, M. Determination of phenolic constituents in citrus juices: Method of high performance liquid chromatography. *Food Chem.* **2004**, *86*, 339–343. [CrossRef]
35. Tu, X.; Yang, S.; Wu, Z.; Wu, C.; Zhang, L.; Lü, X. Simultaneous determination of eleven flavonoids in different lemon (*Citrus limon*) varieties by HPLC. *J. Hunan Agric. Univ.* **2016**, *42*, 543–548.
36. Zheng, J.; Zhao, Q.Y.; Zhang, Y.H.; Jiao, B.N. Simultaneous determination of main flavonoids and phenolic acids in citrus fruit by ultra performance liquid chromatography. *Sci. Agric. Sin.* **2014**, *47*, 4706–4717.

Raman Spectroscopy and Chemometric Modeling to Predict Physical-Chemical Honey Properties from Campeche, Mexico

F. Anguebes-Franseschi [1], M. Abatal [2], Lucio Pat [3], A. Flores [2], A. V. Córdova Quiroz [1], M. A. Ramírez-Elias [1], L. San Pedro [4], O. May Tzuc [4] and A. Bassam [4,*]

[1] Faculty of Chemistry, Autonomous University of Carmen, Street 56 No. 4 Esq. Av. Concordia, Col. Benito Juárez, Z. C. 24180 Ciudad del Carmen, Campeche, Mexico; fanguebes@pampano.unacar.mx (F.A.-F.); acordova@delfin.unacar.mx (A.V.C.Q.); mramirez@pampano.unacar.mx (M.A.R.-E.)
[2] Faculty of Engineering, Autonomous University of Carmen, Campus III, Avenida Central s/n, Esq. Con Fracc. Mundo Maya, C. P. 24115 Ciudad del Carmen, Campeche, Mexico; mabatal@pampano.unacar.mx (M.A.);aflores@pampano.unacar.mx (A.F.);
[3] South Frontier College, Av. Rancho Polígono 2-A, Ciudad Industrial, 24500 Lerma, Campeche, Mexico; lpat@ecosur.mx
[4] Faculty of Engineering, Autonomous University of Yucatan, Av. Industrias no Contaminantes Periférico Norte, Cordemex, Z.C. 97310 Mérida, Yucatan, Mexico; liliana.cedillo@correo.uady.mx (L.S.P.); maytzuc@gmail.com (O.M.T.)
* Correspondence: baali@correo.uady.mx

Academic Editors: Marcello Locatelli and Angela Tartaglia

Abstract: In this work, 10 chemometric models based on Raman spectroscopy were constructed to predict the physicochemical properties of honey produced in the state of Campeche, Mexico. The properties of honey studied were pH, moisture, total soluble solids (TSS), free acidity, lactonic acidity, total acidity, electrical conductivity, Redox potential, hydroxymethylfurfural (HMF), and ash content. These proprieties were obtained according to the methods described by the Association of Official Analytical Chemists, Codex Alimentarius, and the International Honey Commission. For the construction of the chemometric models, 189 honey samples were collected and analyzed in triplicate using Raman spectroscopy to generate the matrix data [X], which were correlated with each of the physicochemical properties [Y]. The predictive capacity of each model was determined by cross validation and external validation, using the statistical parameters: standard error of calibration (SEC), standard error of prediction (SEP), coefficient of determination of cross-validation (R^2_{cal}), coefficient of determination for external validation (R^2_{val}), and Student's *t*-test. The statistical results indicated that the chemometric models satisfactorily predict the humidity, TSS, free acidity, lactonic acidity, total acidity, and Redox potential. However, the models for electric conductivity and pH presented an acceptable prediction capacity but not adequate to supply the conventional processes, while the models for predicting ash content and HMF were not satisfactory. The developed models represent a low-cost tool to analyze the quality of honey, and contribute significantly to increasing the honey distribution and subsequently the economy of the region.

Keywords: quality control; Raman spectroscopy; honey; PLS regression models; physicochemical parameters

1. Introduction

Honey is a natural product, and a complex solution elaborated by honey bees. It is mainly composed of sugars (70–80%) and water (10–20%), and in minor quantities contains flavonoids, phenolic acids, vitamins, proteins, organic acids, lipids, carotenoids, minerals, and enzymes [1]. Honey has been used since ancient times as a food supplement for humans. Additionally, due to its content of phenolic compounds and flavones, it also has several beneficial health effects, which include prebiotic, antimicrobial, anticarcinogenic, antioxidant, antihypertensive, antibacterial, antifungal, anti-inflammatory, and analgesic effects [2–4]. The physical, chemical, and biological properties of honey depend on the type of flowers visited by the honey bees, and the soil where the nectar and pollen are collected. Other influences on its quality are the environmental and storage conditions, as well as the processing for its commercialization [5]. Therefore, quality control of honey represents an important concern for the beekeeping industry, since, on the one hand, it allows tracing of the geographical and botanical origin of the pollen (designation of origin), and, on the other hand, it allows identification of its possible adulteration during processing [6,7].

To classify and determine the honey's quality, standards and methods have been established in the Codex Alimentarious [8], International Honey Commission (IHC) [9], and the Association of Official Analytical Chemists (AOAC) [10]. These standards specify the physical and chemical properties that must be evaluated to determine the honey's quality. The traditional method to perform quality tests on honey involves the analysis of pollen grains contained in its sediments by light microscopy (melissopalynology) [6]. Other methods reported in the literature include chromatography techniques, stable carbon isotope radio analysis, and nuclear magnetic resonance [7,11]. The main drawbacks of all of them are their high cost, time consuming nature, requirement for specialists, and furthermore the fact that many of them are destructive. This has led to the development of analytical methods for the authentication of honey. In this sense, spectroscopy technology combined with chemometric tools represents a good alternative for the fast, reliable, and environmentally friendly quality control of honey samples. The above is due to the development of calibration models that can determine the concentration of a specific chemical species in a mixture of several components [12]. Among the most common chemometric techniques used in honey analysis are Principal Component Analysis (PCA), Hierarchical Clustering Analysis (HCA), Linear Discriminant Analysis (LDA), Partial Least Square (PLS), and Principal Component Regression (PCR) [13].

From the spectrometric techniques available, Raman spectrometry has suitable characteristics for food analysis, such as non-interference from water present in the sample with the Raman measurement, ease of sampling and measurement, and minimal fluorescence interference of the sample matrix variation. In recent years, analytic methods based on Raman spectrometry have been explored as an economic and rapid option to determine honey's destination of origin [14–17]. Corvucci et al. [14] contrasted the ability to identify honey's botanic origin using the melissopalynology technique compared to Raman spectroscopy coupled with multivariable analysis (PCA). The study considered honey samples from Italy, Eastern Europe, and Spain. According to the results, the discrimination of honey origin given by the two first principal components was improved from 85% to 99% using the analytical method. Frausto-Reyes et al. [15] determined the floral origin of honey produced by *Apis Mellifera*, applying Raman spectroscopy together with PCA. The study used 66 samples of both monofloral and polifloral honey collected from several regions of Mexico with different climate types. The use of the chemometric approach was adequate to classify the origin of the sample and the purity of the pollen with 90% accuracy. Jandrić et al. [16] presented a method for the authentication of floral origin honey produced in New Zealand. They combined Raman spectrometry, near infrared spectrometry, and Fourier-transform infrared spectroscopy for the analytical study of honey samples in the range between 200 to 12,000 cm^{-1}. This approach was completed with the use of PLS for the development of chemometric models. The results showed a model fit (R^2), a standard error of calibration (SEC), and standard error of prediction (SEP) of 85.0%, 0.219 and 0.315, respectively. Oroian and Ropciuc [17] applied Raman spectra analysis for the botanical authentication of 76 samples of honey from Romania. The use of this analytic method combined with LDA proved to be an excellent authentication tool, achieving 83.33% cross validation accuracy.

Similarly, the literature reports the use of Raman spectra analysis coupled with multi-variable modeling for the detection of external agents that affect the quality of honey [18–22]. Raman spectroscopy and chemometric models have been used to predict the concentration of glucose, fructose, sucrose, and maltose present in honey samples from Turkey and Greece [18]. The correlation between quantified sugar levels and Raman spectra was performed using both PLS and artificial neural networks (ANN). The statistical R^2 for glucose, fructose, sucrose, and maltose were high, with 0.929, 0.930, 0.937, and 0.893 for PLS and 0.930, 0.931, 0.956, and 0.913 for ANN, indicating that both chemometric tools are efficient for the rapid analysis of sugar content. Oroian et al. [19] used Raman spectroscopy to detect honey adulterated with sugars (glucose, fructose, inverter sugar, hydrolyzed inulin syrup, and malt must). The study considered 900 samples with adulteration levels of 5, 10, 20, 30, 40, and 50%. Authentication of honey purity concentration was performed using PLS and PCR. The chemometric models developed showed good fit for both the calibration (R^2_{cal} = 0.983) and validation (R^2_{val} = 0.981) dataset, with low statistical errors (SEC = 0.009 and SEP = 0.103). Anjos et al. [20] evaluated the potential of Raman spectroscopy in the prediction of the physicochemical composition of *Lavandula* spp. monofloral honey. PLS models were used for the quantitative estimation, and the results were correlated with the values obtained using reference methods. Chemometric models were used for pH, sugar reduction, electrical conductivity, apparent sucrose, total phenol content, total flavonoid content, proline, and total acids, achieving R^2_{cal} in the range of 0.973–0.99, R^2_{val} in the range of 0.833–0.99, SEC in the range of 2.03–0.01, and SEP in the range of 1.71–0.01. In the study by Tahir et al. [21], Raman spectroscopy combined with PLS were applied to predict phenolic compounds and antioxidant activity in honey. It was found that the developed models based on Raman were superior to those established using NIR spectra, with R^2_{cal} and R^2_{val} > 90%, SEC < 1.2, and SEP < 1.7. Raman spectroscopy, and PLS-LDA modeling have also been used to determine the adulteration of Chinese honey with corn syrup [22]. The analysis considered adulteration samples in the range of 10, 20, and 40%. An accuracy prediction of 84.4% was obtained, indicating that combining PLS-LDA with Raman spectra is a potential technique for the detection of impure agents in honey.

In this paper, a study is presented to determine the physical-chemical properties of honey from the Mexican region of the Yucatan Peninsula. In this zone, beekeeping is an ancient activity, carried out since the pre-Columbian era by Mesoamerican cultures like the Maya, who already produced honey from apiaries with honey bees (*Melipona beecheii*) long before the arrival of the Spaniards [23]. After their conquest, the species *Apis mellifera* was introduced in Mexico, which proliferated and dispersed throughout the country due to its higher yields of honey. Currently, the Yucatan Peninsula (located in the south of the country and composed of the states of Yucatan, Campeche, and Quintana Roo) is one of the most fruitful regions for the development of beekeeping activity. This region is characterized by ecosystems with great flora diversity, producing nectars and pollen—many of them endemic—that produce honey with unique organoleptic, physical, and chemical properties; these characteristics make honey from this region very appreciated in national and international markets [24]. In this sense, Mayan beekeepers from the Yucatan Peninsula contribute approximately 35% of the national production. In the state of Campeche, there are 4030 honey producers that generate on average 5571 metric tons of honey per year; Campeche is the second honey producer region nationwide, only surpassed by Yucatan. Of the total produced in this region, 95% is exported, producing profits of up to 12 million US dollars and contributing to generating economic welfare for Mayan beekeepers [25–27]. Thus, the introduction of fast and low-cost tools to analyze the quality of the honey produced would contribute significantly to increasing distribution of this natural food, benefiting local beekeepers and the local economy.

Therefore, due to the economic importance of honey production in the state of Campeche, Mexico, the objective of this work was to develop chemometric models based on Raman spectroscopy for the quantification of the following physical and chemical properties: pH, moisture, total soluble solids (TSS), free acidity, lactonic acidity, total acidity, electrical conductivity (EC), Redox potential, hydroxymethylfurfural (HMF), and ash content. These chemometric models represent useful tools for the quality control of honey produced in the state of Campeche, by quickly and economically predicting the main physicochemical indicators.

2. Analysis of Results

2.1. Raman Analysis

Figure 1 shows that the Raman spectra obtained from the honey samples have spectral bands which cover the ranges of 330–404, 404–440, 440–510, 510–595, 595–691, 691–752, 770– 820, 820–1024, 1024–1094, 1094–1191, 1191–1262, 1262–1300, and 1300–1460 cm^{-1}:

- Spectral region between 230–510 cm^{-1} are related to stretching and bending vibrations of the C-O, C-C-O and C-C-C that form the molecular structure of sugars [21].
- The region between 595–691 cm^{-1} is attributed to stretching vibrations of unsaturated rings present in HMF, carotenes, flavones, flavonoids, and polyphenols [22].
- The peak found between 691–752 cm^{-1} is assigned to stretching vibrations of C-O and C-C-O, and bending vibrations of O-C-O. On the other hand, the band between 770–917 cm^{-1} is a product of the stretching vibrations of the C-C and C-H present in glucose [28].
- Regarding the bands between 820–1024 cm^{-1}, these correspond to deformation vibrations of C-H and methylene bonds $-CH_2-$, as well as the bending vibrations of C-O-H [29].
- The peak present between 1024–1094 cm^{-1} is attributed to bending vibrations of the C-H and C-O-H bonds of sugars, and bending vibrations of the C-N bonds of amino acids and proteins [30].
- The band between 1094–1191 cm^{-1} is assigned to stretching vibrations of the C-O, C-O-C bonds of sugars, and the C-N bonds of proteins and amino acids [18].
- Finally, the spectral region between 1262–1300 cm^{-1} corresponds to vibrations of C-H and O-C-H, while the spectral bands of 1300–1460 cm^{-1} are due to bending and wobble vibrations of the functional groups CH and –OH [30].

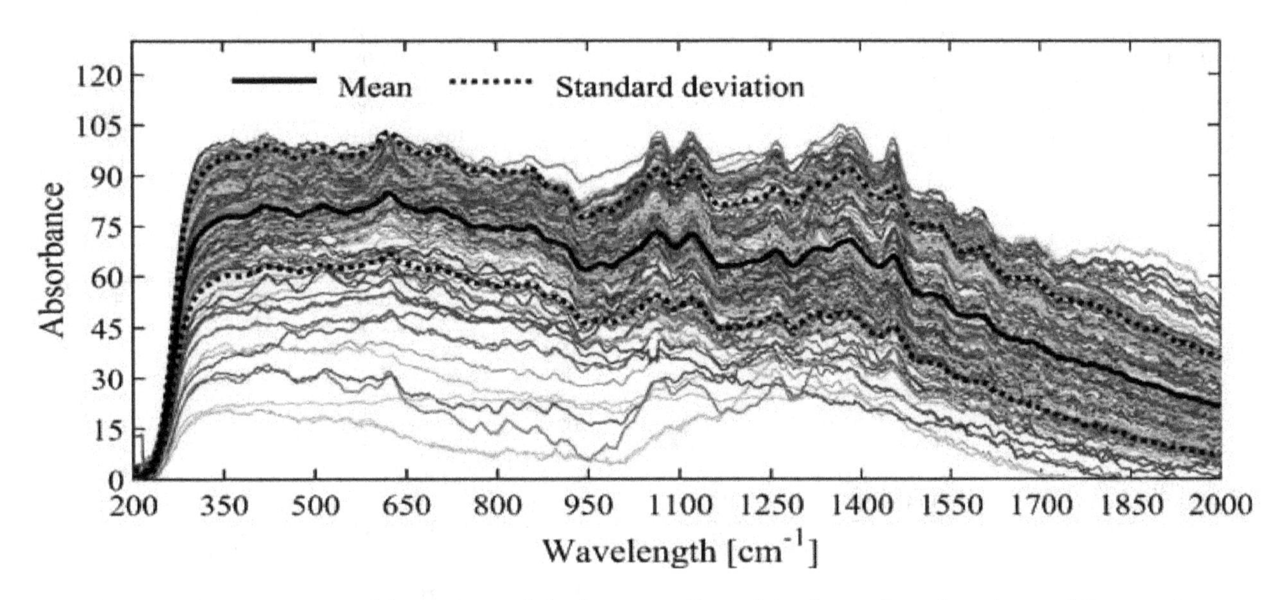

Figure 1. Raman spectral footprints of the honey collected in the various locations of Campeche.

2.2. Chemometric Models

2.2.1. Chemometric Models to Predict pH, Free Acidity, Lactonic Acidity, and Total Acidity

The presence of organic acids, such as gluconic, phenolic, ascorbic, lactic, and metallic ions, causes honey to be slightly acidic by nature. The acidity may be increased due to chemical and biochemical changes that take place in the honey. For example, the glucose oxidase enzyme is capable of transforming glucose into gluconic acid; on the other hand, the ions of the alkaline earth elements can react to form phosphates, sulfates, and chlorides, as well as transform lactone into lactic acid [31]. To measure these chemical changes in honey, in the Codex Alimentarius [8], the pH, free acidity, lactonic acidity, and total acidity were established as quality control criteria. In this sense, free acidity is related to the concentration of organic acids in honey, where a maximum value of 50 meq kg^{-1} is established by the Codex Alimentarius.

Table 1 lists the values of the 10 physicochemical parameters determined for honey samples from the municipalities of the state of Campeche. As reported in the table, the pH of honey samples were in the range of 3.49 to 5.2, within the limit established by the Codex Alimentarius (minimum 3.40 and maximum 6.10). The minimum and maximum values of free acidity were detected between 22.5 and 35.1 meq kg^{-1}, 4.15 y 9.45 meq kg^{-1} for lactonic acidity, and 28.67 a 38.28 meq kg^{-1} for total acidity. According to this, the values of the total acidity present in honey samples agree with the provisions of the Codex Alimentarius, indicating that the honey collected did not show significant degradation.

Table 1. Results obtained for the different physical and chemical parameters of honey from the municipalities of the state of Campeche.

Property	Mean ± σ	Minimum	Maximum	Mean ± σ	Minimum	Maximum
		Calakmul			*Calkini*	
pH	4.01 ± 0.23	3.66	5.11	4.08 ± 0.17	3.80	4.77
Free acidity	21.16 ± 5.03	8.12	32.53	19.79 ± 3.03	15.52	25.51
Lactonic acidity	2.96 ± 1.001	1.23	5.78	2.77 ± 0.84	1.47	4.27
Total acidity	24.17 ± 5.44	11.55	36.78	22.51 ± 3.31	18.25	28.67
Electric conductivity	0.58 ± 0.08	0.35	0.69	0.61 ± 0.05	0.49	0.68
Redox potential	181.94 ± 13.91	133.1	207.2	173.54 ± 8.87	161.6	198.1
Moisture	14.98 ± 1.42	11.81	17.66	12.21 ± 2.27	12.29	16.66
TSS	85.02 ± 1.41	82.37	88.19	85.79 ± 1.09	83.34	87.71
Ash content	0.14 ± 0.06	0.018	0.42	0.143 ± 0.14	0.09	0.21
HMF	2.87 ± 1.33	1.27	5.89	2.31 ± 0.75	1.46	4.35
		Campeche			*Carmen*	
pH	3.95 ± 0.16	3.49	4.18	3.97 ± 0.14	3.64	4.25
Free acidity	17.03 ± 3.52	12.39	26.1	21.22 ± 4.19	8.01	28.53
Lactonic acidity	2.51 ± 0.68	1.47	4.15	3.09 ± 1.08	1.23	5.78
Total acidity	19.53 ± 3.81	14.17	29.65	24.32 ± 4.41	11.45	31.34
Electric conductivity	0.48 ± 0.08	0.28	0.69	0.57 ± 0.08	0.35	0.69
Redox potential	177.49 ± 9.89	151.3	204.2	186.23 ± 8.41	170.1	207.4
Moisture	15.25 ± 3.11	12.76	24.6	15.02 ± 1.53	11.81	17.66
TSS	84.74 ± 3.11	75.42	87.24	84.98 ± 1.53	82.34	88.19
Ash content	0.13 ± 0.018	0.08	0.16	0.14 ± 0.09	0.02	0.88
HMF	2.12 ± 0.46	1.52	3.53	2.98 ± 1.43	1.27	5.89
		Champotón			*Escarcega*	
pH	3.78 ± 0.18	3.55	4.23	3.85 ± 0.17	3.62	4.31
Free acidity	22.81 ± 4.26	11.9	32.5	22.72 ± 5.11	13.5	31.5
Lactonic acidity	3.59 ± 0.78	2.37	5.98	3.51 ± 0.62	1.78	4.37
Total acidity	26.41 ± 4.47	17.01	38.28	26.23 ± 5.13	17.07	35.59
Electric conductivity	0.54 ± 0.11	0.36	0.69	0.58 ± 0.12	0.35	0.755
Redox potential	189.03 ± 11.39	165.4	202.6	172.52 ± 9.38	146.1	185.8
Moisture	16.9 ± 3.11	13.32	25.81	15.16 ± 0.88	13.65	16.89
TSS	83.01 ± 3.11	74.2	86.36	84.83 ± 0.88	83.11	86.35
Ash content	0.14 ± 0.03	0.11	0.17	0.13 ± 0.02	0.068	0.18
HMF	3.34 ± 1.32	1.57	6.39	2.34 ± 1.44	1.57	4.89
		Hecelchacan			*Hopelchén*	
pH	4.09 ± 0.09	3.91	4.21	4.34 ± 0.42	3.51	5.2
Free acidity	17.78 ± 3.06	16.85	22.5	16.64 ± 6.95	6.5	35.1
Lactonic acidity	5.14 ± 2.48	3.19	9.45	3.44 ± 0.91	1.67	5.92
Total acidity	22.93 ± 5.41	21.07	31.95	20.08 ± 6.82	10.41	37.77
Electric conductivity	0.61 ± 0.056	0.51	0.659	0.59 ± 0.08	0.44	0.71
Redox potential	177.49 ± 14.34	167.5	202.1	153.93 ± 22.21	105.6	198.2
Moisture	17.09 ± 3.19	15.17	22.67	14.72 ± 1.23	12.43	17.4
TSS	82.85 ± 3.16	77.33	85.45	85.27 ± 1.23	82.6	87.57
Ash content	0.13 ± 0.015	0.11	0.14	0.14 ± 0.03	0.05	0.21
HMF	2.89 ± 0.265	2.39	3.27	3.18 ± 0.95	1.56	5.78

The variability in the pH, free acidity, lactonic acidity, and total acidity is represented in Table 1 by the standard deviation (σ). In this sense, the honey samples with the highest pH standard deviation were those from the municipalities of Calakmul and Holpechen, with ±0.23 and ±0.42, respectively. This variability is attributed to the diversity of melliferous flora present in the region (Figure 1), which belongs to the Calakmul biosphere reserve and houses more than 150 melliferous flowers, with important differences in their chemical composition [24,25]. On the other hand, honey samples that presented higher pH values (4.18–5.2) correspond to productions from the Tajonal and Mangle Negro plants, characterized by a higher concentration of sodium chloride. The Tajonal is a plant widely distributed in the state of Campeche, which is adapted to alkaline soils and is capable of growing near coastal areas, where a sea breeze is deposited on the flowers. Likewise, Mangle Negro grow in the coastal zone, on the banks of lagoons and estuaries that contain waters with high salinity; this contributes to the fresh honey from these flowers having low acidity due to the presence of sodium chloride.

Regarding total acidity, this presents standard deviations of ±5.44 meq kg^{-1} for honey samples from Calakmul and ±6.82 meq kg^{-1} in honey from Hopelchen. The free acidity for honey from the municipalities of Carmen has standard deviations ±4.41 meq kg^{-1} and ±4.47 meq kg^{-1} for those of Champotón, and ±5.13 meq kg^{-1} for Escarcega. The municipalities of Carmen, Champotón, and Escarcega are geographically are located in the west of the state of Campeche, a region characterized by lagoons, wetlands, rivers and estuaries that are conducive to the growth of melliferous plants such as Arbol de tinto, Pucté, Mangle, Cascarillo, and Ja'abin, among others. The honey of these floral species has a higher moisture content, which favors honey fermentation. On the other hand, the Hecelchacan honey samples showed a standard deviation of ±5.41 meq·kg^{-1}. This variability is attributed to the predominance in this region of melipona honey, which by its nature usually contains water concentrations above 20%, favoring the formation of organic acids by biochemical reactions.

Based on the measurements obtained, chemometric models were created to predict pH, free acidity, lactonic acidity, and total acidity. Figure 2 shows the predictive behavior of the models, while Table 2 contains their statistical performances. The calibration model to predict the pH in honey of the state of Campeche exhibits a standard error of calibration SEC = 0.86 and standard error of prediction SEP = 0.18; likewise, it presents acceptable values for the coefficient correlation of calibration ($R^2_{cal} = 0.92$) and the coefficient correlation of validation ($R^2_{val} = 0.74$). These statistical values show that the chemometric model has an acceptable ability to predict the pH in honey. On the other hand, Student's *t*-test with paired data at 95% confidence obtained $t_c = 0.95$, within the established confidence interval ($t_v = \pm 1.65$). Therefore, the chemometric model based on Near Infrared Spectroscopy (NIRS) has a good reliability but not enough to substitute the standardized method. The statistical values obtained in this work are similar to those reported by Cozzolino et al. [32], who obtained a chemometric model using Vis-NIRS spectroscopy to predict the pH of honey in Uruguay. They also reported values of SEC = 0.13, SEP = 0.21, $R^2_{cal} = 0.88$, and $R^2_{val} = 0.70$. On the other hand, Anjos et al. [20] reported statistical values of SEC = 0.12, SEP = 0.09, $R^2_{val} = 0.83$, and $R^2_{cal} = 0.98$ for a calibration model based on the FT-Raman spectroscopy used to predict the humidity percentage in Portuguese honey.

The chemometric model for predicting free acidity presented a standard error of calibration (SEC = 1.02), a standard error of prediction (SEP = 1.47), coefficient correlation of calibration ($R^2_{cal} = 0.98$, and coefficient correlation of validation ($R^2_{val} = 0.94$). These results indicate that the chemometric model successfully predicts the concentration of honey's free acidity. The Student's *t*-test of paired data ($t_c = 0.64$) for free acidity is within the confidence interval ($t_v = \pm 1.65$), indicating that there are no differences in the prediction capacity of the developed chemometric model with respect to the standard method established in the Codex Alimentarius [8]. In previous studies, such as the one carried out by Ruoff et al. [33], the following statistical values were reported for a chemometric model based on NIRS spectroscopy to predict free acidity in Swiss honey: a standard error of calibration (SEC = 2.01), standard error of prediction (SEP = 2.0), and coefficient correlation of validation (R^2_{val} = 0.737).

Table 2. Values of the statistical parameters obtained in cross-validation and external validation to determine the capacity predictability of each chemometric model.

Properties	Units	Calibration LVs	SEC	R^2_{cal}	Validation LVs	SEP	R^2_{val}
pH	-	5	0.86	0.92	4	0.18	0.743
Free acidity	meq kg^{-1}	6	1.02	0.98	6	1.47	0.935
Lactonic acidity	meq kg^{-1}	6	0.37	0.94	7	0.41	0.911
Total acidity	Meq kg^{-1}	6	1.08	0.98	4	1.23	0.897
Electrical conductivity	mS cm^{-1}	6	0.46	0.87	4	0.85	0.79
Redox potential	*mV*	7	1.06	0.99	8	1.48	0.95
Moisture	%	6	0.42	0.98	9	0.52	0.95
TSS	%	6	0.58	0.92	6	1.32	0.87
Ash content	%	6	1.21	0.78	6	2.54	0.21
HMF	mg kg^{-1}	7	0.76	0.82	8	1.73	0.63

With regards to the chemometric model for predicting lactonic acidity in Campechean honey, it showed good predictive capacity, since the values of cross-validation and external validation, along with the standard error of calibration and standard error of prediction, were small (SEC = 0.37; SEP = 0.41), with the following coefficient correlation of calibration and coefficient correlation of validation (R^2_{cal} = 0.94; R^2_{val} = 0.91). For the Student's *t*-test of paired data (t_c = 0.69) at 95% confidence, the value obtained is in the confidence interval (t_v = ±1.65), so there are no differences in the prediction capacity of lactonic acidity between the obtained chemometric model and the standard method [8].

Finally, the chemometric model to predict total acidity in Campeche honey showed a high coefficient correlation in the cross-validation (R^2_{cal} = 0.98) and coefficient correlation in the external validation (R^2_{val} = 0.89), as well as low values of standard error of calibration (SEC = 1.18) and standard error of external validation (SEP = 1.23). Moreover, the Student's *t*-test of paired data (t_c = 0.75) is in the confidence interval (t_v = ± 1.65), which demonstrates that the chemometric model is as reliable as the standardized method. Comparing the obtained results with those reported by Anjos et al. [20] for an FT-Raman spectroscopy calibration model to predict the acidity total in Portuguese honey, similar values were observed (SEC = 0.22; SEP = 0.28; R^2_{cal} = 0.99; R^2_{val} = 0.99).

In Figure 2, it can be seen that the experimental data of the pH, free acidity, lactonic acidity, and total acidity of the honey samples show a certain degree of dispersion compared to the chemometric model predictions. This can be attributed to the following factors: first, in the state of Campeche, several tropical forests are located that give rise to a great diversity of honey blooms; previous works have identified more than 150 blooms in the area of study [24–26]. Thus, the honeys produced in the region are multifloral, giving rise to a wide variety of physical and chemical properties. Second, the geographical origins where the honey samples were collected—specifically in the east of the state of Campeche, in the municipalities of Carmen, Palizada, Escarcega, and Champotón—are characterized by the presence of rivers, lagoons, wetlands, and swamps. These soils are rich in organic matter and have an acidic pH, which contribute to the development of a great diversity of melliferous flora, such as: Tahonal, Ja'abin, Pukte, huano, Xtabentum, Palo Tinto, hulub, Suuk chak lol, Box káatsim, Bohom, Susuk, cascarillo and mangle negro. Flowers from these botanical origins produce nectar with high concentrations of moisture, which is transferred to the honey [26]. The presence of a high percentage of moisture in honey favors biochemical and chemical reactions—for example, the formation of gluconic acid from glucose and the formation of inorganic acids due to the reaction of water with anions and cations present in honey. This means that honey samples collected in these locations show greater variability in pH, free acidity, lactonic acidity, and total acidity [34,35].

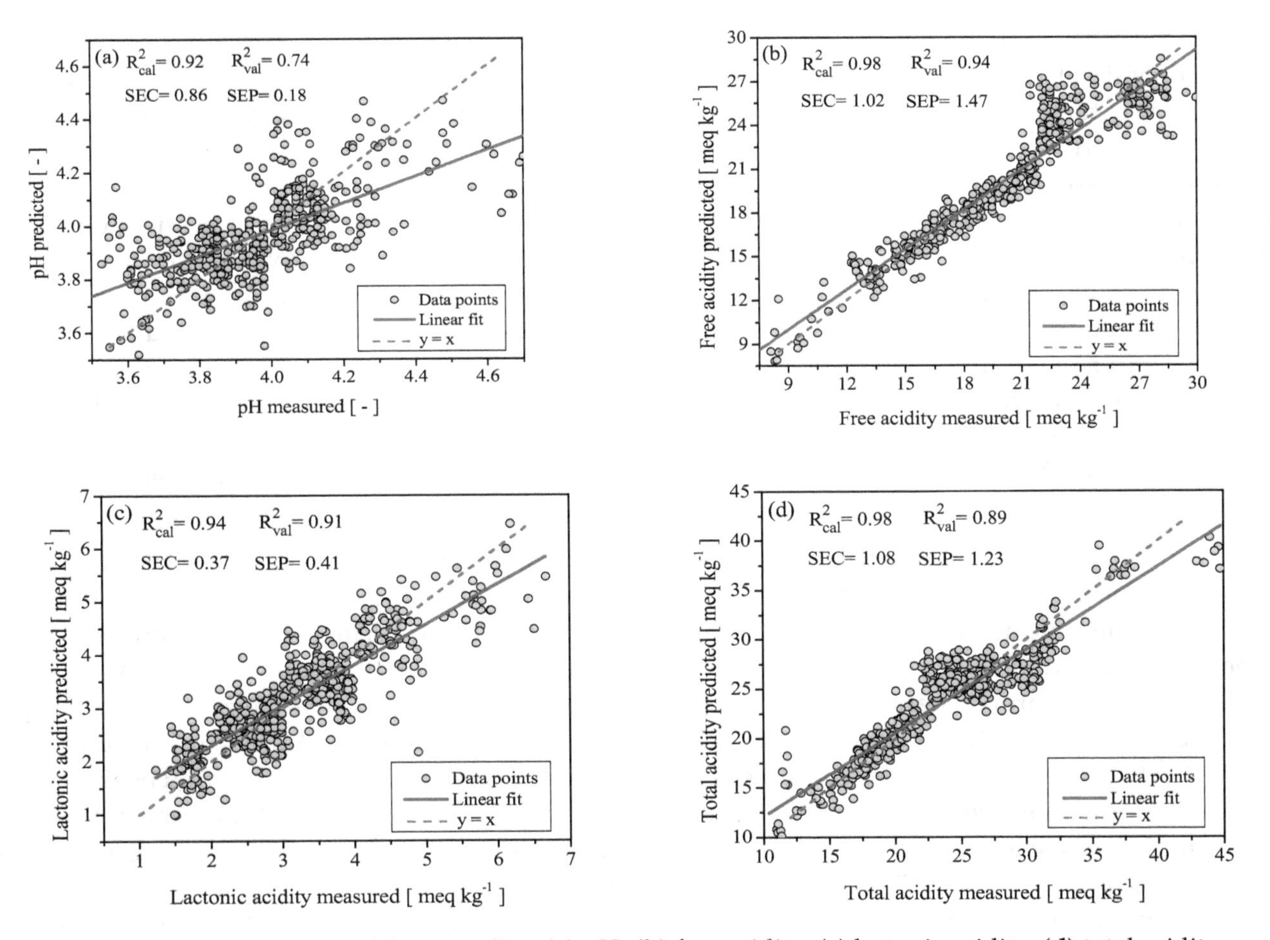

Figure 2. Chemometric models to predict: (**a**) pH; (**b**) free acidity; (**c**) lactonic acidity; (**d**) total acidity.

2.2.2. Chemometric Model to Predict Electrical Conductivity, Redox Potential, Moisture, and TSS

Electrical conductivity is a parameter used to determine the geographical origin of honey. This is related to the content of ashes, organic acids, and dissolved mineral salts; the higher the concentration of these compounds in honey, the greater the value of the electrical conductivity [36]. In this sense, the diverse honey samples from the state of Campeche presented values between 0.28–0.75 mS cm^{-1}, which is below the maximum allowed limit (0.80 mS cm^{-1} [8]). The chemometric model for this physicochemical property had a standard cross-validation error and an external validation error of 0.46 and 0.85, respectively. Moreover, the regression coefficients obtained were R^2_{cal} = 0.87 and R^2_{val} = 0.79. Nevertheless, the R^2_{val} value indicates an acceptable model fit, but are not significant for our propose. In Figure 3a, a noticeable dispersion between the experimental data of the electrical conductivity with respect to the chemometric model is observed. This is attributed to the significant differences in organic matter, salinity content, and carbonates in the soils of the locations where the honey samples were collected. Another cause is the diversity of the honey flora, which contributed to the variation in the content of organic acids in the honeys [24,31].

Comparing the results obtained with previous works, these present slightly lower values than those reported by Anjos et al. [20], who built a chemometric model based on FT-Raman for Portuguese honey. They reported values of calibration errors and external validation of (SEC = 0.01; SEP = 0.01), and coefficients of determination (R^2_{cal} = 0.92.8; R^2_{rval} = 0.938). Nonetheless, the results obtained in our study are similar to those reported by Ruoff et al. [33] for a chemometric model based on NIRS spectroscopy to predict electrical conductivity in Swiss honeys (R^2_{cal} = 0.794 and R^2_{rval} = 0.87); and with the data reported by Cozzolino et al. [32] for a calibration model of Uruguayan honey (R^2_{cal} = 0.83 and R^2_{rval} = 0.80).

On the other hand, honey contains chemical substances dissolved in low concentrations of organic acids, mineral salts, and polyphenols; polyphenols are molecules that contain unsaturated bonds in their chemical structure, and develop a very important function since they are antioxidants; these substances have the property of trapping free radicals generated in biochemical reactions. When honey undergoes cooking processes or remains stored for a long period, the aforementioned substances may undergo oxide–reduction reactions, causing changes in their molecular structure and modifications in the properties of honey. These chemical changes can be monitored using the Redox potential to determine the degree of oxidation. Because the Redox potential can be used as a quality control parameter, it was analyzed in Campeche honeys. The results indicate Redox potential values with a minimum of 133.1 mV and a maximum of 207.2 mV; the difference in these results is attributed to the composition of each bloom. The chemometric model had calibration and validation errors (SEC = 1.06; SEP = 1.48), and high values in the calibration and external validation coefficients ($R^2_{cal} = 0.99$; $R^2_{rval} = 0.95$). The reliability of the model was also confirmed by a Student's *t*-test of paired data, with a value of $t_c = 0.545$ between $t_v = \pm 1.65$ at 95% confidence, so the model has a good predictive capacity.

With regards to moisture, a maximum content of 20% was defined in the Codex Alimentarious [8]. This is because an excess of moisture favors the fermentation of sugars, causing the formation of undesirable organic acids that affect the organoleptic properties [37]. The moisture content in honey depends on several factors, such as floral origin, harvest time, climate change, maturity degree of the honey, and improper handling of the honey by beekeepers [38]. The analyzed honey samples presented humidity values between 11.81–25.81%; some samples showed humidity concentrations above 20% because the honey came from tree blooms located in wetlands, near rivers, and near estuaries. In addition, some samples were from melipona honey, that, by nature, contains high concentrations of moisture [39]. The chemometric model to predict moisture in honey presented SEC = 0.42, SEP = 0.52, $R^2_{cal} = 0.98$, and $R^2_{val} = 0.97$. Additionally, $t_c = 0.41$ was obtained in the Student's *t*-test, which indicates that the chemometric model correctly predicts moisture in the honey. The presented results are similar to those reported by Lichtenberg et al. [40] for a predictive model of moisture in German honeys ($R^2_{cal} = 0.73$ and $\sigma = \pm 1.22$). Likewise, it is consistent with what was reported by García et al. (2000) regarding a chemometric model to predict moisture in honey from the region of Galicia, Spain (SEC = 0.12, SEP = 0.15, and $R^2_{cal} = 0.98$); and with Cozzolino et al. [32], who reported values of (SEC = 2.7, SEP = 3.1, $R^2_{cal} = 0.96$, and $R^2_{val} = 0.94$) for a calibration model focused on predicting moisture content in Uruguayan honey.

The principal component of honey is sugar; honey contains a mixture of sugars, mainly fructose, glucose, sucrose, maltose, and melezitose. Glucose and fructose are the ones that are found in the highest proportion and can represent up to 95% of the sugar content [41]. The honey samples collected in the state of Campeche exhibited total sugar concentrations between 74.19–88.19% w; some samples presented concentrations below 80 ° Brix [8] due to a higher moisture concentration. The chemometric model developed to predict TSS showed the following statistical results: SEC = 0.58; SEP = 1.32; $R^2_{cal} = 0.92$; $R^2_{val} = 0.87$. A Student's *t*-test with a value of $t_c = 0.28$ was in the range $t_v = \pm 1.65$ at 95% reliability, which shows that the model for predicting TSS has an acceptable prediction capacity but not adequate to supply the referenced method. The results obtained in this work were similar to those reported by Mignani et al. [42], who built chemometric models based on Raman spectroscopy to predict glucose and fructose concentrations in Italian honeys (SEC = 7.3; SEP = 11; $R^2_{cal} = 0.96$, $R^2_{val} = 0.92$) and (SEC = 5.5; SEP = 7.6; $R^2_{cal} = 0.89$, $R^2_{val} = 0.82$). Likewise, Özbalci et al. [18] developed calibration models based on Raman spectroscopy to predict glucose and fructose concentrations in Turkish honeys, reporting the following values (SEC = 0.51; SEP = 2.75; $R^2_{cal} = 0.98$; $R^2_{val} = 0.96$). Complementing this, Anjos et al. [20] reported statistical results (SEC = 0.34, SEP = 0.39, $R^2_{cal} = 0.99$, $R^2_{val} = 0.99$) for a predictive calibration model of reducing sugars in Portuguese honey. The comparisons between values predicted by the chemometric models presented in this section and their respective experimental values are shown in Figure 3.

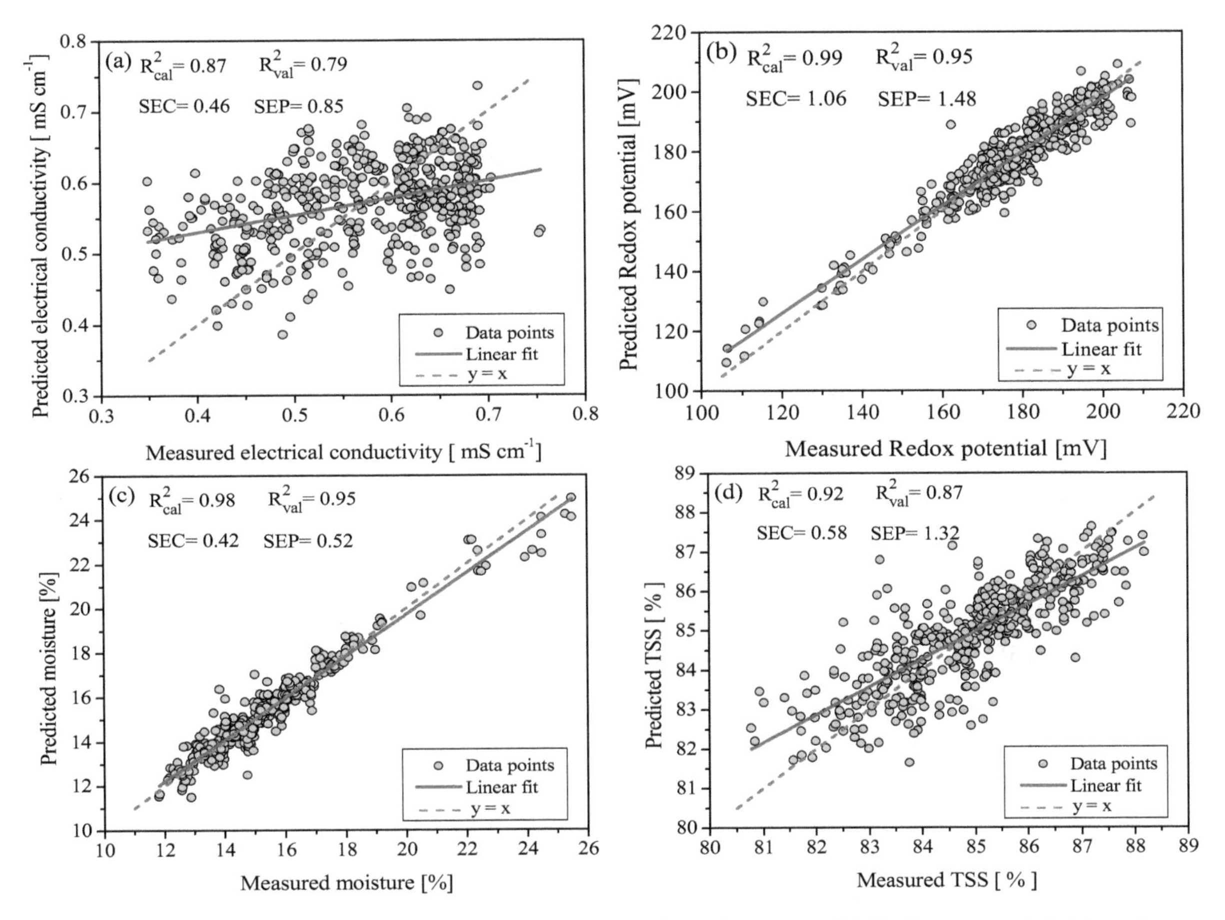

Figure 3. Chemometric models to predict: (**a**) electrical conductivity; (**b**) Redox potential; (**c**) moisture; (**d**) TSS.

2.2.3. Chemometric Model to Predict Content of HMF and Ashes

As presented in Table 2, the chemometric models to predict ash percentage and HMF content presented low coefficients of determination in cross-validation and external validation (R^2_{cal} = 0.78, R^2_{val} = 0.21; and R^2_{cal} = 0.82, R^2_{val} = 0.56). The above indicates that the models are not suitable for the prediction of these chemometric properties.

2.3. Analysis of the PLS loadings

The PLS loading for total acidity, electrical conductivity, Redox potential, humidity and TSS (Figure 4) present six spectral regions (200–600 cm^{-1}, 630–790 cm^{-1}, 870–1000 cm^{-1}, 1080 –1200 cm^{-1}, 1400–1570 cm^{-1}, and 1750–1880 cm^{-1}) that provide useful chemical information for the development of their respective predictive chemometric models. The first spectral band (between 200–400 cm^{-1}) has a positive and negative contribution in the PLS loading. The chemical information provided by this region is related to stretching, bending, and deformation vibrations of C-O, C-C-O, C-C-C and C=O, which form the skeleton of sugar molecules, organic acids, phenolic compounds, and flavonoids. Here, breaks of functional groups and of the sugar backbone can occur due to oxide–reduction reactions; for example, in the transformation of glucose into gluconic acid, fermentation reactions for the production of alcohols and carboxylic acids and the cyclization of fructose produce HMF. These chemical changes in the honey collected would reflect variations in total acidity, pH, Redox potential, and electrical conductivity with respect to time. The band at 630–790 cm^{-1} provides chemical information of the cyclic and alicyclic rings that make up the molecules of HMF, carotenes, flavonols, flavanones, and flavones, among other phenolic compounds. The chemical information related to the band between 870–1000 cm^{-1} is attributed to stretching, bending and deformation vibrations of the C-C, C-H, C-H, –CH_2–, and C-O-H bonds present in the sugars. The Raman region between 1080–1200 cm^{-1} provides information on protein and carbohydrate content in honey, due to stretching vibrations of C-O, C-O-C, C-N carbohydrate, and protein bonds. The region between 1400–1570 cm^{-1} provides chemical information due to bending and wobble vibrations of CH, O-C-H and –OH functional groups present in sugar molecules, and –OH in the water molecules. Finally, the concentrations of moisture, fructose, glucose and moisture in honey are related to stretching vibrations of the unsaturated bonds C=O in fructose and CH=O in glucose, and deformation vibrations –OH of water, which are present in the Raman spectrum between 1750–1880 cm^{-1}.

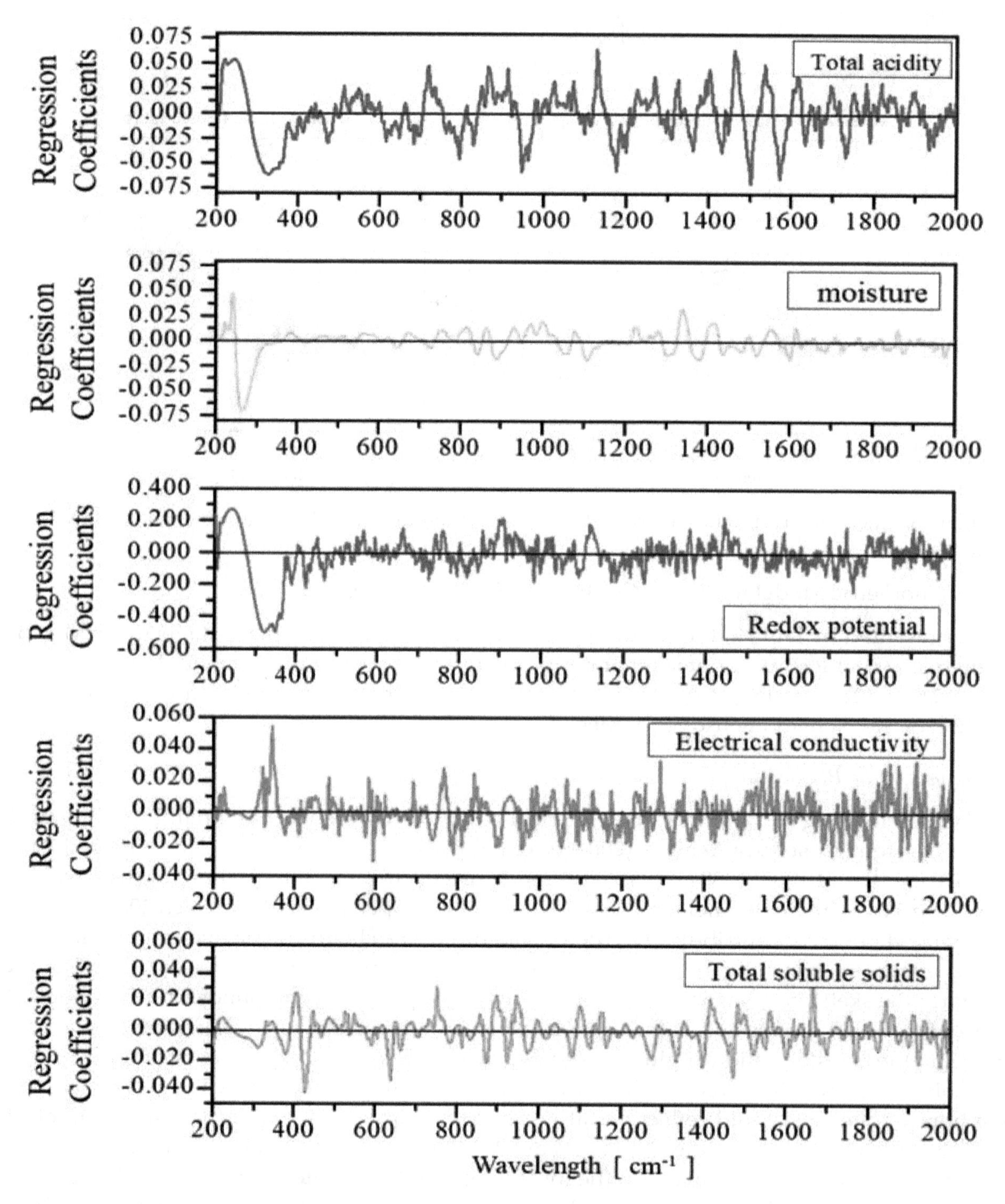

Figure 4. Regression models in the Raman region obtained to predict physical and chemical properties of honey from the state of Campeche.

3. Materials and Methods

3.1. Honey Samples

A total of 189 honey samples were supplied directly from Mayan beekeepers of the state of Campeche, Mexico. The samples were collected between February and June of 2014 and 2015. From the total samples, 175 corresponded to *Apis mellifera* and 14 to *Melipona beecheii*. Figure 5 illustrates the geographical region where the honey samples were collected, which includes the locations of Calakmul (40 samples), Calkiní (14 samples), Campeche (26 samples), Champotón (34 samples), Escárcega (20 samples), Hecelchakán (4 samples), Hopelchén (22 samples), and Sabancuy (29 samples). The predominant floral origin of the honeys was determined according to information provided by the Mayan beekeepers, and included the following: Tahonal (*Viguiera dentata*), Tsíitsilche (*Gymnopodium floribundum*), Ja'abin (*Piscidia piscipula*), Tzalam (*Lysiloma latisiliquum*), Pukte (*Bucida buceras*), Xa'an, huano (*Sabal yapa*), Xtabentum (*Turbina corymbosa*), Palo Tinto (*Haematoxylum campechianum*), Chéechem (*Metopium brownei*), Hulub (*Bravaisia berlandieriana*), Chakàah (*Bursera simaruba*), Suuk, chak lol (*Salvia coccínea*), Box káatsim (*Acacia gaumeri*), Bohom (*Cordia gerascanthus*), Kitim che' (*Caesalpinia gaumeri*), Susuk (*Dyphisa carthagenensis*), Cascarillo (*Erythroxylum confusum*), Machiche (*Lonchocarpus castilloi*) and Mangle negro (*Avicennia germinans*).

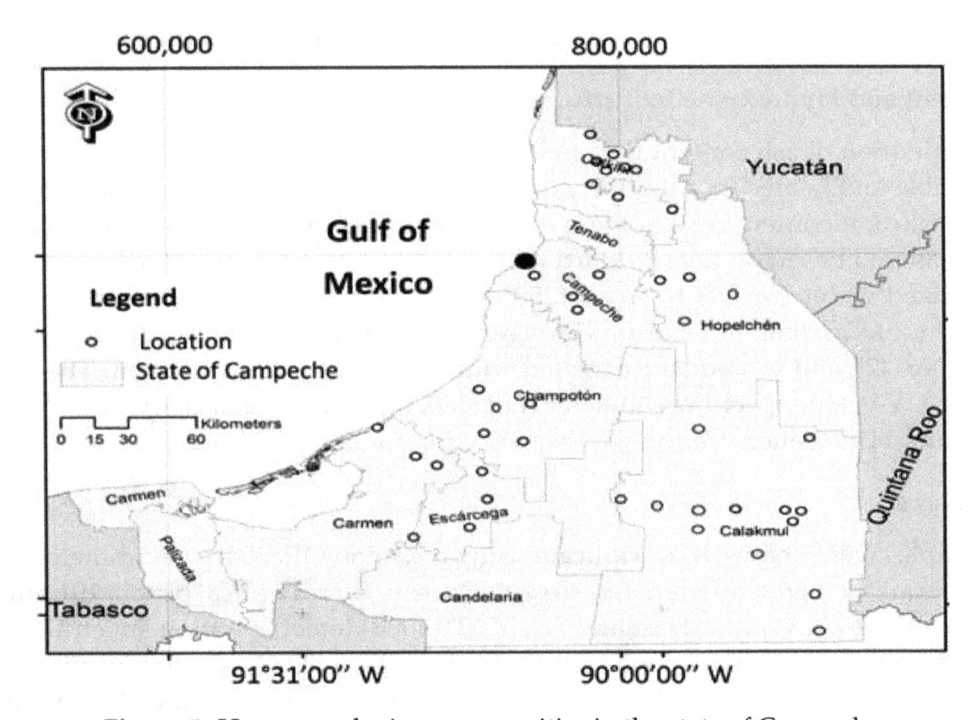

Figure 5. Honey producing communities in the state of Campeche.

3.2. Physicochemical Analysis

The physical-chemical properties of the honey samples were determined according to standards and methods established by Codex Alimentarious [8], International Honey Commission [9], and the Association of Official Analytical Chemists [10]. The honey properties studied were pH, moisture, TSS, free acidity, lactonic acidity, total acidity, EC, Redox potential, HMF, and ash content. The chemical reagents used were standard hydrochloric acid (HCl) solution at 0.05 N, standard sodium hydroxide (NaOH) solution at 0.05 N, deionized water, acetone, buffer solutions, sodium bisulfite (Fermont, Canada), and the reagents Carrez I and Carrez II (Sigma-Aldrich, Saint Louis, MO, USA); all of them of analytical grade. A detailed description of the procedure for obtaining each physical-chemical property is given below.

3.2.1. Moisture and Total Soluble Solid

Moisture and TSS were measured by the refractometric method. One gram of honey was analyzed in an Atago refractometer model PAL-22S (Atago, Tokio, Japan) at 25 °C; TSS was expressed in Brix°, whereas moisture percentage (g/100 g honey) was given according to the method established in [43].

3.2.2. pH, Free Acidity, Lactonic Acidity, and Total Acidity

To determine the pH, 10.0 g of honey was dissolved in 75 mL of deionized water (free CO_2). The solution was analyzed by using a Thermo Scientific brand pH meter (Orion Star A211, Waltham, MA, USA), previously calibrated with standard buffer solutions at pH values of 4–7 and 7–10, respectively. The honey solution was titrated with 0.05 N NaOH until it reached a pH of 8.5 to obtain the free acidity value. Lactonic acidity was determined by adding 10 mL of 0.05 N NaOH to the sample, and then titrating with 0.05 N HCl to return the pH to 8.3. Finally, the total acidity was obtained as the sum of the free acidity and lactonic acidity values, expressed in meq·kg^{-1} [44].

3.2.3. Electrical Conductivity and Redox Potential

The electrical conductivity and Redox potential were measured using a conductivity meter (Thermo Scientific, Waltham, MA, USA), which analyzed a solution composed of 20 g of honey dissolved in 100 mL of deionized water (free CO_2). Measurements were made at 20 °C and the results were expressed in mS·cm^{-1} for electrical conductivity and mV for Redox potential [45,46].

3.2.4. Ash Content and Hydroxymethylfurfural

The determination of ash content was conducted by incineration [47]. Two grams of honey was placed in a crucible and heated in a Lindberg/Blue muffle furnace (Thermo Fisher Scientific, USA) at 650 °C for 6 h. Carbon content results were expressed in g/100 g honey. On the other hand, HMF content was measured based on the standard method [10]. Five grams of honey was dissolved with 25 mL of deionized water (free CO_2) in an (250 mL) Erlenmeyer flask. The solution was clarified by adding 0.5 mL of Carrez I and Carrez II reagents, up to 50 mL. The solution was filtered using Watman paper (No. 42), and subsequently treated with a sodium bisulfite solution. The absorbance was determined on a UV-visible spectrophotometer (DR6000, HACH, Loveland, CO, USA) at wavelengths of 284 and 338 nm. HMF concentration was expressed in mg·kg^{-1}.

3.3. Raman Analysis

Honey samples were analyzed in triplicate using a Raman QE65000 spectrometer (Ocean Optics, Edinburgh, UK) equipped with a symmetric crossed Czerny-Turner optical bench, 101 mm focal length, an RPB 785 fiber optic prove, and Hamamatsu S7031-1006 detector with a spectral range between 780–940 nm. The spectrometer was operated with the SPECTRA SUIT software (version 2.0.162, Ocean Optics, Edinburgh, UK) to establish the interface between the computer and the Raman equipment. To perform the analysis of the samples, 30 mL of honey was deposited in an amber glass bottle and subsequently a laser beam was applied at 785 nm with a power of 20 mW for 10 s. All Raman spectra were collected in the range of 0 to 2200 cm^{-1} at 25 °C with a spectral resolution of 1.55 cm^{-1}. The data between 0–200 cm^{-1} and 2001–2200 cm^{-1} were omitted because they had higher spectral noise. Therefore, the spectral data between 201–2000 cm^{-1} was used.

3.4. Chemometric Model Development

For the development of the chemometric models, an experimental database was created employing the Raman absorbance (matrix X) and the physical-chemical properties of the analyzed honey samples (vectors Y). Raman analysis results were converted into a data matrix using Microsoft Excel 2013 (Microsoft, Redmond, WA, USA) composed of 900 wavelength values and 567 honey samples (510,300 absorbance samples). The data matrix was transposed and exported to the Pirouette V. 4.5

Software (Infometrix, Bogota, Colombia) to be correlated with each of the physicochemical properties. For the construction of chemometric models, partial least square (PLS) regression was used. In order to minimize spectral noise and errors in the development of the chemometric models, the following mathematical and statistical treatments were applied: auto-scaling or centering and subsequently the treatments baseline correction, smoothing, data normalization, first-order derivation, alignment, Log10 analysis, and Standard Normal Variate (SNV) were performed. To determine the predictability of the models developed, a cross-validation was performed (five out) using 90% of the data. Subsequently, an external validation was carried out with the remaining 10% of the data, which were not used in the construction of the chemometric models. The division of the database for the external calibration and validation processes was carried out by the software Pirouette, implementing the Kennard–Stone selection algorithm [33]. The statistical indicators used during the validation phase were: standard error of calibration (SEC), standard error of prediction (SEP), coefficient correlation of calibration (R^2_{cal}) and coefficient correlation of validation (R^2_{val}), and Student's *t*-test of paired data [33,34]. Figure 6 illustrates the computational procedure for the development of the chemometric models.

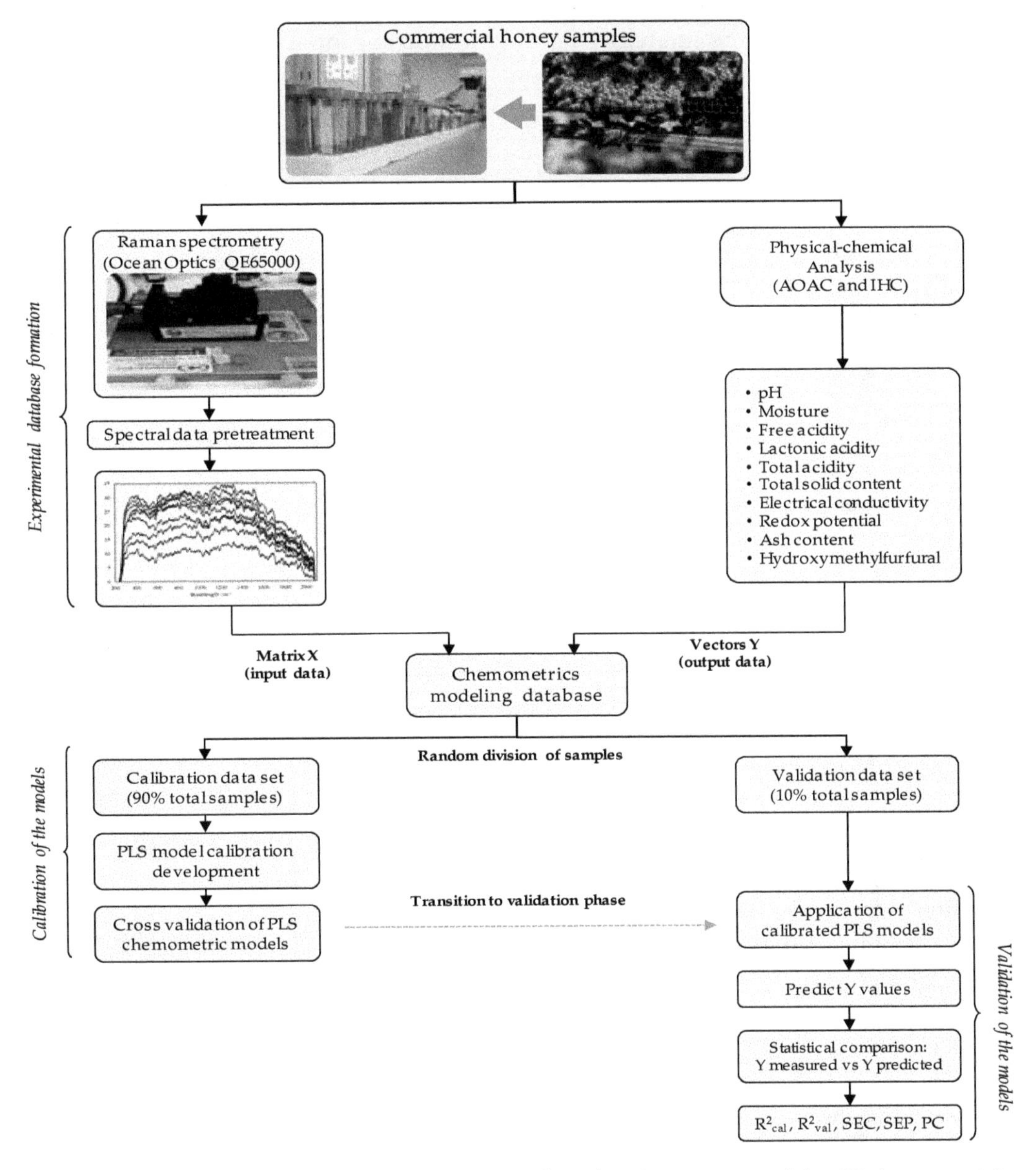

Figure 6. Schematic diagram of the development and evaluation process of the 10 chemometric models for the estimation of physicochemical properties of honey produced in the region of Campeche, Mexico.

4. Conclusions

In this work, it has been demonstrated that the Raman technique is an analytical tool that has advantages over other conventional techniques for the analysis of honey, since it is friendly to the environment and does not use chemical reagents, obtaining results in less time. Furthermore, it has been demonstrated that chemometric modeling based on Raman technology allows the development of numerical models and good capacity of predicting humidity, free acidity, lactonic acidity, total acidity, and Redox potential for Campechean honeys. The statistical parameters used to evaluate the predictability of each chemometric model show an accuracy similar to the conventional methods established in the standards, with the advantage that they are faster and do not use chemical reagents, so they are more environmentally friendly. Chemometric models to predict the content of HMF and ashes did not achieve good predictive capacity, which can be attributed to the fact that these chemical components are at very low concentrations in honey.

According to the study, the chemometric models that presented adequate prediction results represent an interesting alternative to be used in the development of intelligent portable laboratories that facilitate beekeepers in the region to analyze said chemometric properties at the site. Thus, the models presented represent a low-cost option to contribute significantly to the economic development of the honey industry in the region.

Author Contributions: All authors contributed equally to the design and performance of the experiments, the analysis of the data, and the writing and revision of the manuscript.

Acknowledgments: The authors thank the Faculty of Chemistry of the Autonomous University of Carmen for the support provided to perform the physical and chemical analysis of honey samples in the laboratories of instrumental analysis and Environmental Engineering. We also thank PROMEP for the financial support granted through Project DSA/103.5/14/10634 (UNACAR-EXB076) and the Mayan beekeepers of the State of Campeche for donating honey samples.

Abbreviations

σ	Standard deviation
R^2_{cal}	Coefficient of determination of cross-validation
R^2_{val}	Coefficient of determination for external validation
t_c	Student's *t*-test confidence value
t_v	Student's *t*-test external validation value
X	Raman spectroscopy matrix data
Y	Vector value of a physicochemical property
ANN	Artificial Neural Networks
AOAC	Association of Official Analytical Chemists
EC	Electrical conductivity
FT	Fourier transform
HCA	Hierarchical Clustering Analysis
HMF	hydroxymethylfurfural
IHC	International Honey Commission
LDA	Linear Discriminant Analysis
LVs	Latent variables
NIRS	Near infrared spectroscopy
PCA	Principal Component Analysis
PCR	Principal Component Regression
PLS	Partial Least Square
SEC	Standard error of calibration
SEP	Standard error of prediction
SNV	Standard Normal Variate
TSS	Total soluble solids

References

1. Cianciosi, D.; Forbes-Hernández, T.; Afrin, S.; Gasparrini, M.; Reboredo-Rodriguez, P.; Manna, P.; Zhang, J.; Bravo Lamas, L.; Martínez Flórez, S.; Agudo Toyos, P.; et al. Phenolic Compounds in Honey and Their Associated Health Benefits: A Review. *Molecules* **2018**, *23*, 2322. [CrossRef] [PubMed]
2. Ramón-Sierra, J.; Peraza-López, E.; Rodríguez-Borges, R.; Yam-Puc, A.; Madera-Santana, T.; Ortiz-Vázquez, E. Partial characterization of ethanolic extract of Melipona beecheii propolis and in vitro evaluation of its antifungal activity. *Rev. Bras. Farmacogn.* **2019**, *29*, 319–324. [CrossRef]
3. Duca, A.; Sturza, A.; Moacă, E.-A.; Negrea, M.; Lalescu, V.-D.; Lungeanu, D.; Dehelean, C.-A.; Muntean, D.-M.; Alexa, E. Identification of Resveratrol as Bioactive Compound of Propolis from Western Romania and Characterization of Phenolic Profile and Antioxidant Activity of Ethanolic Extracts. *Molecules* **2019**, *24*, 3368. [CrossRef] [PubMed]
4. Przybyłek, I.; Karpiński, T.M. Antibacterial Properties of Propolis. *Molecules* **2019**, *24*, 2047. [CrossRef]
5. Guerrini, A.; Bruni, R.; Maietti, S.; Poli, F.; Rossi, D.; Paganetto, G.; Muzzoli, M.; Scalvenzi, L.; Sacchetti, G. Ecuadorian stingless bee (Meliponinae) honey: A chemical and functional profile of an ancient health product. *Food Chem.* **2009**, *114*, 1413–1420. [CrossRef]
6. Maione, C.; Barbosa, F.; Barbosa, R.M. Predicting the botanical and geographical origin of honey with multivariate data analysis and machine learning techniques: A review. *Comput. Electron. Agric.* **2019**, *157*, 436–446. [CrossRef]
7. Se, K.W.; Wahab, R.A.; Syed Yaacob, S.N.; Ghoshal, S.K. Detection techniques for adulterants in honey: Challenges and recent trends. *J. Food Compos. Anal.* **2019**, *80*, 16–32. [CrossRef]
8. Codex Alimentarius Commission Codex Standard for Honey, CODEX STAN 12-1981. Available online: http://www.fao.org/3/w0076e/w0076e30.htm (accessed on 1 November 2019).
9. Bogdanov, S. Harmonised methods of the International Honey Commission. Available online: http://www.ihc-platform.net/ (accessed on 1 November 2019).
10. AOAC. *Official Methods of Analysis of AOAC International*; Association of Official Analysis Chemists International: Rockville, MD, USA, 2005; ISBN 0935584544.
11. Das, C.; Chakraborty, S.; Acharya, K.; Bera, N.K.; Chattopadhyay, D.; Karmakar, A.; Chattopadhyay, S. FT-MIR supported Electrical Impedance Spectroscopy based study of sugar adulterated honeys from different floral origin. *Talanta* **2017**, *171*, 327–334. [CrossRef]
12. Brereton, R.G. Introduction to multivariate calibration in analytical chemistry. *Analyst* **2000**, *125*, 2125–2154. [CrossRef]
13. Aliaño-González, M.J.; Ferreiro-González, M.; Espada-Bellido, E.; Palma, M.; Barbero, G.F. A screening method based on Visible-NIR spectroscopy for the identification and quantification of different adulterants in high-quality honey. *Talanta* **2019**, *203*, 235–241. [CrossRef]
14. Corvucci, F.; Nobili, L.; Melucci, D.; Grillenzoni, F.-V. The discrimination of honey origin using melissopalynology and Raman spectroscopy techniques coupled with multivariate analysis. *Food Chem.* **2015**, *169*, 297–304. [CrossRef] [PubMed]
15. Frausto-Reyes, C.; Casillas-Peñuelas, R.; Quintanar-Stephano, J.; Macías-López, E.; Bujdud-Pérez, J.; Medina-Ramírez, I. Spectroscopic study of honey from Apis mellifera from different regions in Mexico. *Spectrochim. Acta Part A Mol. Biomol. Spectrosc.* **2017**, *178*, 212–217. [CrossRef] [PubMed]
16. Jandrić, Z.; Haughey, S.A.; Frew, R.D.; McComb, K.; Galvin-King, P.; Elliott, C.T.; Cannavan, A. Discrimination of honey of different floral origins by a combination of various chemical parameters. *Food Chem.* **2015**, *189*, 52–59. [CrossRef] [PubMed]
17. Oroian, M.; Ropciuc, S. Botanical authentication of honeys based on Raman spectra. *J. Food Meas. Charact.* **2018**, *12*, 545–554. [CrossRef]
18. Özbalci, B.; Boyaci, İ.H.; Topcu, A.; Kadılar, C.; Tamer, U. Rapid analysis of sugars in honey by processing Raman spectrum using chemometric methods and artificial neural networks. *Food Chem.* **2013**, *136*, 1444–1452. [CrossRef]
19. Oroian, M.; Ropciuc, S.; Paduret, S. Honey Adulteration Detection Using Raman Spectroscopy. *Food Anal. Methods* **2018**, *11*, 959–968. [CrossRef]
20. Anjos, O.; Santos, A.J.A.; Paixão, V.; Estevinho, L.M. Physicochemical characterization of Lavandula spp. honey with FT-Raman spectroscopy. *Talanta* **2018**, *178*, 43–48. [CrossRef]

21. Tahir, H.E.; Xiaobo, Z.; Zhihua, L.; Jiyong, S.; Zhai, X.; Wang, S.; Mariod, A.A. Rapid prediction of phenolic compounds and antioxidant activity of Sudanese honey using Raman and Fourier transform infrared (FT-IR) spectroscopy. *Food Chem.* **2017**, *226*, 202–211. [CrossRef]
22. Li, S.; Shan, Y.; Zhu, X.; Zhang, X.; Ling, G. Detection of honey adulteration by high fructose corn syrup and maltose syrup using Raman spectroscopy. *J. Food Compos. Anal.* **2012**, *28*, 69–74. [CrossRef]
23. Yam-Puc, A.; Santana-Hernández, A.A.; Yah-Nahuat, P.N.; Ramón-Sierra, J.M.; Cáceres-Farfán, M.R.; Borges-Argáez, R.L.; Ortiz-Vázquez, E. Pentacyclic triterpenes and other constituents in propolis extract from Melipona beecheii collected in Yucatan, México. *Rev. Bras. Farmacogn.* **2019**, *29*, 358–363. [CrossRef]
24. Porter-Bolland, L.; Ellis, E.A.; Guariguata, M.R.; Ruiz-Mallén, I.; Negrete-Yankelevich, S.; Reyes-García, V. Community managed forests and forest protected areas: An assessment of their conservation effectiveness across the tropics. *For. Ecol. Manage.* **2012**, *268*, 6–17. [CrossRef]
25. Güemes-Ricalde, F.J.; Villanueva-G, R.; Eaton, K.D. Honey production by the Mayans in the Yucatan peninsula. *Bee World* **2003**, *84*, 144–154. [CrossRef]
26. Villanueva-Gutiérrez, R.; Moguel-Ordóñez, Y.B.; Echazarreta-González, C.M.; Arana-López, G. Monofloral honeys in the Yucatán Peninsula, Mexico. *Grana* **2009**, *48*, 214–223. [CrossRef]
27. Mondragón-Cortez, P.; Ulloa, J.A.; Rosas-Ulloa, P.; Rodríguez-Rodríguez, R.; Resendiz Vázquez, J.A. Physicochemical characterization of honey from the West region of México. *CyTA—J. Food* **2013**, *11*, 7–13. [CrossRef]
28. White, J.W. Moisture in Honey Review of Chemical and Physical Methods. *J. Assoc. Off. Anal. Chem.* **1969**, *52*, 729–737.
29. Kek, S.P.; Chin, N.L.; Yusof, Y.A.; Tan, S.W.; Chua, L.S. Classification of entomological origin of honey based on its physicochemical and antioxidant properties. *Int. J. Food Prop.* **2017**, *20*, S2723–S2738. [CrossRef]
30. Belay, A.; Solomon, W.K.; Bultossa, G.; Adgaba, N.; Melaku, S. Physicochemical properties of the Harenna forest honey, Bale, Ethiopia. *Food Chem.* **2013**, *141*, 3386–3392. [CrossRef]
31. Jimenez, M.; Beristain, C.I.; Azuara, E.; Mendoza, M.R.; Pascual, L.A. Physicochemical and antioxidant properties of honey from Scaptotrigona mexicana bee. *J. Apic. Res.* **2016**, *55*, 151–160. [CrossRef]
32. Sereia, M.J.; Março, P.H.; Perdoncini, M.R.G.; Parpinelli, R.S.; de Lima, E.G.; Anjo, F.A. Techniques for the Evaluation of Physicochemical Quality and Bioactive Compounds in Honey. In *Honey Analysis*; InTech: London, UK, 2017.
33. Kizil, R.; Irudayaraj, J.; Seetharaman, K. Characterization of Irradiated Starches by Using FT-Raman and FTIR Spectroscopy. *J. Agric. Food Chem.* **2002**, *50*, 3912–3918. [CrossRef]
34. Zhu, X.; Li, S.; Shan, Y.; Zhang, Z.; Li, G.; Su, D.; Liu, F. Detection of adulterants such as sweeteners materials in honey using near-infrared spectroscopy and chemometrics. *J. Food Eng.* **2010**, *101*, 92–97. [CrossRef]
35. Salvador, L.; Guijarro, M.; Rubio, D.; Aucatoma, B.; Guillén, T.; Vargas Jentzsch, P.; Ciobotă, V.; Stolker, L.; Ulic, S.; Vásquez, L.; et al. Exploratory Monitoring of the Quality and Authenticity of Commercial Honey in Ecuador. *Foods* **2019**, *8*, 105. [CrossRef] [PubMed]
36. El Sohaimy, S.A.; Masry, S.H.D.; Shehata, M.G. Physicochemical characteristics of honey from different origins. *Ann. Agric. Sci.* **2015**, *60*, 279–287. [CrossRef]
37. Cozzolino, D.; Corbella, E.; Smyth, H.E. Quality Control of Honey Using Infrared Spectroscopy: A Review. *Appl. Spectrosc. Rev.* **2011**, *46*, 523–538. [CrossRef]
38. Ruoff, K.; Luginbühl, W.; Bogdanov, S.; Bosset, J.-O.; Estermann, B.; Ziolko, T.; Kheradmandan, S.; Amad/, R. Quantitative determination of physical and chemical measurands in honey by near-infrared spectrometry. *Eur. Food Res. Technol.* **2007**, *225*, 415–423. [CrossRef]
39. Silva, L.R.; Videira, R.; Monteiro, A.P.; Valentão, P.; Andrade, P.B. Honey from Luso region (Portugal): Physicochemical characteristics and mineral contents. *Microchem. J.* **2009**, *93*, 73–77. [CrossRef]
40. Karabagias, I.K.; Badeka, A.; Kontakos, S.; Karabournioti, S.; Kontominas, M.G. Characterisation and classification of Greek pine honeys according to their geographical origin based on volatiles, physicochemical parameters and chemometrics. *Food Chem.* **2014**, *146*, 548–557. [CrossRef]
41. Khalil, M.I.; Moniruzzaman, M.; Boukraâ, L.; Benhanifia, M.; Islam, M.A.; Islam, M.N.; Sulaiman, S.A.; Gan, S.H. Physicochemical and Antioxidant Properties of Algerian Honey. *Molecules* **2012**, *17*, 11199–11215. [CrossRef]
42. Siddiqui, A.J.; Musharraf, S.G.; Choudhary, M.I.; Rahman, A.-. Application of analytical methods in authentication and adulteration of honey. *Food Chem.* **2017**, *217*, 687–698. [CrossRef]

43. Chen, C. Relationship between Water Activity and Moisture Content in Floral Honey. *Foods* **2019**, *8*, 30. [CrossRef]
44. Lemos, M.S.; Venturieri, G.C.; Dantas Filho, H.A.; Dantas, K.G.F. Evaluation of the physicochemical parameters and inorganic constituents of honeys from the Amazon region. *J. Apic. Res.* **2018**, *57*, 135–144. [CrossRef]
45. Lichtenberg-Kraag, B.; Hedtke, C.; Bienefeld, K. Infrared spectroscopy in routine quality analysis of honey. *Apidologie* **2002**, *33*, 327–337. [CrossRef]
46. Yücel, Y.; Sultanog˘lu, P. Characterization of honeys from Hatay Region by their physicochemical properties combined with chemometrics. *Food Biosci.* **2013**, *1*, 16–25. [CrossRef]
47. Grazia Mignani, A.; Ciaccheri, L.; Mencaglia, A.A.; Di Sanzo, R.; Carabetta, S.; Russo, M. Dispersive Raman Spectroscopy for the Nondestructive and Rapid Assessment of the Quality of Southern Italian Honey Types. *J. Light. Technol.* **2016**, *34*, 4479–4485. [CrossRef]

8

Assessing Geographical Origin of *Gentiana Rigescens* using Untargeted Chromatographic Fingerprint, Data Fusion and Chemometrics

Tao Shen [1,2,3], Hong Yu [1,2,*] and Yuan-Zhong Wang [4]

1 Yunnan Herbal Laboratory, Institute of Herb Biotic Resources, School of Life and Sciences, Yunnan University, Kunming 650091, China
2 The International Joint Research Center for Sustainable Utilization of Cordyceps Bioresouces in China and Southeast Asia, Yunnan University, Kunming 650091, China
3 College of Chemistry, Biological and Environment, Yuxi Normal University, Yu'xi 653100, China
4 College of Traditional Chinese Medicine, Yunnan University of Chinese Medicine, Kunming 650500, China
* Correspondence: hongyu@ynu.edu.cn or herbfish@163.com

Academic Editor: Marcello Locatelli

Abstract: *Gentiana rigescens* Franchet, which is famous for its bitter properties, is a traditional drug of chronic hepatitis and important raw materials for the pharmaceutical industry in China. In the study, high-performance liquid chromatography (HPLC), coupled with diode array detector (DAD) and chemometrics, were used to investigate the chemical geographical variation of *G. rigescens* and to classify medicinal materials, according to their grown latitudes. The chromatographic fingerprints of 280 individuals and 840 samples from rhizomes, stems, and leaves of four different latitude areas were recorded and analyzed for tracing the geographical origin of medicinal materials. At first, HPLC fingerprints of underground and aerial parts were generated while using reversed-phase liquid chromatography. After the preliminary data exploration, two supervised pattern recognition techniques, random forest (RF) and orthogonal partial least-squares discriminant analysis (OPLS-DA), were applied to the three HPLC fingerprint data sets of rhizomes, stems, and leaves, respectively. Furthermore, fingerprint data sets of aerial and underground parts were separately processed and joined while using two data fusion strategies ("low-level" and "mid-level"). The results showed that classification models that are based OPLS-DA were more efficient than RF models. The classification models using low-level data fusion method built showed considerably good recognition and prediction abilities (the accuracy is higher than 99% and sensibility, specificity, Matthews correlation coefficient, and efficiency range from 0.95 to 1.00). Low-level data fusion strategy combined with OPLS-DA could provide the best discrimination result. In summary, this study explored the latitude variation of phytochemical of *G. rigescens* and developed a reliable and accurate identification method for *G. rigescens* that were grown at different latitudes based on untargeted HPLC fingerprint, data fusion, and chemometrics. The study results are meaningful for authentication and the quality control of Chinese medicinal materials.

Keywords: authentication; liquid chromatography fingerprint; chemometrics; random forest; OPLS-DA; data fusion; *Gentiana rigescens*

---

1. Introduction

Gentiana rigescens Franchet (Dian long dan) is a herbaceous species that grows in mountainous regions of Yunnan-Guizhou Plateau in the southwest of China [1]. Like European traditional medicinal plant yellow gentian (*G. lutea* L), *G. rigescens* is famous for its bitter properties that are due to the bitter active principles (e.g., loganin, gentiopicroside, swertiamarin, sweroside, etc.) [2–4]. Those compounds have pharmacological effects of anti-inflammation, antioxidant, anti-cancer, antiviral, cholagogic agent, hepatoprotective, wound-healing activities, and so forth [3,5]. Additionally, they are used to stimulate appetite and improve digestion [5–7]. In addition, a series of neuritogenic compounds had been isolated from the aerial and underground parts of *G. rigescens*, which could be used as raw material for the preparation of functional food and a therapeutic drug for Alzheimer's disease [8–11]. Now, *G. rigescens* have been the official drug of Chinese pharmacopoeia (2015 edition) for chronic hepatitis and important raw materials for the pharmaceutical industry in China [12].

G. rigescens were usually collected from different regions of Yunnan-Guizhou Plateau in order to provide satisfaction of continuously increasing industrial demands for raw materials. However, some of the researchers had reported that chemical constitutions of underground part of *G. rigescens* were extremely variable and diverse according to plant grown location or producing area [13–15]. Quantitative analysis of bioactivity compounds (such as gentiopicroside, sweroside, swertiamarin, isoorientin, and other compounds) from rhizomes, stems, leaves, and flowers indicated that northwest of Yunnan-Guizhou Plateau was suitable for chemical compounds accumulation [13–16]. Additionally, conversion and transport of those compounds might be influenced by climatic conditions in the plant habitat [14,17].

Latitude has a strong impact on the local climate environment in southwest China [18,19]. As the main distribution area of *G. rigescens*, Yunnan-Guizhou Plateau is characterized by very complex topography and it displays a wide variety of micro-climates [18–21]. There are six climatic zones from the north towards the south [20]. Especially, in the higher latitude areas, such as northwest Yunnan or south of the Hengduan Mountains (26–28° N), the temperature gradients are more abrupt than in the other regions [19]. Furthermore, precipitation and temperature in the Yunnan-Guizhou Plateau also show clear variations along the latitude gradients [19,21]. Therefore, it is necessary to explore the variation of phytochemical and medicinal material quality of *G. rigescens* that were grown in different latitudes and build a classification model for tracing producing areas of medicinal materials.

As we know, the contents of bioactive compounds and quality of medicinal materials have a close relationship with the environment of producing area [22–25]. Quality control and geographical indication of medicinal materials raise many concerns by pharmaceutical industries with the expansion in the use of herbal medicines. However, using few marker compounds could not reflect the chemical complexity of herbs and this method is hard to effectively authenticate the origin of herbal medicines [26,27]. Chemical fingerprints, as a comprehensive evaluation methodology, have been widely used to deal with the problem [26,28,29]. In recent years, infrared spectroscopy (IR), UV-Vis spectroscopy (UV-Vis), and other spectral fingerprints have been well-established analytical techniques for geographical traceability studies of *G. rigescens* and other medicinal plants in the worldwide [30–34]. In contrast, there were limited reports on the use of chromatographic fingerprint to identify the producing regions of herbal materials [30–35]. Although there were many reports about discrimination of herbs according to their producing areas while using liquid chromatography technology, most of them are based on the information of limited chemical markers or chromatographic profiles [36–39]. The potential of chromatographic fingerprints for herbs authentication needs to be further explored.

When compared with chemical marker or chromatographic profile (targeted), chromatographic fingerprint (untargeted) contains unspecific and non-evident information and chemometric tools should extract chemical information [40]. Recently, literature reported some successful studies applying chromatographic fingerprint, together with chemometric methodology, to discriminate herbs and food samples of different origin or cultivars [41–44]. All of those studies suggested that it is possible to develop a reliable and accurate method for the geographical tracing of *G. rigescens* by applying the chromatographic fingerprint methodology.

In the progression of improving geographical authentication of food and drugs, one of the important goals is building discrimination models with a less error rate and reducing the uncertainty of the prediction results [33,44]. Data fusion strategy has been widely used in the last years in the field of food authentication in order to improve class discrimination techniques [45]. Some reports about *Panax notoginseng*, *Paris Polyphylla* var. *yunnanensis* and other herb materials also showed the huge potential of this strategy in the discrimination of medicinal materials producing areas [46–48]. Today, most of the fused data come from spectral fingerprint and very few studies report the data fusion of chromatographic fingerprint [42,43]. Furthermore, data fusion studies are mostly based on the fusion of multivariate instrumental techniques [42,43], while reports of *P. Polyphylla* var. *yunnanensis*, *Macrohyporia cocos*, and other species indicated that reliable classification results were also available by the fusion analysis of chemical fingerprint data collected from different medicinal parts of herbs [35,49]. Accumulation and distribution of metabolites in the different parts of plants were different because of the differential response of root, stem, flower and other organs to the environment variation of producing area [17,50]. Therefore, fingerprint data fusion of multi-medicinal parts may provide integrated chemical information for the authentication of medicinal materials. At the same time, this method also contributes to a more comprehensive understanding of the response and adaptation of medicinal plants to complex geographical environments.

The aim of this study is to explore the variation of chromatographic fingerprints of *G. rigescens* along the latitude gradients and to use chemometrics to mine fingerprint chemical information, and to investigate the potential of the untargeted chromatographic fingerprint to trace herbs grown at different latitudes. For this purpose, we developed fingerprint of rhizomes, stems, and leaves of *G. rigescens* by high-performance liquid chromatography with diode array detection (HPLC-DAD) technology. Subsequently, classification models for the identification of different producing areas were built by HPLC fingerprint combined with RF (random forest algorithm) and OPLS-DA (orthogonal partial least-squares discriminant analysis). At last, two types of data fusion strategies, "low- level" and "mid-level" data fusion, were studied in order to improve the model performances.

2. Results and Discussion

2.1. Chromatographic Fingerprints Variation Along the Latitude Gradients

Figure 1 displays the representative chromatographic fingerprints of rhizome, stem, and leaf. From HPLC fingerprints, it can be found that the five marker compounds of iridoids were eluted before 15 min. The retention times (*t*/min) of loganin (1), 6′-*O*-β-D-glucopyranosylgentiopicroside (2), swertiamarine (3), gentiopicroside (4), and sweroside (5) were 7.279, 9.213, 9.573, 11.376, and 11.622 min, respectively. Loganin and gentiopicroside were mainly accumulation in the underground part and sweroside accumulated more in the overground parts. Furthermore, differences in the chemical composition of rhizome, stem, and leaf can also be visually observed through chromatographic fingerprints. For facilitating subsequent data exploration and modeling analysis, the retention time of fingerprints signal was replaced by variables (Figure 1d–f). As a result, there were 3839, 4140, and 4140 variables of rhizome, stem, and leaf fingerprints, respectively.

Principal component analysis (PCA) and two-dimensional score plots visualized the differences and variation trends of three medicinal parts. Figure 2 shows that the rhizomes and stems of *G. rigescens* tended to cluster to the left part, while the leaves data scattered to the right.

Although the fingerprints between the aboveground and underground medicinal parts were obvious differences, an interesting result is that a trend of separation according to product region latitude was observed from the PCA and score plots of samples of three medicinal parts. For example, two-dimensional score plots of chromatographic fingerprint of rhizomes showed that the samples separation trend increases with an increase in geographical distance and a clear separation between samples that were collected from lower latitude and higher latitude regions (Figure 3). In contrast to this, when considering the separation between samples with product regions geographically close to each other, we observed that the rhizome samples separation trend decreases with a decrease in the geographical distance (Figure 4). The PCA score plots of stems and leaves changed in the same trend as rhizomes (Figures S1–S4).

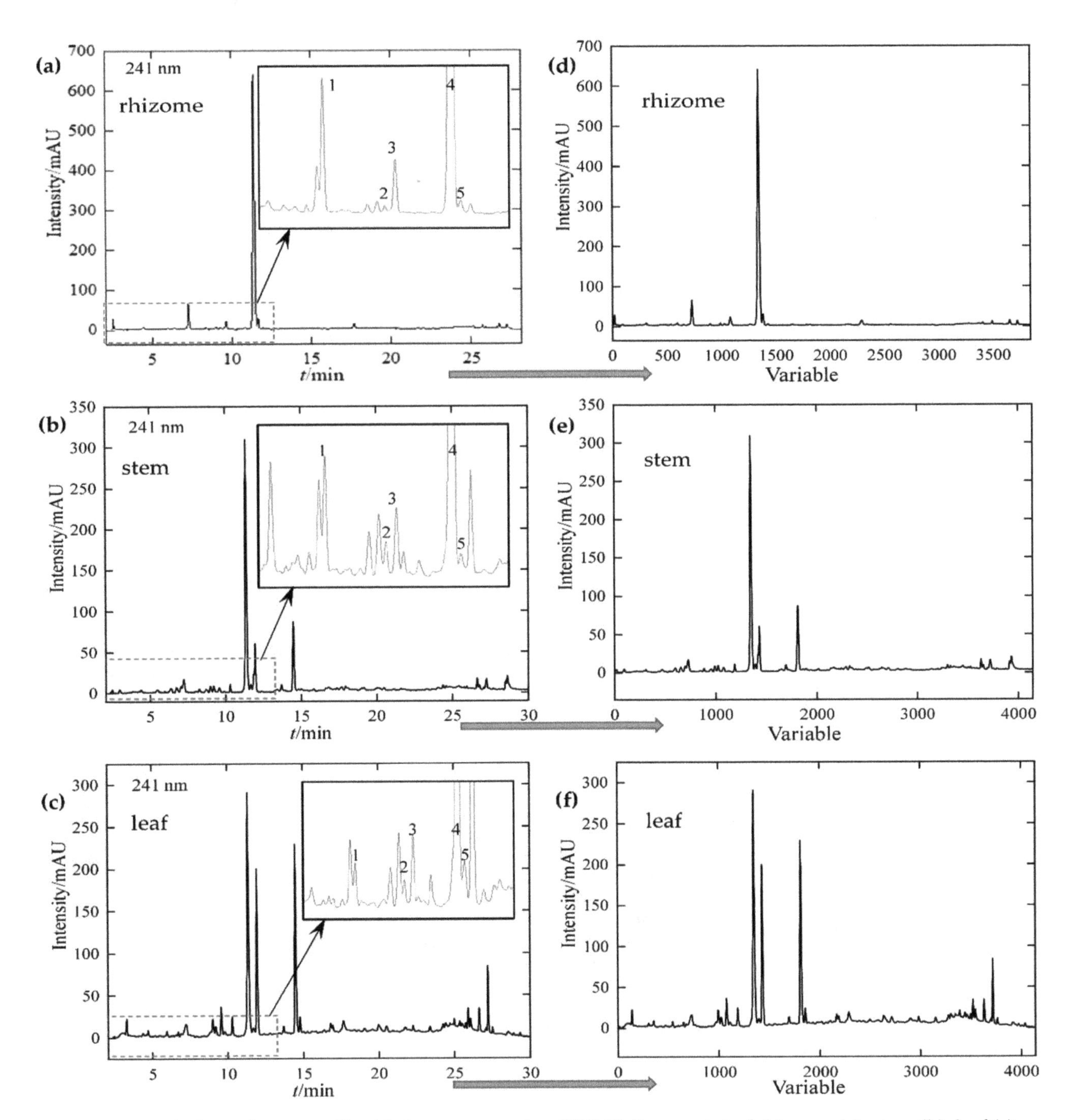

Figure 1. High-performance liquid chromatography (HPLC) fingerprint of rhizome (**a**), stem (**b**), leaf (**c**) and fingerprints after variable transformation (**d–f**). (1) loganin, (2) 6′-*O*-β-D-glucopyranosylgentiopicroside, (3) swertiamarine, (4) gentiopicroside, and (5) sweroside.

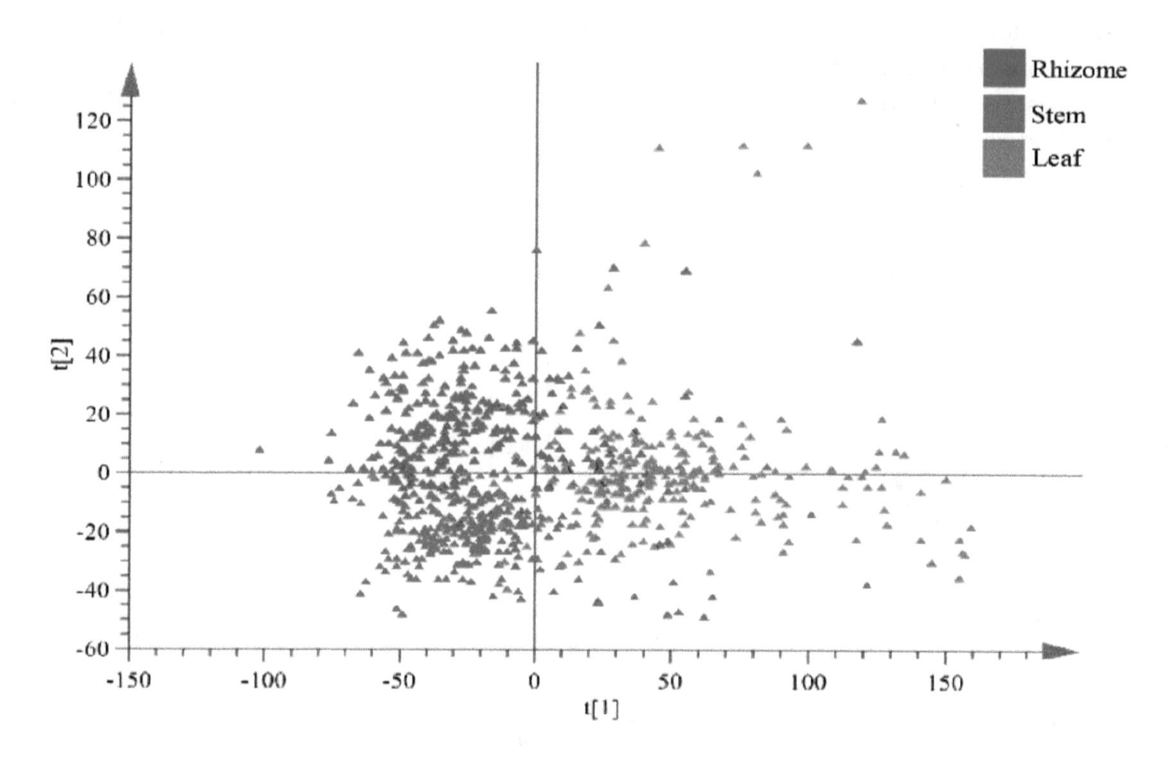

Figure 2. Two-dimensional principal component score plot of rhizomes, stems, and leaves samples based on chromatographic fingerprint data.

The results of PCA highlighted that the chromatographic fingerprints of *G. rigescens* were different among rhizomes, stems, and leaves, and were affected by latitude gradients of the production regions. Especially between lower latitudes and higher latitudes, the samples seem to be clearly distinguishable. Based on PCA exploratory analysis (unsupervised methods), supervised pattern recognition (OPLS-DA) should be applied to gain better classification results for samples that were grown in different latitudes (Figures 5 and 6), and OPLS-DA and variable importance in the projection (VIP) analysis were used to further investigate the fingerprint variables of *G. rigescens* that were sensitive to latitude changes.

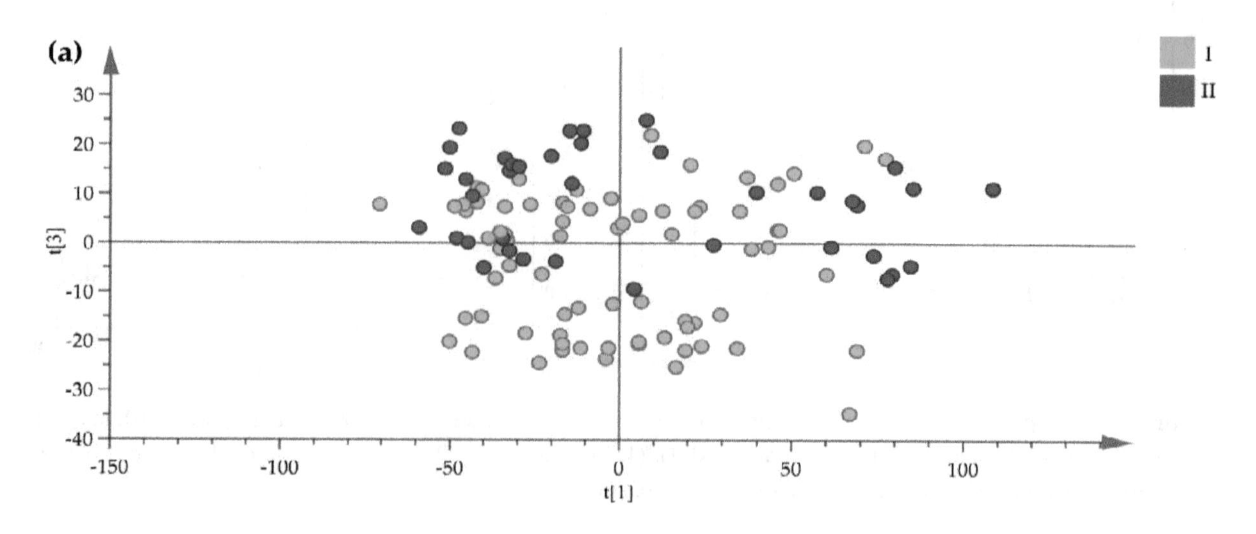

Figure 3. *Cont.*

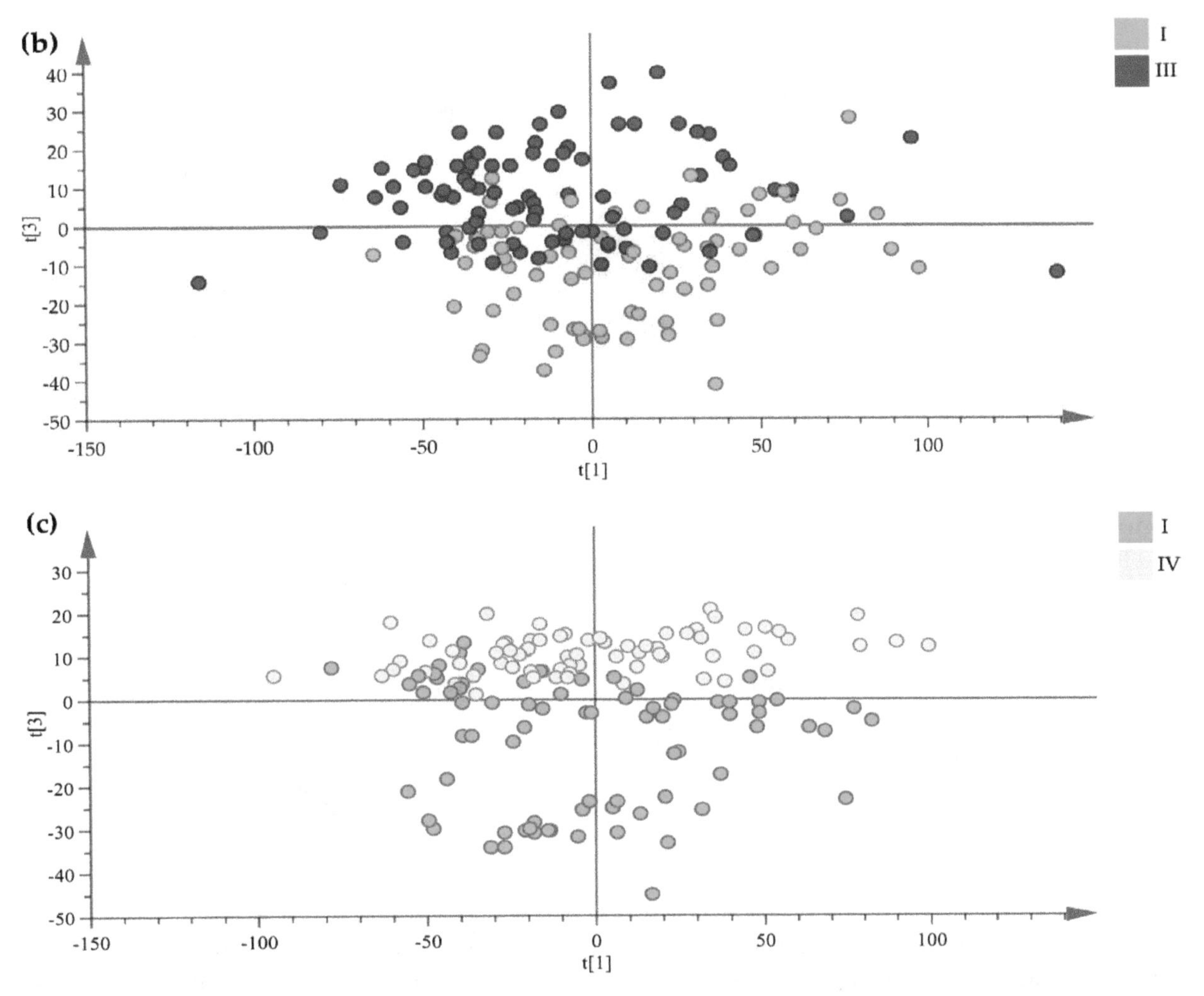

Figure 3. Variation of rhizomes score plots along the latitude gradients. (**a**) is low latitude and mid-latitude, (**b**) is low latitude and mid-high latitude and (**c**) is low latitude and high latitude (green circles = low latitudes area, 23.92–23.66° N, blue circles = mid-latitude area, 24.95–25.06° N, red circles = mid-high latitude area, 26.49–26.64° N, yellow circles = high latitude area, 27.34–28.52° N).

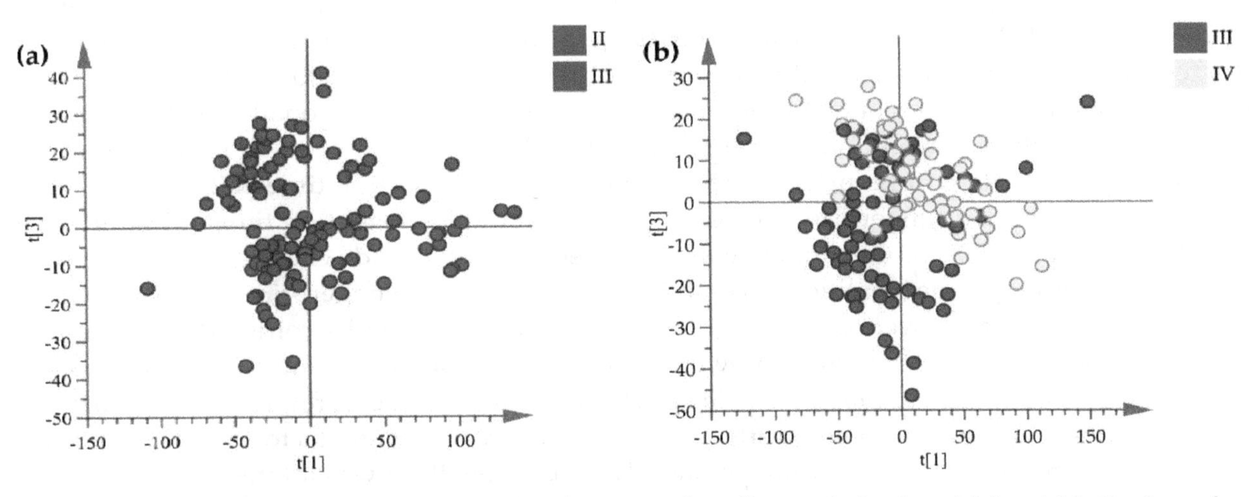

Figure 4. Variation of rhizomes score plots between the adjacent latitudes. (**a**) is mid-latitude and mid-high latitude and (**b**) is mid-high latitude and high- latitude (blue circles = mid-latitude area, 24.95–25.06° N, red circles = mid-high latitude area, 26.49–26.64° N, yellow circles = high latitude area, 27.34–28.52° N).

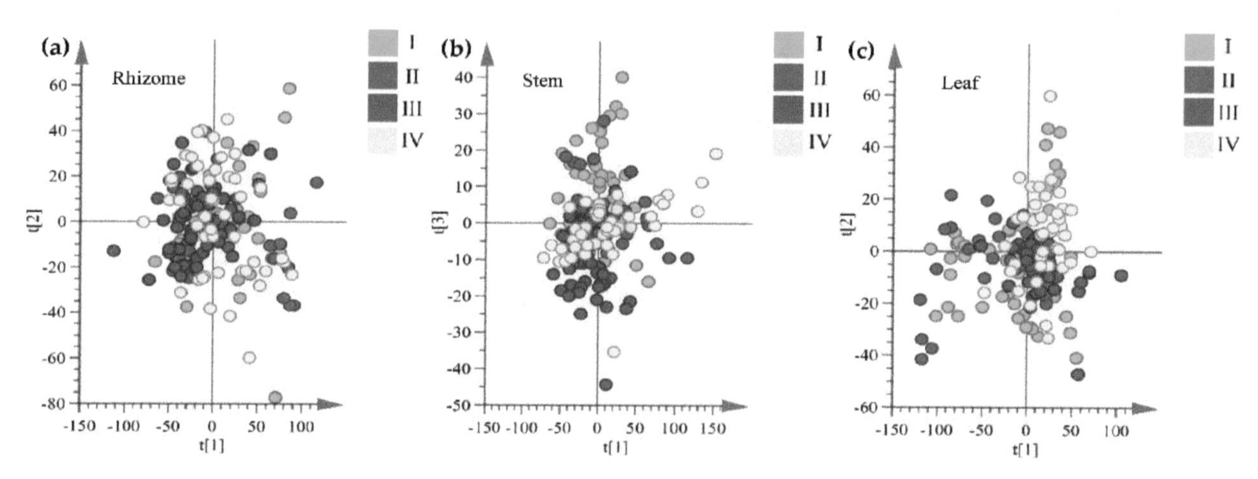

Figure 5. Two-dimensional principal component score plots for samples of rhizomes (**a**), stems (**b**), and leaves (**c**) of *G. rigescens* grown at four latitudes.

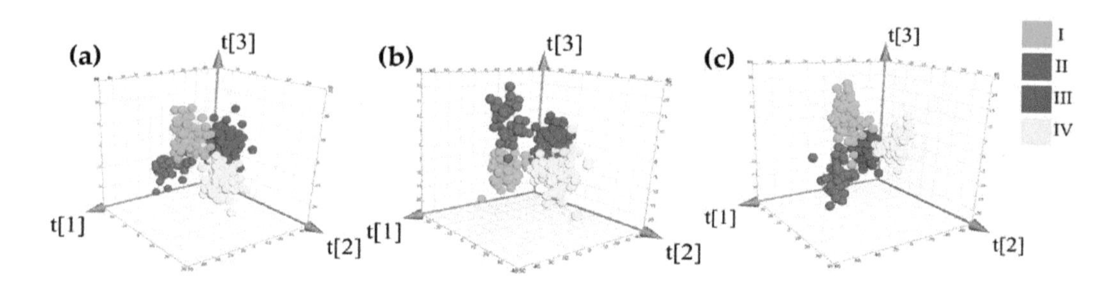

Figure 6. Three-dimensional (3D) Scores-plot diagram of rhizomes (**a**), stems (**b**), and leaves (**c**) orthogonal partial least-squares discriminant analysis (OPLS-DA) analysis among four different latitudes (OPLS-DA model (**a**) $R^2 = 0.74$ and $Q^2 = 0.68$, model (**b**) $R^2 = 0.75$ and $Q^2 = 0.68$, model (**c**) $R^2 = 0.72$ and $Q^2 = 0.71$, permutation plot of three models were shown in Figures S5–S7).

The variable's VIP value was greater than 1.00, which indicates that the variable was obviously affected by the change of the latitude of the producing areas. From Figure 7a, it could be found that the change of three ranges of rhizome's fingerprint was closely related to producing areas latitude. The first range was related to variables of retention time at 2.00–13.00 min. The second range was related to variables of retention time at 15.00–20.00 min. Additionally, the third range was related to the variables of retention time after 25.00 min. Figure 7b showed that important variables (VIP value > 1.00) of stem fingerprint relate to the variables of retention time at 2.00–20.00 min and 25.00–30.00 min. For leaf fingerprint, chromatographic variables, retention time at 2.00–15.00 min, 17.00–19.00 min and 25.00–30.00 min, were the most sensitive to latitude changes of producing areas (Figure 7c). According to the identification of the major compounds in fingerprint, it showed that many of these important variables were chromatographic signals of iridoids and secoiridoids, such as loganin, 6′-*O*-β-ᴅ-glucopyranosylgentiopicroside, swertiamarine, gentiopicroside, and sweroside. A previous study regarding the spatial profiling of iridoids phytochemical constituents found that the geographical variation of those compounds could be attributed to some environmental factors [13,17], for example, the difference of precipitation of natural habitats [17]. Additionally, it was interesting to note that the number of important variables after 25 min is gradually increasing from the rhizome to the leaves. The results suggested that, in addition to iridoids, other low polarity products in *G. rigescens* have implications for the differentiation of different geographical origins.

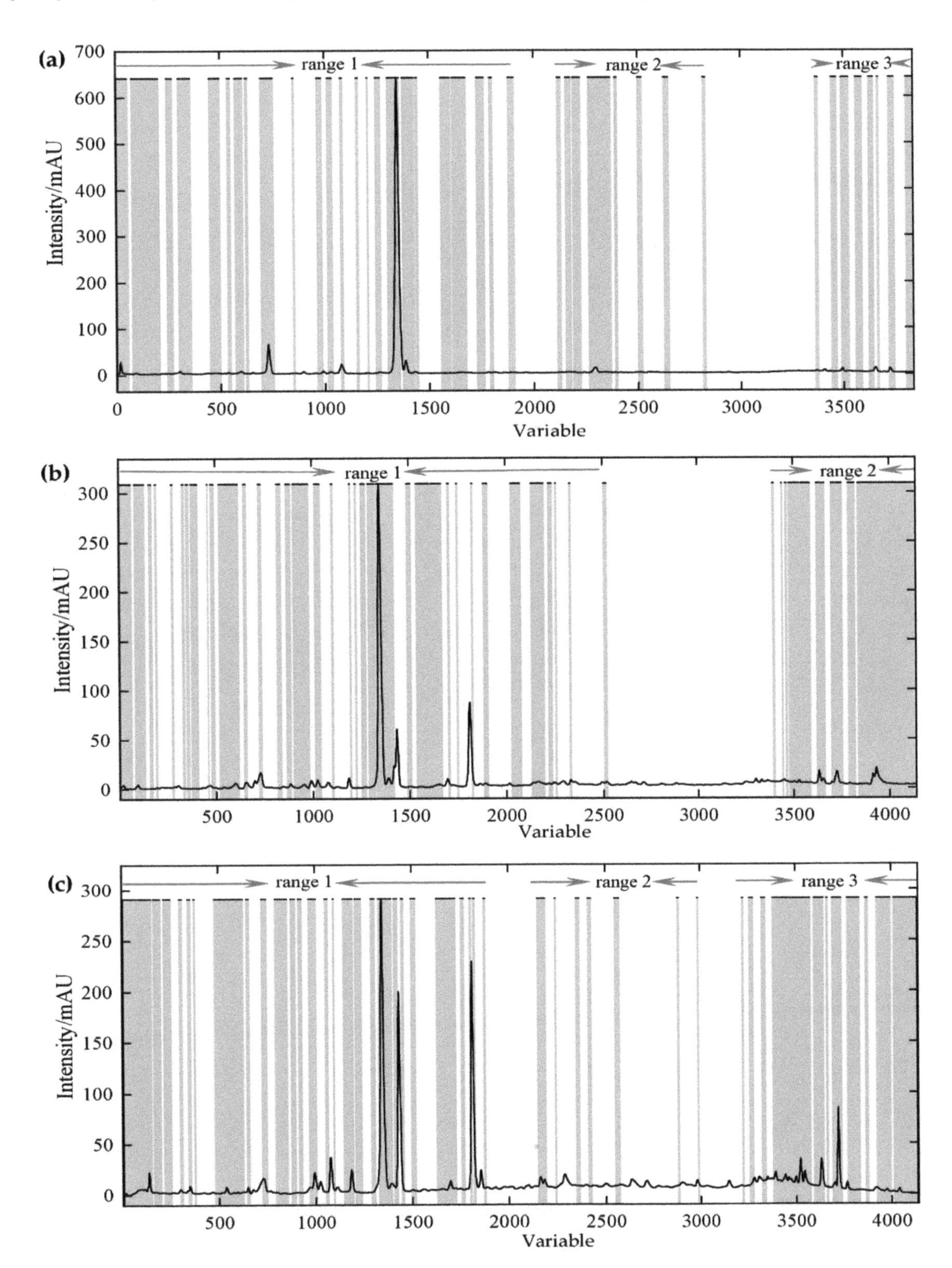

Figure 7. Important variables of fingerprint (purple = variable VIP value > 1) (**a**) rhizome, (**b**) stem, and (**c**) leaf.

In a word, current research indicated that the chemical composition of *G. rigescens* changes with the grown latitude in a way that could be traced with the chromatographic fingerprint. Furthermore, three-dimensional (3D) score plots and VIP analysis showed a difference of phytochemical geographic variation for overground and underground parts. Those differences might affect the result of geographical origin traceability of samples.

2.2. *Geographic Authentication Based on Fingerprints of Different Medicinal Parts*

In recent years, literature had already reported satisfying classification results that were obtained by RF or OPLS-DA models [51–54]. As an ensemble learning method, the RF algorithm could correct for decision trees' habit of overfitting to their training set [55]. Additionally, OPLS could help to overcome these obstacles by separating useful information from noise and improve complex chemical data features and interpretability [56,57]. In this work, we tested RF and OPLS-DA models, combined with rhizome, stem, and leaf fingerprint data in order to classify *G. rigescens* according to their grown latitude.

2.2.1. RF Classification

In the beginning, samples from the data set of rhizomes (280 samples and 3839 variables) were separated into a calibration set (186 samples) and a validation set (94 samples) by the Kennard-Stone algorithm. Subsequently, 186 rhizome samples that were collected from four latitude gradients were used to establish the calibration model (R_RF). During the modeling process, the initial value of n_{tree} (needs to be optimized) was defined as 2000, the initial value of m_{try} was defined as the square root of the number of variables, and the rest of the parameters were defined as the default value. Subsequently, OOB errors were calculated and the value of the best n_{tree} was obtained according to the lowest OOB error. Figure 8 shows that the minimum error and the standard error are the lowest, with 663 trees. Based on the optimal number of trees, m_{try} was re-selected by searching the values ranged from 50 to 75. The calculation results found that the m_{try} value should be defined as 61, because of the model had the lowest OOB classification error. Finally, a final classification model was established based on optimum n_{tree} and m_{try} values.

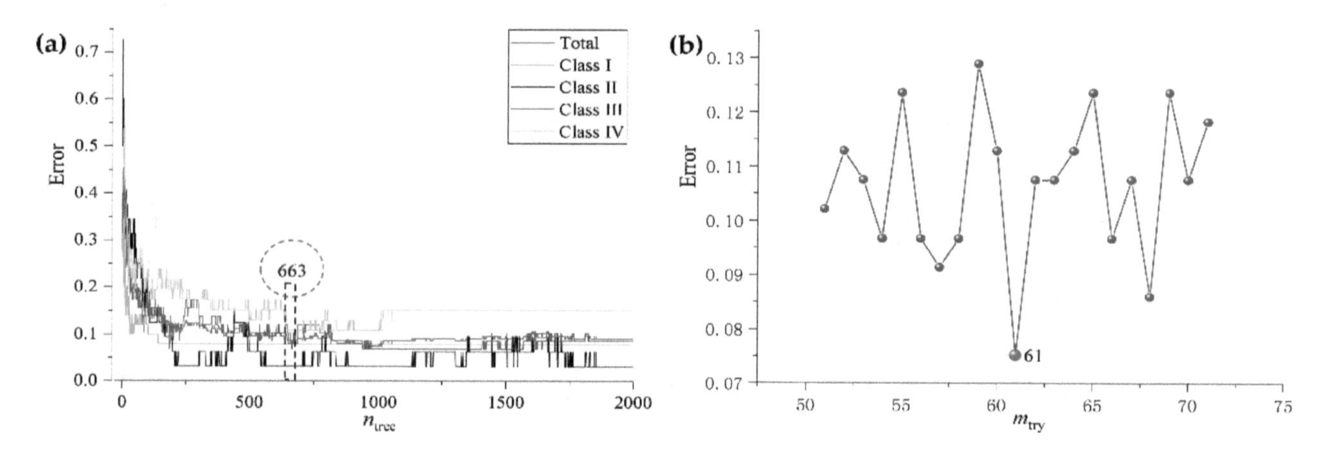

Figure 8. The n_{tree} (**a**) and m_{try} (**b**) screening of RF models based on rhizomes fingerprints.

Table 1 shows that the accuracies for samples of calibration set were 96.77% for low latitude samples, 99.46% for mid-latitude samples, 94.62% for mid-high latitude samples, and 94.09% for high latitude samples. Additionally, the accuracies of samples of validation set were 91.49%, 95.74%, 94.68%, and 98.94% for four different latitudes samples, respectively.

Table 1. The major parameters of random forest (RF) model based on rhizomes data set.

Model	Performance	Calibration Set				Validation Set			
		I	II	III	IV	I	II	III	IV
R_RF	ACC (%)	96.77	99.46	94.62	94.09	91.49	95.74	94.68	98.94
	SE	0.92	0.97	0.93	0.89	0.92	0.75	0.93	0.96
	SP	0.99	1.00	0.95	0.96	0.91	1.00	0.95	1.00
	MCC	0.92	0.98	0.88	0.84	0.80	0.84	0.88	0.97
	EFF	0.95	0.98	0.94	0.92	0.92	0.87	0.94	0.98

Like previous investigations of the rhizome model, the data set of stems (280 samples and 4140 variables) and leaves (280 samples and 4140 variables) were separated into calibration sets and validation sets, respectively. Subsequently, RF calibration modes of stems (S_RF) and leaves (L_RF) were built. The optimum n_{tree} and m_{try} could be found in Figures 9 and 10.

For the RF model of the stem, the accuracies of samples of calibration set of 92.47%, 94.62%, 93.01%, and 93.01% were achieved for low latitudes, mid-latitudes, mid-high latitudes, and high latitudes. Additionally, the accuracies of samples of validation set were 98.94%, 97.87%, 96.81%, and 97.87%, respectively (Table 2).

For RF model of the leaf, accuracies of 92.47%, 96.24%, 93.01%, and 94.62% were achieved for the calibration set. Additionally, accuracies of 85.11%, 93.62%, 89.36%, and 93.62% for the validation set (Table 3).

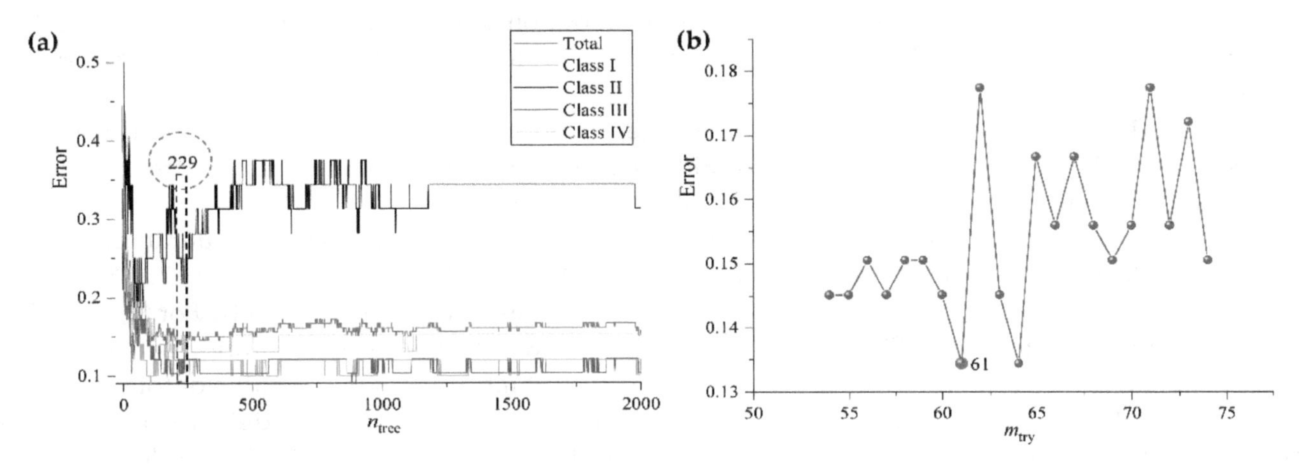

Figure 9. The n_{tree} (**a**) and m_{try} (**b**) screening of RF models based on stems fingerprints.

Table 2. The major parameters of RF model based on stems data set.

Model	Performance	Calibration Set				Validation Set			
		I	II	III	IV	I	II	III	IV
S_RF	ACC (%)	92.47	94.62	93.01	93.01	98.94	97.87	96.81	97.87
	SE	0.92	0.69	0.91	0.87	1.00	0.88	1.00	0.91
	SP	0.93	1.00	0.94	0.95	0.99	1.00	0.95	1.00
	MCC	0.82	0.80	0.84	0.81	0.97	0.92	0.93	0.94
	EFF	0.92	0.83	0.93	0.91	0.99	0.94	0.98	0.96

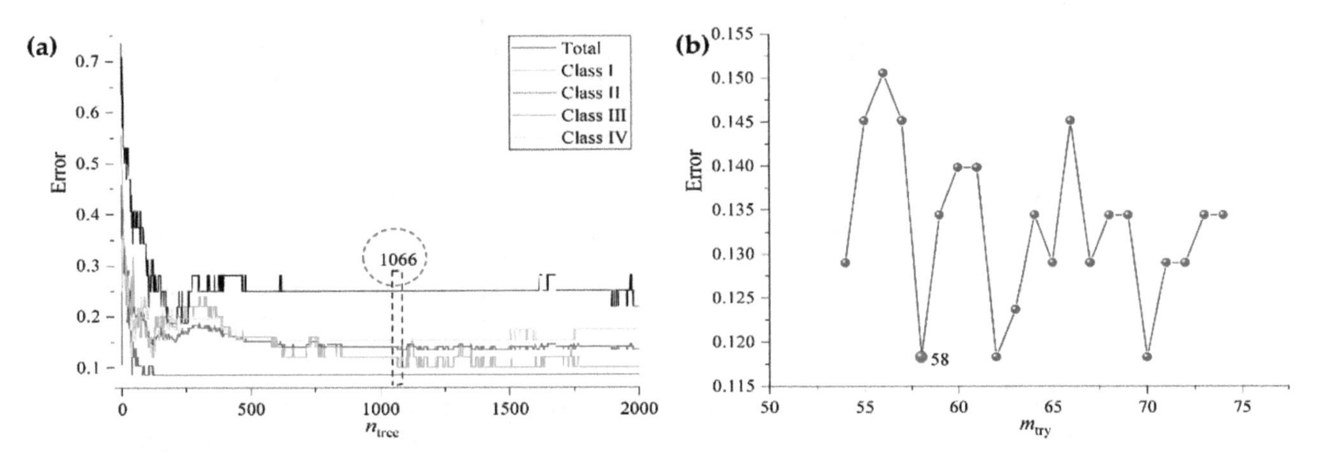

Figure 10. The n_{tree} (**a**) and m_{try} (**b**) screening of RF models based on leaves fingerprints.

Table 3. The major parameters of RF model based on leaves data set.

Model	Performance	Calibration Set				Validation Set			
		I	II	III	IV	I	II	III	IV
L_RF	ACC (%)	92.47	96.24	93.01	94.62	85.11	93.62	89.36	93.62
	SE	0.94	0.78	0.91	0.85	0.88	0.69	0.86	0.74
	SP	0.92	1.00	0.94	0.98	0.84	0.99	0.91	1.00
	MCC	0.82	0.86	0.84	0.85	0.67	0.76	0.76	0.83
	EFF	0.93	0.88	0.93	0.91	0.86	0.82	0.88	0.86

2.2.2. OPLS-DA Classification

The OPLS-DA models of rhizomes (R_OPLS-DA), stems (S_OPLS-DA), and leaves (L_OPLS-DA) were constructed based on the same calibration and validation sets that were used in RF models. All of the models were constructed based on the internal seven-fold cross-validation and permutation plot could be found in Supplementary Materials.

Table S1 showed that the R^2 of models ranged from 0.77 to 0.82 and the Q^2 of models were larger than 0.50, which indicated that the OPLS-DA models were well fitted and better predictive. The permutation test results could be found in Figures S14–S16.

The classification results of R_OPLS-DA model showed (Table 4) accuracies of calibration set were 98.92% for all classes. Accuracies of validation set were 95.47%, 98.94%, 94.86%, and 97.87% for low latitudes, mid-latitudes, mid-high latitudes, and high latitudes samples, respectively. For S_OPLS-DA models (Table 4), although 98.92%, 99.46%, 98.92%, and 98.39% values of calibration set accuracies were obtained for samples that were grown in four different latitudes, a lower value of total accuracy rate of validation set was obtained (93.62%). Parameters of L_OPLS-DA model showed (Table 4) that the accuracies of the calibration set were 97.31%, 99.46%, 97.31%, and 98.39% for low latitude, mid-latitude, mid-high latitude, and high latitude samples, respectively. However, the total accuracy of the validation set was lower than the calibration set. Especially, for samples of class 1, the accuracy was only 88.30%.

Table 4. The major parameters of OPLS-DA models.

Model	Performance	Calibration Set				Validation Set			
		I	II	III	IV	I	II	III	IV
R_OPLS-DA	ACC (%)	98.92	98.92	98.92	98.92	95.74	98.94	94.68	97.87
	SE	0.98	0.97	0.98	0.98	0.92	0.94	0.93	0.96
	SP	0.99	0.99	0.99	0.99	0.97	1.00	0.95	0.99
	MCC	0.97	0.96	0.97	0.97	0.89	0.96	0.88	0.94
	EFF	0.99	0.98	0.99	0.99	0.95	0.97	0.94	0.97
S_OPLS-DA	ACC (%)	98.92	99.46	98.92	98.39	91.49	93.62	91.49	97.87
	SE	1.00	0.97	1.00	0.93	0.92	0.81	0.83	0.91
	SP	0.99	1.00	0.98	1.00	0.91	0.96	0.95	1.00
	MCC	0.97	0.98	0.98	0.96	0.80	0.77	0.80	0.94
	EFF	0.99	0.98	0.99	0.97	0.92	0.88	0.89	0.96
L_OPLS-DA	ACC (%)	97.31	99.46	97.31	98.39	88.30	95.74	92.55	93.62
	SE	0.94	1.00	0.93	1.00	0.81	0.88	0.90	0.83
	SP	0.99	0.99	0.99	0.98	0.91	0.97	0.94	0.97
	MCC	0.93	0.98	0.94	0.96	0.71	0.85	0.83	0.82
	EFF	0.96	1.00	0.96	0.99	0.86	0.92	0.92	0.90

Finally, we made a comprehensive comparison to the six models' classification performance superiority on the basis of the above analysis. For the RF model, the order of calibration total accuracy was as follows: R_RF (96.24%) > L_RF (94.09%) > S_RF (93.28%). The order of validation total accuracy was as follows: S_RF (97.87%) > R_RF (95.21%) > L_RF (90.43%). For the OPL-DA model, the order of calibration total accuracy was as follows: R_OPL-DA (98.92%) and S_OPLS-DA (98.92%) > L_OPLS-DA (98.12%). The order of validation total accuracy was as follows: R_OPL-DA (96.81%) > S_OPLS-DA (93.62%) > L_OPLS-DA (92.55%). Classification models that were built by using leaf data set presented the worst performance from the accuracy point of view. Additionally, validation sets of the L_RF and L_OPL-DA model had lower Matthews correlation coefficient (MCC) values. By contrast, all of the models based on rhizome data set presented a better classification performance (total accuracy ranged from 95.21% to 98.92%). The best total accuracy occurred when rhizome data combined with the OPLS algorithm. We could find that phenomenon of imbalance category recognition in R_OPLS-DA model was better than other models from SE values, SP values, MCC values, and EFF value.

Although the classification performance for OPLS-DA and RF models on the basis of rhizome data set was good, the model classification ability, accuracy, sensitivity (SE), specificity (SP), MCC, and efficiency (EFF), need to be enhanced. In a further step, the feasibility of combining the information from rhizome, stem, and leaf fingerprint data for samples geographical traceability was investigated by low-level and mid-level data fusion strategies.

2.3. *Geographic Authentication Based on Data Fusion Strategy*

2.3.1. Low-Level Data Fusion

According to the method that was described in data preprocessing (Figure 11), fingerprint data sets of overground and underground organs as subsets were used to concatenate into a single data block (a new data set). In the case of the low-level strategy, four data sets, rhizome combined with stem (RS), rhizome combined with leaf (RL), stem combined with leaf (SL), and all data combined (RSL), were used to build RF (RS_RF, RL_RF, SL_RF, and RSL_RF) and OPLS-DA (RS_OPLS-DA, RL_OPLS-DA, SL_OPLS-DA, and RSL_OPLS-DA) models. For every data set, the samples were randomly selected as a calibration set and the rest of the samples were used as a validation set (finished by Kennard-Stone algorithm).

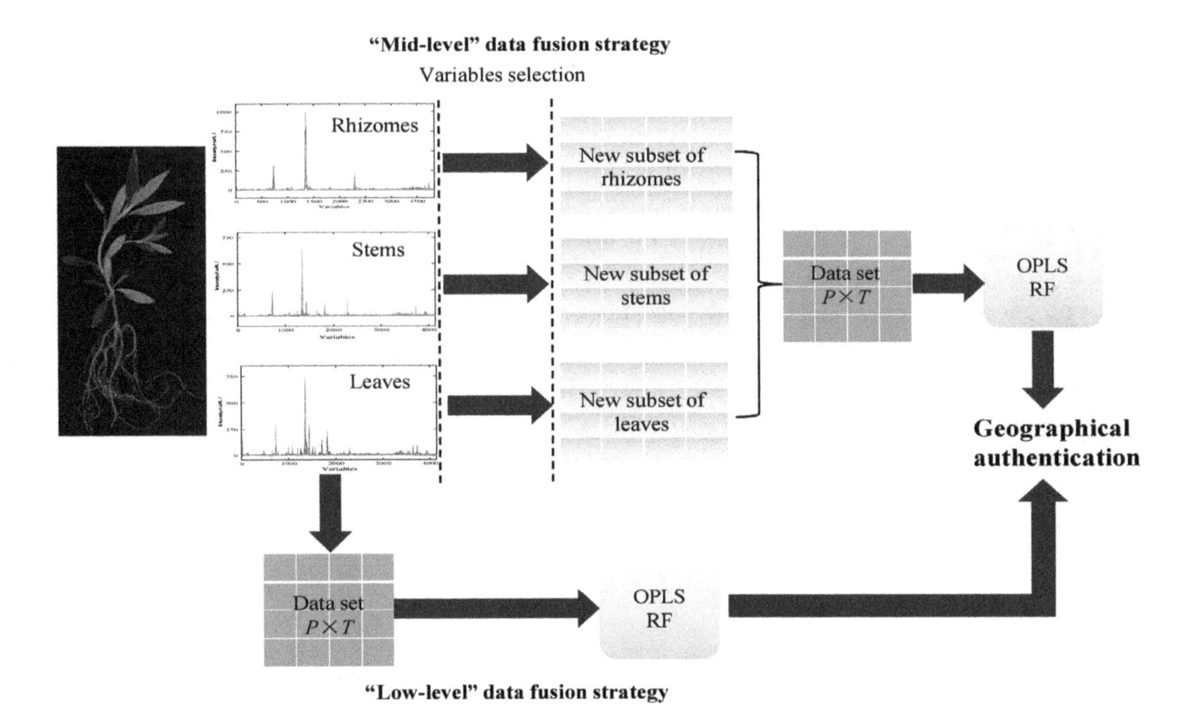

Figure 11. The workflow of geographical authentication of *G. rigescens* grown at different latitudes using data fusion strategy.

The optimum n_{tree} and m_{try} values were selected at first (Figure S8). Afterwards, final classification models were established based on the best values of arguments. From Table 5, it could be seen that the samples collected from four different latitudes were better discriminated by using RS data set and RSL data set. RS_RF model achieved 95.43% total accuracy for the calibration set and achieved 96.81% total accuracy for calibration set. RSL_RF model achieved 94.89% correctly for the calibration set and achieved 97.37% correctly for the calibration set. From a comparison with SE, SP, MCC, and EFF values of S_RF and L_RF models (Tables 1 and 3), we found that the low-level data fusion strategy improved the phenomenon of imbalance category recognition in the RF model (Table 5). However, the total accuracy of models was not obviously improved.

Table 5. The major parameters of RF models based on low-level data fusion strategy.

Model	Class	Calibration Set				Validation Set			
		I	II	III	IV	I	II	III	IV
RS_RF	ACC (%)	96.77	98.92	92.47	93.55	95.74	97.87	95.74	97.87
	SE	0.92	0.94	0.91	0.87	0.92	0.88	0.97	0.96
	SP	0.99	1.00	0.93	0.96	0.97	1.00	0.95	0.99
	MCC	0.92	0.96	0.83	0.83	0.89	0.92	0.90	0.94
	EFF	0.95	0.97	0.92	0.91	0.95	0.94	0.96	0.97
RL_RF	ACC (%)	94.09	98.39	93.01	96.24	87.23	94.68	91.49	92.55
	SE	0.90	0.91	0.91	0.91	0.96	0.75	0.86	0.70
	SP	0.96	1.00	0.94	0.98	0.84	0.99	0.94	1.00
	MCC	0.85	0.94	0.84	0.90	0.74	0.80	0.80	0.80
	EFF	0.93	0.95	0.93	0.95	0.90	0.86	0.90	0.83
SL_RF	ACC (%)	93.55	95.70	92.47	93.55	90.43	96.81	96.81	94.68
	SE	0.94	0.75	0.91	0.85	0.92	0.88	0.93	0.83
	SP	0.93	1.00	0.93	0.96	0.90	0.99	0.98	0.99
	MCC	0.84	0.84	0.83	0.82	0.78	0.88	0.92	0.85
	EFF	0.94	0.87	0.92	0.90	0.91	0.93	0.96	0.90
RSL_RF	ACC (%)	95.70	99.46	91.94	92.47	94.68	96.81	100.00	97.87
	SE	0.94	0.97	0.86	0.85	0.86	1.00	1.00	0.96
	SP	0.96	1.00	0.95	0.95	0.98	0.96	1.00	0.99
	MCC	0.89	0.98	0.81	0.80	0.87	0.88	1.00	0.94
	EFF	0.95	0.98	0.90	0.90	0.92	0.98	1.00	0.97

The permutation plot of all models could be found in Supplementary Materials (Figures S17–S20). The classification results of OPLS-DA models based on low-level data fusion showed models' R^2 values ranged from 0.86 to 0.90 and Q^2 values ranged from 0.74 to 0.80 (Table S2). Total accuracy rates of the calibration set of RS_OPLS-DA, RL_OPLS-DA, SL_OPLS-DA, and RSL_OPLS-DA were 99.46%, 99.73, 100.00%, and 99.73%, respectively (Table 6). Additionally, correct classification rates of validation sets varied from 97.34% to 98.40% (Table 6). The comparison parameters for SE, SP, MCC, and EFF (Tables 4 and 6), the results highlight classification abilities of data fusion OPLS-DA models were better than the individual data set models. What is more, the RS_OPLS-DA model was the optimum classification model when using low-level data fusion strategy (Tables 5 and 6).

Table 6. The major parameters of OPLS-DA models based on low-level data fusion strategy.

Model	Class	Calibration Set				Validation Set			
		I	II	III	IV	I	II	III	IV
RS_OPLS-DA	ACC (%)	99.46	100.00	99.46	98.92	97.87	98.94	97.87	98.94
	SE	1.00	1.00	1.00	0.96	0.96	1.00	0.97	0.96
	SP	0.99	1.00	0.99	1.00	0.99	0.99	0.98	1.00
	MCC	0.99	1.00	0.99	0.97	0.95	0.96	0.95	0.97
	EFF	1.00	1.00	1.00	0.98	0.97	0.99	0.98	0.98

Table 6. *Cont.*

RL_OPLS-DA	ACC (%)	99.46	100.00	100.00	99.46	95.74	97.87	97.87	97.87
	SE	1.00	1.00	1.00	0.98	0.88	1.00	0.97	0.96
	SP	0.99	1.00	1.00	1.00	0.99	0.97	0.98	0.99
	MCC	0.99	1.00	1.00	0.99	0.89	0.93	0.95	0.94
	EFF	1.00	1.00	1.00	0.99	0.93	0.99	0.98	0.97
SL_OPLS-DA	ACC (%)	100.00	100.00	100.00	100.00	94.68	98.94	97.87	97.87
	SE	1.00	1.00	1.00	1.00	1.00	0.94	0.93	0.91
	SP	1.00	1.00	1.00	1.00	0.93	1.00	1.00	1.00
	MCC	1.00	1.00	1.00	1.00	0.88	0.96	0.95	0.94
	EFF	1.00	1.00	1.00	1.00	0.96	0.97	0.96	0.96
RSL_OPLS-DA	ACC (%)	99.46	100.00	100.00	99.46	96.81	98.94	97.87	97.87
	SE	1.00	1.00	1.00	0.98	0.92	1.00	0.97	0.96
	SP	0.99	1.00	1.00	1.00	0.99	0.99	0.98	0.99
	MCC	0.99	1.00	1.00	0.99	0.92	0.96	0.95	0.94
	EFF	1.00	1.00	1.00	0.99	0.95	0.99	0.98	0.97

2.3.2. Mid-Level Data Fusion

At the end of the research, the feasibility for further optimizing the model parameters by feature subset selection and data fusion was investigated (Figure 11). Variables selection was one of the steps of the mid-level data fusion strategy. For the RF model, the "Boruta" algorithm was used to identify important chromatographic signal variables that significantly contributed to the classification performance. "Boruta" selection was finished based on three RM models that were built while using data sets of rhizomes (3839 variables), stems (4140 variables), and leaves (4140 variables), respectively. After comparing original attributes' importance with importance achievable at random, 200 variables of rhizome data set, 305 variables of stem data set, and 359 of variables for leaf data set were retained as relevant features variables for sample discrimination (Figures S9–S11). Subsequently, those feature subsets were combined as a new data block and the fused data set (505 variables for RS, 559 variables for RL, 664 variables for SL, and 864 variables for RSL) was used to establish final classification models. The optimum n_{tree} and m_{try} values of RS_RF, RL_RF, SL_RF, and RSL_RF model could be found in Figure S12.

Table 7 lists the statistical results for the classification ability of the four RF models based on mid-level data fusion. The average accuracies of the calibration set and validation set were achieved for 96.44% and 97.21% by using RF algorithm. It is notable that the RL_RF model had accuracies that ranged from 94.09% to 99.46% in the calibration set and accuracy ranging from 96.81% to 100% in the validation set. In addition, parameters of SE (0.87–1.00), SP (0.94–1.00), MCC (0.87–1.00), and EFF (0.92–1.00) for each class of RL_RF model were higher than most RF classification models. As a result, mid-level data fusion strategy could eliminate the unnecessary variables, enhance model classification ability, and improve the phenomenon of imbalance category recognition in the RF model relative to low-level data fusion strategy.

For the OPLS-DA model, in front of all, three independent classification models were built while using original data sets of rhizome, stem, and leaf, respectively. Subsequently, the VIP value of variables in different classification models was calculated by SIMCA software. The results showed (Figure S12) that a total of 4486 variables (1309 variables selected from rhizome data set, 1538 variables selected from stem data set and 1639 variables selected from leaf data set) VIP values were greater than 1. Those variables with large importance for the geographical traceability of samples were combined into a new data set (2847 variables for RS, 2948 variables for RL, 3177 variables for SL, and 4486 variables for RSL) for final classification model building. The R^2 and Q^2 values and the permutation plot of RS_OPLS-DA, RL_OPLS-DA, SL_OPLS-DA, and RSL_OPLS-DA model were shown in Table S2 and Figures S21–S24.

Table 7. The major parameters of RF models based on mid-level data fusion strategy.

Model	Class	Calibration Set				Validation Set			
		I	II	III	IV	I	II	III	IV
RS_RF	ACC (%)	99.46	99.46	94.09	95.16	98.94	100.00	96.81	97.87
	SE	0.98	0.97	0.95	0.87	1.00	1.00	0.97	0.91
	SP	1.00	1.00	0.94	0.98	0.99	1.00	0.97	1.00
	MCC	0.99	0.98	0.87	0.87	0.97	1.00	0.93	0.94
	EFF	0.99	0.98	0.94	0.92	0.99	1.00	0.97	0.96
RL_RF	ACC (%)	95.70	96.77	96.24	97.31	91.49	98.94	91.49	94.68
	SE	0.92	0.88	0.97	0.93	0.92	1.00	0.86	0.78
	SP	0.97	0.99	0.96	0.99	0.91	0.99	0.94	1.00
	MCC	0.89	0.88	0.91	0.93	0.80	0.96	0.80	0.86
	EFF	0.94	0.93	0.96	0.96	0.92	0.99	0.90	0.88
SL_RF	ACC (%)	95.16	96.77	93.55	96.24	97.87	100.00	98.94	96.81
	SE	0.94	0.81	0.95	0.89	0.96	1.00	1.00	0.91
	SP	0.96	1.00	0.93	0.99	0.99	1.00	0.98	0.99
	MCC	0.88	0.88	0.86	0.90	0.95	1.00	0.98	0.91
	EFF	0.95	0.90	0.94	0.94	0.97	1.00	0.99	0.95
RSL_RF	ACC (%)	97.85	99.46	94.09	95.70	96.81	100.00	95.74	98.94
	SE	0.94	0.97	0.93	0.91	0.96	1.00	0.93	0.96
	SP	0.99	1.00	0.95	0.97	0.97	1.00	0.97	1.00
	MCC	0.94	0.98	0.86	0.88	0.92	1.00	0.90	0.97
	EFF	0.97	0.98	0.94	0.94	0.97	1.00	0.95	0.98

The classification results showed that average accuracies of calibration and validation sets were achieved for 99.66% and 96.81%, respectively (Table 8). The four models exhibit good performances (MCC values ranged from 0.96 to 1.00 and EFF values ranged from 0.92 to 1.00 (Table 8). OPLS-DA models based on mid-level data fusion and low-level data fusion showed similar accuracy and model performance although feature selection was useful for reducing irrelevant variable when classifying samples.

Table 8. The major parameters of OPLS-DA models based on mid-level data fusion strategy.

Model	Class	Calibration Set				Validation Set			
		I	II	III	IV	I	II	III	IV
RS_OPLS-DA	ACC (%)	100.00	100.00	99.46	99.46	93.62	97.87	94.68	98.94
	SE	1.00	1.00	1.00	0.98	0.88	1.00	0.90	0.96
	SP	1.00	1.00	0.99	1.00	0.96	0.97	0.97	1.00
	MCC	1.00	1.00	0.99	0.99	0.84	0.93	0.87	0.97
	EFF	1.00	1.00	1.00	0.99	0.92	0.99	0.93	0.98
RL_OPLS-DA	ACC (%)	100.00	100.00	99.46	99.46	96.81	97.87	97.87	98.94
	SE	1.00	1.00	1.00	0.98	0.92	1.00	0.97	0.96
	SP	1.00	1.00	0.99	1.00	0.99	0.97	0.98	1.00
	MCC	1.00	1.00	0.99	0.99	0.92	0.93	0.95	0.97
	EFF	1.00	1.00	1.00	0.99	0.95	0.99	0.98	0.98
SL_OPLS-DA	ACC (%)	100.00	100.00	98.92	98.92	93.62	97.87	94.68	96.81
	SE	1.00	1.00	0.98	0.98	0.92	0.94	0.90	0.91
	SP	1.00	1.00	0.99	0.99	0.94	0.99	0.97	0.99
	MCC	1.00	1.00	0.97	0.97	0.85	0.92	0.87	0.91
	EFF	1.00	1.00	0.99	0.99	0.93	0.96	0.93	0.95
RSL_OPLS-DA	ACC (%)	100.00	100.00	99.46	99.46	95.74	98.94	96.81	97.87
	SE	1.00	1.00	1.00	0.98	0.92	1.00	0.93	0.96
	SP	1.00	1.00	0.99	1.00	0.97	0.99	0.98	0.99
	MCC	1.00	1.00	0.99	0.99	0.89	0.96	0.92	0.94
	EFF	1.00	1.00	1.00	0.99	0.95	0.99	0.96	0.97

Overall, it can be seen that there is an improvement in the results that were provided by data fusion when compared with performances of models based on independent data sets. When considering the similar accuracy and a higher SE, SP, MCC, and EFF values between calibration set and validation set, the RS_OPLS-DA models that were based on low-level data fusion strategy was the best performance.

3. Materials and Methods

3.1. Plant Material Collection

Plant materials (29 population and 280 individuals) of *G. rigescens* were collected in the fall of 2012 and 2013 at the time of local traditional harvest period, at the different location of Yunnan, Guizhou, and Sichuan (Figure 12). Four producing areas were divided according to the location of population. (I) low latitudes area, with latitudes ranging from 23.92–23.66° N, South of Yunnan (eight population and 76 individuals), (II) mid-latitude area, with latitudes ranges from 24.95–25.06° N, Middle of Yunnan (five population and 48 individuals), (III) mid-high latitude area, with latitudes ranges from 26.49–26.64° N, Northwest of Yunnan and West of Guizhou (nine population and 76 individuals 87), and (IV) high latitude area, with latitudes ranges from 27.34–28.52° N, Hengduan Mountains Region of Yunnan and mountainous regions of Southwest of Sichuan (seven population and 69 individuals). The fresh materials were authenticated and transported to the laboratory of Yuxi normal University. Subsequently, samples were wash cleaning and dried at 50 °C as soon as possible. At last, all samples (rhizomes, stems and leaves) were stored in a relatively dry environment prior to the extraction procedure.

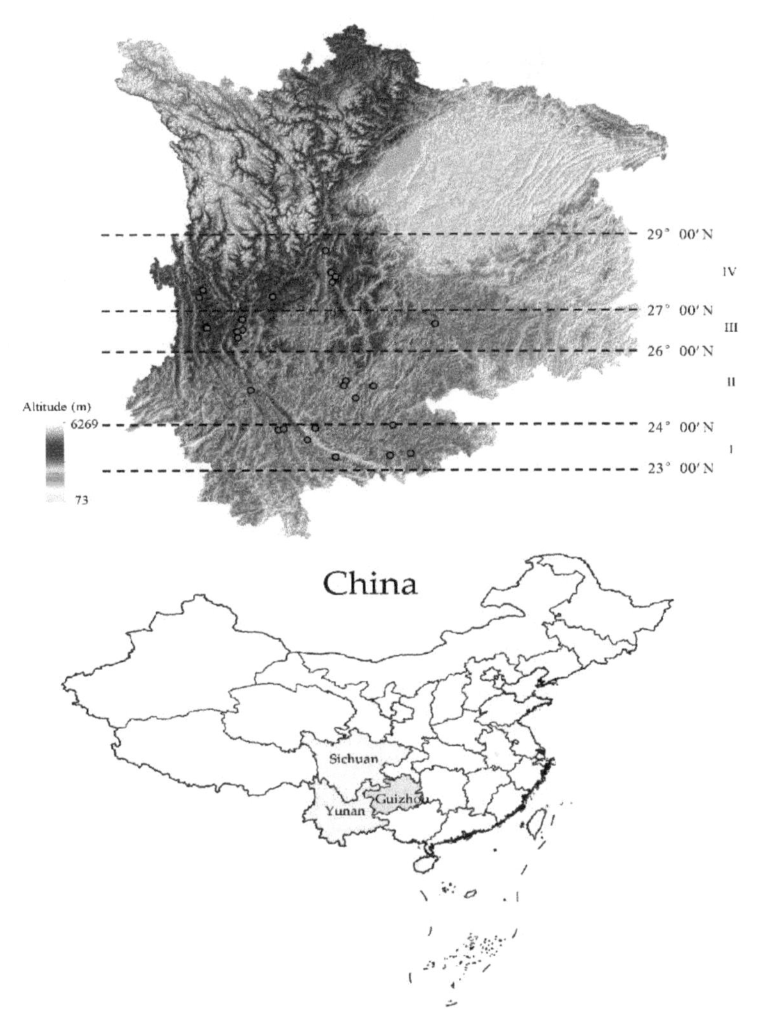

Figure 12. Geographical distribution of sample information.

3.2. Chemicals and Reagents

HPLC-grade acetonitrile, methanol (MeOH) were supplied by Thermo Fisher Scientific (Waltham, MA, USA). HPLC-grade formic acid was purchased from Sigma-Aldrich (Steinheim, Germany). Deionized water was obtained from Wahaha Group Co., Ltd. (Hangzhou, Zhejiang, China). The primary grade reference standards loganin (purity: ≥98%), 6′-O-β-D-glucopyranosylgentiopicroside (purity: ≥98%), swertiamarine (purity: ≥98%), gentiopicroside (purity: ≥98%), and sweroside (purity: ≥98%) were purchased from the Chinese National Institute for Food and Drug Control (Beijing, China), Shanghai Shifeng Biological Technology (Shanghai, China), respectively.

3.3. Sample Preparation

The dried samples (rhizomes, stems, and leaves) were ground and then passed through a 100 mesh sieves. Each sample powder (25 mg) was accurately weighed and extracted while using 1.5 mL 80% methanol-water solution, at 25 °C. The samples were extracted while using an Ultrasonic extractor for 40 min. The final extract was filtered with a 0.22 μm syringe filter into an HPLC vial and then subjected to HPLC analysis [16,58].

3.4. Instrumentation and HPLC Analysis

Chromatographic analyses were performed with an Agilent 1260 Infinity LC system (Agilent Technologies, Santa Clara, CA, USA), which was equipped with a G1315D diode-array detector, a G1329B ALS autosampler, and a thermostated column compartment. The HPLC fingerprint was recorded by Chemstation software (Agilent Technologies, Waldbron, Germany).

The analytical separation was adopted from a published method for chemical fingerprinting analysis [16]. The separation was achieved on a reversed phase C18 (Agilent Intersil, 5 μm, 4.6 × 150 mm) column (Agilent, Santa Clara, CA, USA). The composition of the mobile phase was: (A) 0.1% phosphoric acid in water and (B) 100% acetonitrile. The separation was as follows: 0.00–2.50 min: 7–10% B, 2.50–20.00 min: 10–26% B, 20.00–29.02 min: 26–58.3% B, 29.02–30.00 min: 58.3–90% B. The column was subsequently washed with 90% B and re-equilibrated with 7% B prior to injection of the next sample. The flow rate was 1.0 mL/min and the column temperature was 30 °C. The injection volume was 5 μL and the detective wavelength of UV spectra was set at 241 nm. Chromatographic data was processed while using OpenLab software (Agilent Technologies) [16,58].

3.5. Data Analysis

HPLC fingerprints from the 280 rhizome samples, 280 stem samples, and 280 leaf samples, a total of 840 fingerprint data was exported in CSV format and imported to MATLAB R2018b (The MathWorks, Inc., Natick, MA, USA), which was used for correlation optimized warping (COW) alignment preprocessing of chromatographic fingerprint. MATLAB code of COW is freely available from www.models.kvl.dk. The preprocessing fingerprint was analyzed in the following work [59].

Exploratory data analysis (EDA) is necessary for building predictive models [60,61]. It can help in determining interesting correlations among all of the samples or variables and summarize data sets main characteristics [60]. Principal component analysis (PCA) is a popular primary tool in EDA [61,62]. It is often used to visualize the relatedness between samples and explains the variance in the data. Hence, PCA, as an unsupervised pattern recognition technique, was widely used to extract key information from chemical fingerprint for geographical origin or Modelling Research [61].

Unlike PCA, orthogonal partial least squares discriminant analysis (OPLS-DA) is a supervised pattern recognition technique. As an extension of PLS, an inbuilt orthogonal signal correction filter was incorporated in the OPLS-DA model [56]. This algorithm effectively divides the X variable into two parts: one part that is related to class information (Y-predictive) and the other is orthogonal or unrelated to class information (Y-uncorrelated). Therefore, interpretability and prediction performance of the model was enhanced [56].

Random forest (RF) is another supervised pattern recognition technique utilized in the study. RF is an ensemble learning method [55]. A large number of trees were produced by RF algorithm in order to improve model predictive ability, and trees' decision results were combined as final decision results. In other words, the more trees built in the random forest classifier, the higher accuracy could be achieved. However, many researches showed that an optimum tree number was of great importance in modeling classification performance [33,46].

In this work, exploratory data analysis of HPLC fingerprints of *G. rigescens* grown in four different latitudes was finished with PCA. Two supervised pattern recognition techniques, OPLS-DA and RF, were applied to build classification models for *G. rigescens* producing areas. SIMCA 14.1 software managed PCA and OPLS-DA (Umetrics AB, Umea, Sweden). RF classification models were established with R 3.5.1 program and package randomForest (Version 4.6-14) [63].

Data Fusion Strategy

In the case of low-level fusion strategy (Figure 11), different subsets HPLC fingerprint data matrix of rhizomes, stems, and leaves) are straightforwardly concatenated and compiled into a new chromatographic data matrix for subsequent classification model construction [45,46]. Furthermore, each subset must be totally aligned and keep all the variables on the same scale before subsets reconnection [45,46].

In the case of mid-level fusion (Figure 11), the first step of data treatment is feature selection that is based on rhizomes, stems, or leaves classification models. When compared with the raw data sets, feature selection of subsets minimizes the data content and reduces data dimensions. Subsequently, new subsets of rhizomes, stems, and leaves were rebuilt while using variables of feature selection [45]. At last, those subsets are concatenated and compiled into a final data matrix for model construction [45].

In the research, relevant variables of RF classification models were determined by the R software package Boruta [64], and VIP was used for important variables selection of OPLS-DA [65].

3.6. Model Evaluation

Five parameters, including accuracy (ACC), sensitivity (SE), specificity (SP), efficiency (EFF), and Matthews correlation coefficient (MCC) were applied to evaluate the identification ability of RF and OPLS-DA models. The ruggedness of OPLS-DA model was investigated through 200 times permutation tests. Furthermore, cumulative prediction ability (Q^2), cumulative interpretation ability (R^2), root mean square error of estimation (RMSEE), root mean square error of cross-validation (RMSECV), and root mean square error of prediction (RMSEP) were important evaluation indexes for the predictive power of OPLS-DA model [33,66].

Values of TP (Correctly identified samples of positive class), TN (correctly identified samples of negative class), FN (incorrectly identified samples of positive class), and FP (incorrectly identified samples of negative class) were calculated according to confusion matrixes of classification models. Subsequently, ACC, SE, SP, EFF, and MCC were calculated while using Equations (1)–(5) and values of Q^2, R^2, RMSEE, RMSECV, and RMSEP computed by software SIMCA 14.1.

$$\mathrm{ACC} = \frac{(\mathrm{TN} + \mathrm{TP})}{(\mathrm{TP} + \mathrm{TN} + \mathrm{FP} + \mathrm{FN})} \tag{1}$$

$$\mathrm{SE} = \frac{\mathrm{TP}}{(\mathrm{TP} + \mathrm{FN})} \tag{2}$$

$$\mathrm{SP} = \frac{\mathrm{TN}}{(\mathrm{TN} + \mathrm{FP})} \tag{3}$$

$$\mathrm{EFF} = \sqrt{\mathrm{SE} \times \mathrm{SP}} \tag{4}$$

$$\mathrm{MCC} = \frac{(\mathrm{TP} \times \mathrm{TN} - \mathrm{FP} \times \mathrm{FN})}{\sqrt{(\mathrm{TP} + \mathrm{FP})(\mathrm{TP} + \mathrm{FN})(\mathrm{TN} + \mathrm{FP})(\mathrm{TN} + \mathrm{FN})}} \tag{5}$$

For model performance, lower values of RMSEE, RMSECV, and RMSEP mean better predictive ability for the models. Conversely, the closer that values of ACC, SE, SP, EFF, MCC, and Q^2, R^2 are to 1, the more well performance the model is.

4. Conclusions

The findings in this study showed that *G. rigescens* chemical profiles were influenced by the latitude gradients of producing areas and lower latitudes and higher latitudes samples seemed to be clearly distinguishable. According to the score plots of PCA and OPLS-DA, the phytochemical geographic variation of the overground and underground part along the latitude gradients was visualized. Subsequently, the potential of fingerprint data obtained while using HPLC-DAD to discriminate and classify *G. rigescens* grown in four different latitudes was investigated. Additionally, RF and OPLS-DA models were used to develop an effective way for geographical traceability of the *G. rigescens* that were grown in four different latitudes. When using independent data sets to build models, rhizomes data set combined with OPLS-DA presented the best performance with a classification accuracy of calibration and validation set varied from 94.68% to 98.94%. In a further step, the feasibility of combining the chromatographic fingerprint data from overground and underground organs was investigated based on two kinds of data fusion strategies in order to improve the performance of classification models: low-level and mid-level. Notably, classification performances of OPLS-DA models were efficiently improved by low-level data fusion strategy and better performances of RF models appeared to be achieved by mid-level data fusion strategy. Although satisfactory results were obtained with both RF and OPLS-DA based on two kinds of data fusion strategies, OPLS-DA combined with rhizome-stem fusion data set was the optimum model for discriminating *G. rigescens* samples according to their grown latitudes, with an accuracy of (97.87–100.00%), SE of (0.96–1.00), SP of (0.98–1.00), MCC of (0.95–1.00), and EFF of (0.97–1.00).

Supplementary Materials: The following are available online. Figure S1: Variation of stems score plots along the latitude gradients, Figure S2: Variation of stems score plots between the adjacent latitudes, Figure S3: Variation of leaves score plots along the latitude gradients, Figure S4: Variation of leaves score plots between the adjacent latitudes, Figure S5: Permutation plot of the OPLS-DA of rhizome samples, Figure S6: Permutation plot of the OPLS-DA of stem samples, Figure S7: Permutation plot of the OPLS-DA of leaf samples, Figure S8: The n_{tree} and m_{try} screening of RF models based on low-level data fusion strategy, Figure S9: Result of variables selection of rhizome fingerprint data based on "Boruta" algorithm, Figure S10. Result of variables selection of stem fingerprint data based on "Boruta" algorithm, Figure S11: Result of variables selection of leaf fingerprint data based on "Boruta" algorithm, Figure S12: The n_{tree} and m_{try} screening of RF models based on mid-level data fusion strategy, Figure S13: The importance variables of OPLS-DA models of rhizomes, stems and leaves fingerprints data, Figure S14: Permutation testing (200 times) of the R_OPLS-DA model, Figure S15: Permutation testing (200 times) of the S_OPLS-DA model, Figure S16: Permutation testing (200 times) of the L_OPLS-DA model, Figure S17: Permutation testing (200 times) of the RS_OPLS-DA model based on low-level data fusion, Figure S18: Permutation testing (200 times) of the RL_OPLS-DA model based on low-level data fusion, Figure S19: Permutation testing (200 times) of the SL_OPLS-DA model based on low-level data fusion, Figure S20: Permutation testing (200 times) of the RSL_OPLS-DA model based on low-level data fusion, Figure S21: Permutation testing (200 times) of the RS_OPLS-DA model based on mid-level data fusion, Figure S22: Permutation testing (200 times) of the RL_OPLS-DA model based on mid-level data fusion, Figure S23: Permutation testing (200 times) of the SL_OPLS-DA model based on mid-level data fusion, Figure S24: Permutation testing (200 times) of the RSL_OPLS-DA model based on mid-level data fusion, Table S1: The evaluation indexes for predictive power of OPLS-DA model of rhizome, stem and leaf, Table S2: The evaluation indexes for predictive power of OPLS-DA models based on low-level and mid-level data fusion strategies.

Author Contributions: H.Y. and Y.-Z.W. designed the project and revised the manuscript. T.S. performed the experiments, analyzed the data and wrote the manuscript.

References

1. Flora of China Editorial Committee. *Flora of China*; Science Press and Missouri Botanical Garden Press: Beijing, China, 1995; Volume 22.
2. Mustafa, A.M.; Caprioli, G.; Ricciutelli, M.; Maggi, F.; Marín, R.; Vittori, S.; Sagratini, G. Comparative HPLC/ESI-MS and HPLC/DAD study of different populations of cultivated, wild and commercial

Gentiana lutea L. *Food Chem.* **2015**, *174*, 426–433. [CrossRef] [PubMed]

3. Pan, Y.; Zhao, Y.L.; Zhang, J.; Li, W.Y.; Wang, Y.Z. Phytochemistry and pharmacological activities of the genus *Gentiana* (Gentianaceae). *Chem. Biodivers.* **2016**, *13*, 107–150. [CrossRef] [PubMed]
4. Jiang, R.W.; Wong, K.L.; Chan, Y.M.; Xu, H.X.; But, P.P.H.; Shaw, P.C. Isolation of iridoid and secoiridoid glycosides and comparative study on *Radix gentianae* and related adulterants by HPLC analysis. *Phytochemistry* **2005**, *66*, 2674–2680. [CrossRef] [PubMed]
5. Xu, Y.; Li, Y.; Maffucci, K.; Huang, L.; Zeng, R. Analytical methods of phytochemicals from the genus *Gentiana*. *Molecules* **2017**, *22*, 2080. [CrossRef] [PubMed]
6. Mustafa, A.M.; Ricciutelli, M.; Maggi, F.; Sagratini, G.; Vittori, S.; Caprioli, G. Simultaneous determination of 18 bioactive compounds in Italian bitter liqueurs by reversed-phase high-performance liquid chromatography—Diode array detection. *Food Anal. Method* **2014**, *7*, 697–705. [CrossRef]
7. Mirzaee, F.; Hosseini, A.; Jouybari, H.B.; Davoodi, A.; Azadbakht, M. Medicinal, biological and phytochemical properties of *Gentiana* species. *J. Tradit. Complement. Med.* **2017**, *7*, 400–408. [CrossRef] [PubMed]
8. Gao, L.J.; Li, J.Y.; Qi, J.H. Gentisides A and B, two new neuritogenic compounds from the traditional Chinese medicine *Gentiana rigescens* franch. *Bioorgan. Med. Chem.* **2010**, *18*, 2131–2134. [CrossRef]
9. Gao, L.J.; Xiang, L.; Luo, Y.; Wang, G.F.; Li, J.Y.; Qi, J.H. Gentisides C-K: Nine new neuritogenic compounds from the traditional Chinese medicine *Gentiana rigescens* franch. *Bioorgan. Med. Chem.* **2010**, *18*, 6995–7000. [CrossRef]
10. Li, J.; Gao, L.J.; Sun, K.Y.; Xiao, D.; Li, W.Y.; Xiang, L.; Qi, J.H. Benzoate fraction from *Gentiana rigescens* franch alleviates scopolamine-induced impaired memory in mice model in vivo. *J. Ethnopharmacol.* **2016**, *193*, 107–116. [CrossRef]
11. Mustafa, A.M.; Caprioli, G.; Dikmen, M.; Kaya, E.; Maggi, F.; Sagratini, G.; Vittori, S.; öztürk, Y. Evaluation of neuritogenic activity of cultivated, wild and commercial roots of *Gentiana lutea* L. *J. Funct. Foods* **2015**, *19*, 164–173. [CrossRef]
12. Pharmacopoeia, C.C. *Pharmacopoeia of the People's Republic of China*; China Medicinal Science Press: Beijing, China, 2015.
13. Pan, Y.; Zhang, J.; Shen, T.; Zhao, Y.L.; Zuo, Z.T.; Wang, Y.Z.; Li, W.Y. Investigation of chemical diversity in different parts and origins of ethnomedicine *Gentiana rigescens* franch using targeted metabolite profiling and multivariate statistical analysis. *Biomed. Chromatogr.* **2016**, *30*, 232–240. [CrossRef] [PubMed]
14. Li, J.; Zhang, J.; Zhao, Y.L.; Huang, H.Y.; Wang, Y.Z. Comprehensive quality assessment based specific chemical profiles for geographic and tissue variation in *Gentiana rigescens* using HPLC and FTIR method combined with principal component analysis. *Front. Chem.* **2017**, *5*, 125. [CrossRef] [PubMed]
15. Wu, Z.; Zhao, Y.L.; Zhang, J.; Wang, Y.Z. Quality assessment of *Gentiana rigescens* from different geographical origins using FT-IR spectroscopy combined with HPLC. *Molecules* **2017**, *22*, 1238. [CrossRef] [PubMed]
16. Wang, Y.; Shen, T.; Zhang, J.; Huang, H.Y.; Wang, Y.Z. Geographical authentication of *Gentiana rigescens* by high-performance liquid chromatography and infrared spectroscopy. *Anal. Lett.* **2018**, *51*, 2173–2191. [CrossRef]
17. Ailer, B.; Avramov, S.; Banjanac, T.; Cvetković, J.; Nestorović Živković, J.; Patenković, A.; Mišić, D. Secoiridoid glycosides as a marker system in chemical variability estimation and chemotype assignment of *Centaurium erythraea* Rafn from the Balkan Peninsula. *Ind. Crop. Prod.* **2012**, *40*, 336–344.
18. Yang, Y.M.; Tian, K.; Hao, J.M.; Pei, S.J.; Yang, Y.X. Biodiversity and biodiversity conservation in Yunnan, China. *Biodivers. Conserv.* **2004**, *13*, 813–826. [CrossRef]
19. Fan, Z.X.; Thomas, A. Spatiotemporal variability of reference evapotranspiration and its contributing climatic factors in Yunnan Province, SW China, 1961–2004. *Clim. Chang.* **2013**, *116*, 309–325. [CrossRef]
20. Liu, M.X.; Xu, X.L.; Sun, A.Y.; Wang, K.L.; Yue, Y.M.; Tong, X.W.; Liu, W. Evaluation of high-resolution satellite rainfall products using rain gauge data over complex terrain in southwest China. *Theor. Appl. Climatol.* **2015**, *119*, 203–219. [CrossRef]
21. Tang, Q.H.; Ge, Q.S. *Atlas of Environmental Risks Facing China under Climate Change*; Springer Verlag: Berlin, Germany, 2018.
22. Zhao, Z.Z.; Guo, P.; Brand, E. The formation of daodi medicinal materials. *J. Ethnopharmacol.* **2012**, *140*, 476–481. [CrossRef]
23. Sun, M.M.; Li, L.; Wang, M.; van Wijk, E.; He, M.; van Wijk, R.; Koval, S.; Hankemeier, T.; van der Greef, J.; Wei, S.L. Effects of growth altitude on chemical constituents and delayed luminescence properties in medicinal

rhubarb. *J. Photoch. Photobio. B* **2016**, *162*, 24–33. [CrossRef]

24. Song, X.Y.; Jin, L.; Shi, Y.P.; Li, Y.D.; Chen, J. Multivariate statistical analysis based on a chromatographic fingerprint for the evaluation of important environmental factors that affect the quality of *Angelica sinensis*. *Anal. Methods* **2014**, *6*, 8268–8276. [CrossRef]
25. Yao, R.Y.; Heinrich, M.; Zou, Y.F.; Reich, E.; Zhang, X.L.; Chen, Y.; Weckerle, C.S. Quality variation of Goji (fruits of *Lycium* spp.) in China: A comparative morphological and metabolomic analysis. *Front. Pharmacol.* **2018**, *9*, 151. [CrossRef] [PubMed]
26. Huang, Y.P.; Wu, Z.W.; Su, R.H.; Ruan, G.H.; Du, F.Y.; Li, G.K. Current application of chemometrics in traditional Chinese herbal medicine research. *J. Chromatogr. B* **2016**, *1026*, 27–35. [CrossRef] [PubMed]
27. Zhang, C.; Zheng, X.; Ni, H.; Li, P.; Li, H.J. Discovery of quality control markers from traditional Chinese medicines by fingerprint-efficacy modeling: Current status and future perspectives. *J. Pharmaceut. Biomed.* **2018**, *159*, 296–304. [CrossRef] [PubMed]
28. Chen, D.D.; Xie, X.F.; Ao, H.; Liu, J.L.; Peng, C. Raman spectroscopy in quality control of Chinese herbal medicine. *J. Chin. Med. Assoc.* **2017**, *80*, 288–296. [CrossRef] [PubMed]
29. Wang, P.; Yu, Z.G. Species authentication and geographical origin discrimination of herbal medicines by near infrared spectroscopy: A review. *J. Pharm. Anal.* **2015**, *5*, 277–284. [CrossRef] [PubMed]
30. Qi, L.M.; Zhang, J.; Zhao, Y.L.; Zuo, Z.T.; Wang, Y.Z.; Jin, H. Characterization of *Gentiana rigescens* by ultraviolet-visible and infrared spectroscopies with chemometrics. *Anal. Lett.* **2017**, *50*, 1497–1511. [CrossRef]
31. Zhao, Y.L.; Zhang, J.; Jin, H.; Zhang, J.Y.; Shen, T.; Wang, Y.Z. Discrimination of *Gentiana rigescens* from different origins by fourier transform infrared spectroscopy combined with chemometric methods. *J. Aoac Int.* **2015**, *98*, 22–26. [CrossRef]
32. Lee, D.Y.; Kang, K.B.; Kim, J.; Kim, H.J.; Sung, S.H. Classficiation of bupleuri radix according to geographical origins using near infrared spectroscopy (NIRS) combined with supervised pattern recognition. *Nat. Prod. Sci.* **2018**, *24*, 164. [CrossRef]
33. Pei, Y.F.; Wu, L.H.; Zhang, Q.Z.; Wang, Y.Z. Geographical traceability of cultivated *Paris polyphylla* var.*yunnanensis* using ATR-FTMIR spectroscopy with three mathematical algorithms. *Anal. Methods* **2019**, *11*, 113–122. [CrossRef]
34. Chen, H.; Lin, Z.; Tan, C. Fast discrimination of the geographical origins of notoginseng by near-infrared spectroscopy and chemometrics. *J. Pharmaceut. Biomed.* **2018**, *161*, 239–245. [CrossRef] [PubMed]
35. Wang, Q.Q.; Huang, H.Y.; Wang, Y.Z. Geographical authentication of *Macrohyporia cocos* by a data fusion method combining ultra-fast liquid chromatography and fourier transform infrared spectroscopy. *Molecules* **2019**, *24*, 1320. [CrossRef] [PubMed]
36. Ma, X.D.; Fan, Y.X.; Jin, C.C.; Wang, F.; Xin, G.Z.; Li, P.; Li, H.J. Specific targeted quantification combined with non-targeted metabolite profiling for quality evaluation of *Gastrodia elata* tubers from different geographical origins and cultivars. *J. Chromatogr. A* **2016**, *1450*, 53–63. [CrossRef] [PubMed]
37. Tang, J.F.; Li, W.X.; Zhang, F.; Li, Y.H.; Cao, Y.J.; Zhao, Y.; Li, X.L.; Ma, Z.J. Discrimination of radix polygoni multiflori from different geographical areas by UPLC-QTOF/MS combined with chemometrics. *Chin. Med.* **2017**, *12*, 1–12. [CrossRef] [PubMed]
38. Zhu, L.X.; Xu, J.; Wang, R.J.; Li, H.X.; Tan, Y.Z.; Chen, H.B.; Dong, X.P.; Zhao, Z.Z. Correlation between quality and geographical origins of *Poria cocos* revealed by qualitative fingerprint profiling and quantitative determination of triterpenoid acids. *Molecules* **2018**, *23*, 2200. [CrossRef] [PubMed]
39. Sun, L.L.; Wang, M.; Zhang, H.J.; Liu, Y.N.; Ren, X.L.; Deng, Y.R.; Qi, A.D. Comprehensive analysis of polygoni multiflori radix of different geographical origins using ultra-high-performance liquid chromatography fingerprints and multivariate chemometric methods. *J. Food Drug Anal.* **2018**, *26*, 90–99. [CrossRef] [PubMed]
40. Cuadros-Rodríguez, L.; Ruiz-Samblás, C.; Valverde-Som, L.; Pérez-Castaño, E.; González-Casado, A. Chromatographic fingerprinting: An innovative approach for food 'identitation' and food authentication—A tutorial. *Anal. Chim. Acta* **2016**, *909*, 9–23. [CrossRef] [PubMed]
41. Lucio-Gutiérrez, J.R.; Coello, J.; Maspoch, S. Enhanced chromatographic fingerprinting of herb materials by multi-wavelength selection and chemometrics. *Anal. Chim. Acta* **2012**, *710*, 40–49. [CrossRef]
42. Zhang, L.L.; Liu, Y.Y.; Liu, Z.L.; Wang, C.; Song, Z.Q.; Liu, Y.X.; Dong, Y.Z.; Ning, Z.C.; Lu, A.P. Comparison of the roots of *Salvia miltiorrhiza* bunge (danshen) and its variety *S. miltiorrhiza* Bge f. Alba (baihua danshen) based on multi-wavelength HPLC-fingerprinting and contents of nine active components. *Anal. Methods* **2016**, *8*, 3171–3182. [CrossRef]

43. Wang, X.; Li, B.Q.; Xu, M.L.; Liu, J.J.; Zhai, H.L. Quality assessment of traditional Chinese medicine using HPLC-PAD combined with tchebichef image moments. *J. Chromatogr. B* **2017**, *1040*, 8–13. [CrossRef]
44. Jiménez-Carvelo, A.M.; Cruz, C.M.; Olivieri, A.C.; González-Casado, A.; Cuadros-Rodríguez, L. Classification of olive oils according to their cultivars based on second-order data using LC-DAD. *Talanta* **2019**, *195*, 69–76. [CrossRef] [PubMed]
45. Borràs, E.; Ferré, J.; Boqué, R.; Mestres, M.; Aceña, L.; Busto, O. Data fusion methodologies for food and beverage authentication and quality assessment—A review. *Anal. Chim. Acta* **2015**, *891*, 1–14. [CrossRef] [PubMed]
46. Li, Y.; Zhang, J.Y.; Wang, Y.Z. FT-MIR and NIR spectral data fusion: A synergetic strategy for the geographical traceability of *Panax notoginseng*. *Anal. Bioanal. Chem.* **2018**, *410*, 91–103. [CrossRef] [PubMed]
47. Wu, X.M.; Zuo, Z.T.; Zhang, Q.Z.; Wang, Y.Z. FT-MIR and UV–vis data fusion strategy for origins discrimination of wild *Paris polyphylla* Smith var. *yunnanensis*. *Vib. Spectrosc.* **2018**, *96*, 125–136. [CrossRef]
48. Wang, H.Y.; Song, C.; Sha, M.; Liu, J.; Li, L.P.; Zhang, Z.Y. Discrimination of medicine radix astragali from different geographic origins using multiple spectroscopies combined with data fusion methods. *J. Appl. Spectrosc.* **2018**, *85*, 313–319. [CrossRef]
49. Pei, Y.F.; Zhang, Q.Z.; Zuo, Z.T.; Wang, Y.Z. Comparison and Identification for rhizomes and leaves of *Paris yunnanensis* based on Fourier transform mid-Infrared spectroscopy combined with chemometrics. *Molecules* **2018**, *23*, 3343. [CrossRef] [PubMed]
50. Yang, H.; Liu, J.; Chen, S.; Hu, F.; Zhou, D. Spatial variation profiling of four phytochemical constituents in *Gentiana straminea* (Gentianaceae). *J. Nat. Med.* **2014**, *68*, 38–45. [CrossRef]
51. Lei, M.; Yu, X.H.; Li, M.; Zhu, W.X. Geographic origin identification of coal using near-infrared spectroscopy combined with improved random forest method. *Infrared Phys. Technol.* **2018**, *92*, 177–182. [CrossRef]
52. Sayago, A.; González-Domínguez, R.; Beltrán, R.; Fernández-Recamales, Á. Combination of complementary data mining methods for geographical characterization of extra virgin olive oils based on mineral composition. *Food Chem.* **2018**, *261*, 42–50. [CrossRef]
53. Jandrić, Z.; Haughey, S.A.; Frew, R.D.; Mccomb, K.; Galvin-King, P.; Elliott, C.T.; Cannavan, A. Discrimination of honey of different floral origins by a combination of various chemical parameters. *Food Chem.* **2015**, *189*, 52–59. [CrossRef]
54. Jolayemi, O.S.; Ajatta, M.A.; Adegeye, A.A. Geographical discrimination of palm oils (*Elaeis guineensis*) using quality characteristics and UV-visible spectroscopy. *Food Sci. Nutr.* **2018**, *6*, 773–782. [CrossRef] [PubMed]
55. Breiman, L. Random forests. *Mach. Learn.* **2001**, *45*, 5–32. [CrossRef]
56. Wold, S.; Antti, H.; Lindgren, F.; öhman, J. Orthogonal signal correction of near-infrared spectra. *Chemometr. Intell. Lab.* **1998**, *44*, 175–185. [CrossRef]
57. Boccard, J.L.; Rutledge, D.N. A consensus orthogonal partial least squares discriminant analysis (OPLS-DA) strategy for multiblock omics data fusion. *Anal. Chim. Acta* **2013**, *769*, 30–39. [CrossRef] [PubMed]
58. Lyv, W.Q.; Zhang, J.; Zuo, Z.T.; Wang, Y.Z.; Zhang, Q.Z. Quality evaluation of *Gentiana rigescence* by grey relational analysis method. *Chin. J. Exp. Tradit. Med. Formul.* **2017**, *23*, 66–73.
59. Skov, T.; van den Berg, F.; Tomasi, G.; Bro, R. Automated alignment of chromatographic data. *J. Chemometr.* **2006**, *20*, 484–497. [CrossRef]
60. Kürzl, H. Exploratory data analysis: Recent advances for the interpretation of geochemical data. *J. Geochem. Explor.* **1988**, *30*, 309–322. [CrossRef]
61. Esteki, M.; Shahsavari, Z.; Simal-Gandara, J. Food identification by high performance liquid chromatography fingerprinting and mathematical processing. *Food Res. Int.* **2019**, *122*, 303–317. [CrossRef]
62. Berrueta, L.A.; Alonso-Salces, R.M.; Héberger, K. Supervised pattern recognition in food analysis. *J. Chromatogr. A* **2007**, *1158*, 196–214. [CrossRef]
63. Liaw, A.; Wiener, M. Classification and regression by randomForest. *R News* **2002**, *2*, 18–22.
64. Kursa, M.B.; Rudnicki, W.R. Feature selection with the boruta package. *J. Stat. Softw.* **2010**, *36*, 1–13. [CrossRef]
65. Mehmood, T.; Liland, K.H.; Snipen, L.; Sæbø, S. A review of variable selection methods in partial least squares regression. *Chemometr. Intell. Lab.* **2012**, *118*, 62–69. [CrossRef]
66. Cao, D.S.; Hu, Q.N.; Xu, Q.S.; Yang, Y.N.; Zhao, J.C.; Lu, H.M.; Zhang, L.X.; Liang, Y.Z. In silico classification of human maximum recommended daily dose based on modified random forest and substructure fingerprint. *Anal. Chim. Acta* **2011**, *692*, 50–56. [CrossRef] [PubMed]

9

Identification of Metabolites of Eupatorin In Vivo and In Vitro based on UHPLC-Q-TOF-MS/MS

Luya Li [1], Yuting Chen [1], Xue Feng [1], Jintuo Yin [1], Shenghao Li [2], Yupeng Sun [1] and Lantong Zhang [1,*]

[1] School of Pharmacy, Hebei Medical University, Shijiazhuang 050017, China
[2] School of Pharmacy, Hebei University of Chinese Medicine, Shijiazhuang 050000, China
* Correspondence: zhanglantong@263.net or zhanglantong@hebmu.edu.cn

Academic Editor: Marcello Locatelli

Abstract: Eupatorin is the major bioactive component of Java tea (*Orthosiphon stamineus*), exhibiting strong anticancer and anti-inflammatory activities. However, no research on the metabolism of eupatorin has been reported to date. In the present study, ultra-high-performance liquid chromatography coupled with hybrid triple quadrupole time-of-flight mass spectrometry (UHPLC-Q-TOF-MS) combined with an efficient online data acquisition and a multiple data processing method were developed for metabolite identification in vivo (rat plasma, bile, urine and feces) and in vitro (rat liver microsomes and intestinal flora). A total of 51 metabolites in vivo, 60 metabolites in vitro were structurally characterized. The loss of CH_2, CH_2O, O, CO, oxidation, methylation, glucuronidation, sulfate conjugation, N-acetylation, hydrogenation, ketone formation, glycine conjugation, glutamine conjugation and glucose conjugation were the main metabolic pathways of eupatorin. This was the first identification of metabolites of eupatorin in vivo and in vitro and it will provide reference and valuable evidence for further development of new pharmaceuticals and pharmacological mechanisms.

Keywords: eupatorin; UHPLC-Q-TOF-MS/MS; metabolism; in vivo *and in vitro*; rat liver microsomes; rat intestinal flora

---

1. Introduction

Eupatorin (5,3′-di-hydroxy-6,7,4′-tri-methoxy-flavone, Figure 1), belonging to the natural methoxyflavone compound, is widely found in Java tea (*Orthosiphon stamineus*, OS) which is a popular medicinal herb used in traditional Chinese medicine as a diuretic agent and for renal system disorders in Southeast Asia and European countries [1–3]. OS has gained a great interest nowadays due to its wide range of pharmacological effects such as antibacterial, antioxidant, hepatoprotection, antidiabetic, anti-hypertension, anti-inflammatory and antiproliferative activities [4–9]. Eupatorin, as a major bioactive flavonoid constituent in OS possesses numerous strong biological activities, including anticancer, anti-inflammatory and vasorelaxation activities [10–17]. Its anticancer activities have attracted more and more attention and it was expected to be developed as a cancer chemopreventive and as an adjuvant chemotherapeutic agent. Although there is literature on the qualitative and quantification profile of eupatorin in OS [6], the metabolism study of eupatorin has not been studied to date, which was necessary for the exploration of the biological activity and the clinical therapeutic effect of eupatorin. Thus, an investigation is essential to explore the identification of metabolites of eupatorin for further understanding of its biological activities.

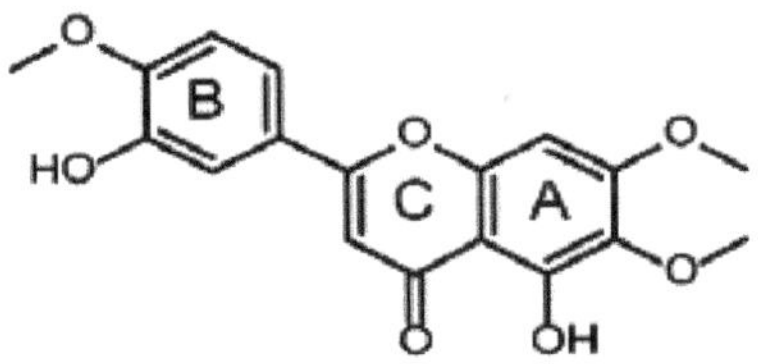

Figure 1. Chemical structure of eupatorin.

To the best of our knowledge, a series of biotransformations will occur when drugs are orally taken into the body, there are four aspects of pharmacological consequences in these biotransformation processes: (1) Transforming into inactive substances; (2) transforming the drug with no pharmacological activity into active metabolites; (3) changing the types of pharmacological actions of drugs; (4) and producing toxic substances [18]. Therefore, it is extremely crucial to study the metabolism of drugs in vivo to make sure of safety of use. In addition, as the main metabolic organ of the human body, the liver is rich in enzymes, especially cytochrome P450 enzymes, which are closely related to the biological transformation of drugs [19]. Furthermore, the gastrointestinal tract is also a vital place for drug metabolism, and its intestinal flora have a significant impact on drug absorption, metabolism and toxicology [20,21]. Hence, in this paper, mass spectrometry was employed to investigate the metabolism of eupatorin in rats, liver microsomes and intestinal flora, in order to characterize the metabolites and structural information of the products, which will lay a foundation for further studies on the safety and efficacy of metabolites and will provide greater possibilities for the development of new drugs.

With the development of technology, a quadrupole time-of-flight mass spectrometry has been widely used as a reliable analytical technique to detect metabolites due to its advantages of high resolution, high sensitivity, high-efficiency separation and accurate quality measurement [22,23]. In this study, high-sensitivity ultra-high-performance liquid chromatography coupled with hybrid triple quadrupole time-of-flight mass spectrometry (UPLC-Q-TOF-MS) full scan mode, electrospray ionization (ESI) source negative ion mode monitoring combined with multiple mass loss (MMDF) and dynamic background subtraction (DBS) were employed to collect data online. Correspondingly, multiple data processing methods were applied by using PeakView 2.0 and MetabolitePilot 2.0.4 software developed by AB SCIEX company, including a variety of data handing functions such as the extraction of ion chromatograms (XIC), mass defect filter (MDF), product ion filter (PIF) and neutral loss filtering (NLF), which provided accurate secondary mass spectral information [24]. Based on the above methods, the metabolic pathways of eupatorin were explored and summarized for the first time and 51 metabolites in vivo and 60 metabolites in vitro were finally identified. These metabolic studies are important parts of drug discovery and development and can also provide a basis for further pharmacological research.

2. Results and Discussion

2.1. Analytical Strategy

In this study, UHPLC-Q-TOF-MS/MS combined with an online data acquisition and multifarious processing methods was adopted to systematically identify the metabolites of eupatorin in vivo and in vitro.

The workflow of the analytic procedure was segmented into three steps. First, an online full-scan data acquisition was performed based on the MMDF and DBS to collect data online and to capture all potential metabolites. Next, a multiple data processing method was employed by using PeakView 2.0 and MetabolitePilot 2.0.4 software, which contained many data-processing tools such as XIC, MDF, PIF and NIF, these provided accurate MS/MS information to determine the metabolites of eupatorin. Finally, plenty of metabolites were identified according to accurate mass datasets, specific secondary mass spectrometry information and so on. With regard to the isomers of metabolites, Clog P values calculated by ChemDraw 14.0 were used to further distinguish them. Generally speaking, the larger the Clog P value, the longer the retention time will be in the reversed-phase chromatography system [25–27].

2.2. Mass Fragmentation Behavior of Eupatorin

In order to identify the metabolites of eupatorin, it is of significance to understand the pyrolysis of parent drug (M0). The chromatographic and mass spectrometric behaviors of eupatorin were explored in the negative ESI scan mode by UHPLC-Q-TOF-MS. Eupatorin ($C_{18}H_{16}O_7$) was eluted at 12.22 min and yielded at 343.0821 $[M-H]^-$. The characteristic fragment ions of M0 at *m/z* 328.0585, 313.0348, 298.0111, 285.0398, 270.0160, 267.0285, 254.0217, 241.0503, 221.0434, 147.0461, 132.0214 were detected according to the MS/MS spectrum. Fragment ions at *m/z* 328.0585, 313.0348, 298.0111, 270.0160 and 254.0217 were generated by M0 through losing CH_3, CH_3, CH_3, CO and O continuously. The ion at *m/z* 343.0821 yielded other representative fragment ions at *m/z* 267.0285, 241.0503 and 221.0434 by loss of CO_2 and 2O, $C_4H_6O_3$, $C_7H_6O_2$, respectively. The product ion at *m/z* 285.0398 was created by dropping CO from the ion at *m/z* 313.0348. Last but not the least, the conspicuous product ion at *m/z* 147.0461 was formed because of the Retro-Diels-Alder (RDA) reaction in ring C of the flavonoid, which gained the ion at *m/z* 132.0214 by loss of CH_3 [28]. The MS/MS spectrum and the fragmentation pathways of eupatorin are shown in Figure 2.

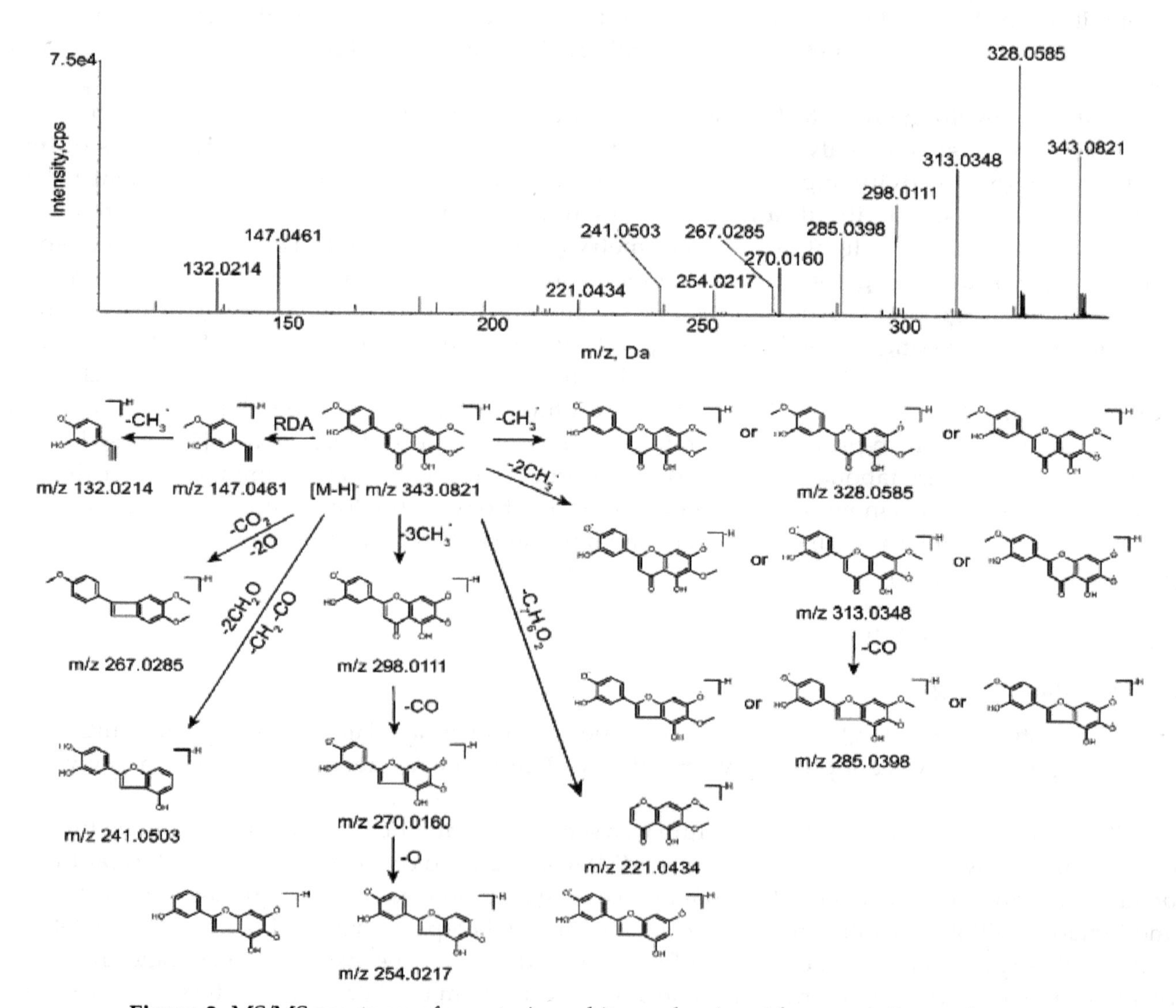

Figure 2. MS/MS spectrum of eupatorin and its predominant fragmentation pathways.

2.3. Identification of Metabolites in Vivo and in Vitro

Metabolites M1, M2 and M3 ($C_{17}H_{14}O_7$) were isomers with the deprotonated molecular ions $[M-H]^-$ at *m/z* 329.0660, 329.0668 and 329.0662, which were 14 Da (CH_2) lower than that of M0. They were eluted at 9.93 min, 10.27 min and 10.79 min, respectively. In the MS/MS spectrum, product ions at *m/z* 314.0427, 313.0384, 299.0188 and 285.0371 were formed after losing CH_3, O, $2CH_3$ and CO_2, respectively. The prominent fragment ion at *m/z* 133.0287 created after the RDA reaction was 14 Da lower than the ion *m/z* 147.0461 of the parent drug, suggested that CH_2 was lost at the methoxy group at 4′position. At the same time, the fragment ions at *m/z* 207.7129 and 207.7166 were 14 Da lower than that of M0, which showed that the loss of CH_2 occurred at the methoxy group at 6 or 7 position of A ring. Additionally, the Clog P values of M1, M2 and M3 were 2.26422, 2.26434, 2.51422, respectively. Therefore, M1–M3 were illustrated according to the above information.

Metabolites M4 and M5 ($C_{16}H_{12}O_7$) were eluted at 7.26 and 8.50 min, with the deprotonated molecular ions $[M-H]^-$ at *m/z* 315.0500 and 315.0504, 28 Da (C_2H_4) lower than that of the parent drug, which indicated that it lost $2CH_2$. Fragment ions at *m/z* 300.0279 and 297.1740 were generated by loss of CH_3 and H_2O, respectively. The product ion at *m/z* 269.1760 was obtained through dropping CO from the ion at *m/z* 297.1740. According to the dominant fragment ion at *m/z* 133.0270 gained by the RDA reaction, loss of CH_2 and CH_2 occurred at the position of 4′, 6 or 4′, 7. In addition, the distinctive ion at *m/z* 147.0821 was similar with that of the parent drug, which implied that the reaction occurred at the position of 6 and 7.

Metabolite M6 ($C_{17}H_{14}O_6$) was obtained with a peak at *m/z* 313.0713 in the UPLC system, which was eluted at 13.86 min, 30 Da (CH_2O) lower than that of eupatorin. Prominent fragment ions at *m/z* 298.0483 and 283.0250 were created by dropping CH_3 and CH_3 successively. In addition, the characteristic fragment ions at *m/z* 117.0364 was produced by RDA reaction, which was 30 Da lower than that of M0, showing that loss of CH_2O occurred at the position of 4′. Similarly, the product ion at *m/z* 147.0078 was consistent with M0, indicating that loss of CH_2O occurred at the position of 6 or 7. Thus, it was speculated that it may have three missing CH_2O sites.

Metabolite M7 ($C_{16}H_{12}O_6$) was detected at 10.10 min and exhibited the molecular ion $[M-H]^-$ at *m/z* 299.0562, which was 44 Da lower than that of M0. Based on the information of chemical elements and software provided, it indicated that M7 lost CH_2O and CH_2. Crucial fragment ions at *m/z* 284.0326 and 251.1281 were obtained by loss of CH_3 and 3O from M7, respectively. Furthermore, M7 had common fragment ion at *m/z* 146.9687 with that of the parent drug, it is equally important that the noteworthy fragment ion at *m/z* 281.1787 was generated by loss of H_2O from M7, which implied that loss of CH_2O and CH_2 occurred at the position of 7 or 6, respectively. Hence, it was identified.

Metabolite M8 ($C_{16}H_{12}O_5$) was eluted at 13.60 min, which displayed deprotonated molecular ion $[M-H]^-$ at *m/z* 283.0614, 60 Da ($C_2H_4O_2$) lower than that of the parent drug. Fragment ions at *m/z* 268.0379 and 240.0428 were produced by dropping CH_3 and CO continuously from *m/z* 283.0614. In addition, the dominant fragment ion at *m/z* 146.9655 was consistent with that of the parent drug, while the diagnostic fragment ion at *m/z* 161.0025 was 60 Da lower than 221.0434 of M0, these suggested that loss of CH_2O and CH_2O reaction happened at C-6 and C-7 of A ring. So, the structure of M8 could be inferred.

Metabolites M9 and M10 ($C_{18}H_{16}O_6$) appeared as deprotonated molecular ions $[M-H]^-$ at *m/z* 327.0882 and 327.0872, together with the retention time of 4.98 min and 7.47 min, respectively, which were 16 Da lower than M0, suggesting they lacked one oxygen atom compared with the parent. The MS/MS spectra showed the fragment ions at *m/z* 309.0800, 299.0957 and 281.2489, which were created by loss of O, CO and C_2H_6O, respectively. In addition, M9 had common fragment ion at *m/z* 146.9380 with that of the parent drug, and meanwhile the characteristic fragment ion at *m/z* 205.0025 was 16 Da lower than 221.0434 of M0, which implied that loss of O occurred at C-5 of A ring. Nevertheless, the ion at *m/z* 130.9716 gained after the RDA cleavage was 16 Da lower than that of the parent drug, showing that loss of O occurred at C-4′ of B ring. Therefore, the structures of metabolites

M9 and M10 were determined. Moreover, they were also validated with the Clog P values of M9 and M10 which were 2.45814 and 3.44497, respectively.

Metabolite M11 ($C_{18}H_{16}O_5$) was turned up in the chromatogram at 9.55 min with the deprotonated molecular ion at *m/z* 311.0930 $[M-H]^-$ and was 32 Da less than that of M0, suggesting that the loss of two oxygen atoms reaction took place. A series of diagnostic product ions at *m/z* 250.9816, 204.9868 and 130.9658 were yielded by loss of $C_2H_4O_2$, C_7H_6O and RDA reaction. In addition, the product ion at *m/z* 174.9556 was obtained through dropping CH_2O from the ion at *m/z* 204.9868. According to the above characteristic fragment ions and analysis, loss of O and O occurred at C-5 and C-3′.

Metabolite M12 ($C_{17}H_{14}O_5$), the deprotonated molecular ion of *m/z* 297.0768 was observed at the retention time of 7.33 min and was 46 Da lower than that of eupatorin. According to its secondary mass spectrum and the information software provided, implying that M12 lost O and CH_2O. Fragment ions at *m/z* 267.1016, 253.0865, 175.0394 and 147.0452 were produced by loss of CH_2O, CO_2, $C_7H_6O_2$ and RDA reaction. It was important that the typical ion at *m/z* 147.0452 was similar with the fragment ion at *m/z* 147.0461 of the parent drug, together with the dominant fragment ion at *m/z* 175.0394, 46 Da lower than that of M0, all of which indicated that the reaction was likely to occur in the A ring. Above all, loss of O happened at the hydroxyl group at the 5 position, while loss of CH_2O occurred at the methoxy group at 6 or 7 position.

Metabolite M13 ($C_{17}H_{16}O_6$) exhibited a sharp peak at an elution time of 12.74 min in the XIC with a deprotonated ion at *m/z* 315.0862 and it was 28 Da (CO) less than eupatorin. Product ions at *m/z* 300.0633, 285.0401 and 270.0144 were formed after dropping CH_3 continuously. In addition, the MS^2 spectrum of M13 presented other vital fragment ions at *m/z* 193.0503 and 147.0445 by losing $C_7H_6O_2$ and undergoing RDA reaction.

Metabolites M14, M15, M16 and M17 ($C_{18}H_{16}O_8$): Four chromatographic peaks were eluted at 10.01 min, 10.50 min, 11.47 min and 12.23 min with deprotonated molecular ions $[M-H]^-$ at *m/z* 359.0772, 359.0768, 359.0767 and 359.0767, which were 16 Da (O) higher than that of eupatorin. Characteristic ions at *m/z* 344.0542, 329.0304, and 314.0064 were obtained by loss of CH_3 successively. Furthermore, noteworthy fragment ions at *m/z* 221.0098 and 163.0368 were produced by loss of $C_7H_6O_3$ and RDA reaction. The ion at *m/z* 163.0368 was 16 Da (O) larger than *m/z* 147.0461, showing that oxidation occurred at C-2′, C-5′ or C-6′ of B ring. However, the prominent ion at *m/z* 147.0130 was similar with the fragment ion at *m/z* 147.0461 of the parent drug, indicating that the reaction happened at the position of 8 in the A ring. The Clog P values of M14-M17 were 1.79518, 1.84518, 1.86518 and 1.87123, respectively. Thus, M14-M17 were characterized by comparing the different values of Clog P.

Metabolite M18 ($C_{18}H_{16}O_9$), the deprotonated molecular ion of *m/z* 375.0709 was observed at the retention time of 9.90 min, which was 32 Da (2O) higher than that of eupatorin. A series of product ions at *m/z* 329.0669, 221.1216 and 178.9947 were detected by loss of CH_2O_2, $C_7H_6O_4$ and RDA reaction in its secondary mass spectrum. Product ions at *m/z* 314.0434 and 299.0191 were produced by losing CH_3 and CH_3 continuously from the ion at *m/z* 329.0669. What's more, the key fragment ions at *m/z* 178.9947 was 32 Da higher than 147.0461 of eupatorin, implying that di-oxidation reaction occurred in the B ring, then M18 was identified.

Metabolite M19 ($C_{18}H_{16}O_{10}$) was detected at 12.26 min and showed the deprotonated molecular ion $[M-H]^-$ at *m/z* 391.0673, 48 Da (3O) higher than that of the parent drug, which contained the fragment ions at *m/z* 345.0869, 330.0636 and 315.0393 by loss of CH_2O_2, CH_3 and CH_3 continuously. More importantly, distinctive fragment ions at *m/z* 221.0399 and 195.0289 were created by loss of $C_7H_6O_5$ and RDA reaction. The pivotal fragment ions at *m/z* 195.0289 was 48 Da higher than 147.0461 of the parent drug, suggesting that tri-oxidation happened at C-2′, C-5′ and C-6′ of B ring. Hence, M19 was recognized.

Metabolites M20 and M21 ($C_{17}H_{14}O_8$) were eluted at 9.43 min and 10.29 min, with the deprotonated molecular ions $[M-H]^-$ at *m/z* 345.0605 and 345.0606 and were increased by 2 Da compared with M0, indicating that it carried out demethylation and oxidation reaction. The representative secondary fragment ions at *m/z* 330.0384, 301.0719, 221.0028, 125.0311 and 149.0234 generated by the loss of CH_3,

CO_2, $C_6H_4O_3$, $C_{11}H_8O_5$ and RDA reaction implied that demethylation and oxidation occurred in ring B. Furthermore, the Clog P values of M20 and M21 were 1.5017 and 1.59734, respectively, so their structures were identified.

Metabolites M22 and M23 ($C_{19}H_{18}O_7$) were obtained in the extracted chromatogram at *m/z* 357.0972 and 357.0969 with the retention time of 10.02 min and 12.86 min, which were 14 Da (CH_2) higher than that of eupatorin. The diagnostic fragment ions at *m/z* 342.0740, 327.0503, 312.0266 and 297.0033 were attributed to the loss of CH_3 successively. In addition, because of the prominent fragment ions at *m/z* 235.0434 and 147.0433 obtained after RDA reaction, it was proposed that methylation happened at hydroxyl group at 5 position. Nevertheless, the fragment ion at *m/z* 161.0269 was 14 Da higher than 147.0461 of eupatorin, indicating that it occurred at C-3′ of B ring. Furthermore, the Clog P values of M22 and M23 were 2.06632 and 3.18323, respectively, so they were verified.

Metabolites M24 and M25 ($C_{19}H_{18}O_6$) were eluted at 7.15 min and 8.79 min, respectively. They had the deprotonated molecular ions $[M-H]^-$ at *m/z* 341.1025 and 341.1027, which were 2 Da lower than that of eupatorin, presumably they occurred a loss of O and a methylation reaction. The distinctive fragment ion at *m/z* 130.9906 was 16 Da (O) lower than 147.0461 of M0, along with fragment ions at *m/z* 235.0607 and 107.0440 produced by loss of C_7H_6O and $C_{12}H_{10}O_5$, implying that the loss of O occurred at the hydroxyl group at C-3′, while methylation happened at the hydroxyl group at C-5. Similarly, according to the representative fragment ion at *m/z* 161.0595, 14 Da (CH_2) lower than 147.0461 of M0 and the diagnostic fragment ions at *m/z* 204.9196 and 137.0553 obtained by loss of $C_8H_8O_2$ and $C_{11}H_8O_4$, the loss of O occurred at the hydroxyl group at 5 position, while methylation took place at the hydroxyl group at 3′ position. In addition, they were also validated with the Clog P values of M24 and M25 which were 2.80306 and 2.9313, respectively.

Metabolite M26 ($C_{15}H_{10}O_5$), displayed a peak at 9.89 min, as well as a deprotonated molecular ion $[M-H]^-$ at *m/z* 269.0459, 14 Da (CH_2) lower than that of M8, suggesting that demethylation occurred on the basis of M8. The fragment ions at *m/z* 253.0124, 241.0500, 225.0555 and 133.0298 were attributed to the loss of O, CO, CO_2 and RDA reaction, which was 14 Da (CH_2) lower than that of M0 and M8, implying that demethylation occurred at the methoxy group at 4′ position. Like the M8, the loss of CH_2O and CH_2O took place at C-6 and C-7 of A ring.

Metabolite M27 ($C_{15}H_{10}O_6$) was detected at a retention time of 8.45 min with the deprotonated molecular ion $[M-H]^-$ at *m/z* 285.0402, 14 Da (CH_2) lower than that of M7, indicating that M27 was demethylated on the basis of M7. Product ions at *m/z* 267.0130, 241.0462, 221.0063, 177.0189 and 133.0307 were produced by loss of H_2O, CO_2, 4O, $C_6H_4O_2$ and RDA reaction which was 14 Da (CH_2) lower than that of M0 and M7, it means demethylation happened at the methoxy group at 4′ position. Similar to M7, loss of CH_2O and CH_2 occurred at C-7 and C-6 of A ring, respectively.

Metabolites M28 and M29 ($C_{24}H_{24}O_{13}$) arose as deprotonated molecular ions $[M-H]^-$ at *m/z* 519.1140 and 519.1151, together with the retention time of 7.10 min and 8.14 min, respectively, which were 176 Da higher than that of eupatorin, suggesting that glucuronidation was carried out. The key product ion at *m/z* 343.0822 was yielded by dropping a glucuronic acid. Moreover, the crucial ion at *m/z* 146.9662 was similar to the fragment ion at *m/z* 147.0461 and while the fragment ion at *m/z* 397.0442 was 176 Da higher than that of the parent drug, indicating that glucuronidation happened at the hydroxyl group at 5 position. Nevertheless, the prominent fragment ions at *m/z* 323.0173 was 176 Da larger than 147.0461 of M0, inferring that the reaction occurred at the hydroxyl group at 3′ position. Furthermore, M28 and M29 were also proved by the different Clog P values of −0.494983 and 0.621934, respectively.

Metabolite M30 ($C_{23}H_{22}O_{13}$) was detected at 7.88 min with the deprotonated molecular ion $[M-H]^-$ at *m/z* 505.0979, 14 Da lower than that of M28 and M29, which suggested that it occurred glucuronide conjugation and demethylation. The characteristic product ion at *m/z* 329.0669 was obtained by losing a glucuronic acid. The distinctive fragment ions at *m/z* 285.0735 and 309.0687 which was 162 Da larger than 147.0461 of M0 were attributed to the loss of $C_{11}H_8O_5$ and RDA reaction, implying glucuronide conjugation and demethylation occurred at the position of 3′ and 4′, respectively.

Metabolites M31 and M32 ($C_{18}H_{16}O_{10}S$) appeared as deprotonated molecular ions $[M-H]^-$ at *m/z* 423.0391 and 423.0387 with the retention time of 8.93 min and 9.20 min. S elemental was found, suggesting that it had been a sulfate bound. The characteristic product ion at *m/z* 343.0830 was created by the loss of SO_3. Remaining ions at *m/z* 328.0593, 313.0355, 285.0413 and 147.0037 were similar to the fragment ions of the parent drug, inferring that sulfate conjugation occurred at the hydroxyl group at 5 position. However, the pivotal fragment ion at *m/z* 227.0084 was 80 Da higher than 147.0461 of eupatorin, implying sulfate conjugation took place at the hydroxyl group at 3′ position of B ring. In addition, the Clog P values of M31 and M32 were 0.270316 and 1.38723, respectively. So, they were also validated.

Metabolite M33 ($C_{17}H_{14}O_{10}S$) was eluted at the retention time of 9.01 min on the UPLC system. Its deprotonated molecular ion $[M-H]^-$ at *m/z* 409.0233 lacked CH_2 compared with M31 and M32. The representative product ion at *m/z* 329.0670 was acquired by dropping SO_3. In addition, M33 created the dominant fragment ion at *m/z* 212.0456 through the RDA reaction, and the product ion at *m/z* 132.0208 was formed by the loss of SO_3 from it. Therefore, it might occur at the methoxy group at 4′ position.

Metabolite M34 ($C_{18}H_{16}O_9S$) was eluted at a retention time of 12.65 min. The MS/MS spectrum of M34 showed the deprotonated molecular ion $[M-H]^-$ at *m/z* 407.0434, lacked one oxygen atom compared with M31 and M32. The crucial fragment ions at *m/z* 327.0826, 301.0034 and 131.0573 were attributed to the loss of SO_3, C_7H_6O and RDA reaction. In addition, the product ion at *m/z* 220.9818 was acquired by the loss of SO_3 from the fragment ions at *m/z* 301.0034. Based on the information above, the loss of O and sulfate conjugation might occur at the hydroxyl group at 3′ and 5 position, respectively.

Metabolites M35 and M36 ($C_{20}H_{18}O_8$): Two isomers were simultaneously extracted in the XIC at 13.36 and 13.90 min and were detected at *m/z* 385.0917 and 385.0925, respectively. The noteworthy ion at *m/z* 343.0846 was yielded by the loss of acetyl. In M36, the diagnostic fragment ion at *m/z* 189.0551 generated by RDA reaction, which was 42 Da higher than 147.0461 of eupatorin and the distinctive fragment ion at *m/z* 221.0781 indicated that acetylation reaction happened at the hydroxyl group at 3′ position. Likewise, according to the prominent fragment ions at *m/z* 263.0551 and 147.0513, the acetylation reaction happened at the hydroxyl group at position 5 of M35. In addition, Clog P values of M35 and M36 were 1.49632 and 2.61323, respectively, which could also support the confirmation of the structures.

Metabolites M37 and M38 ($C_{20}H_{18}O_7$) were observed at 13.02 and 13.90 min in the XIC and were detected at *m/z* 369.0987 and 369.0975 in the mass spectra, respectively, which were decreased by 16 Da (O) compared with M35 and M36. The typical fragment ion at *m/z* 130.9934, 16 Da lower than 147.0461 of eupatorin, together with the representative fragment ion at *m/z* 263.1681, 42 Da higher than that of eupatorin, inferring that loss of O and acetylation reaction occurred at the hydroxyl group at 3′and 5 position, respectively. Similarly, based on the crucial product ions at *m/z* 174.9586 and 164.9289, loss of O and acetylation reaction occurred at the hydroxyl group at 5 and 3′ position, respectively. In addition, M37 and M38 were also verified based on their Clog P values of 2.23306 and 2.3613, respectively.

Metabolite M39 ($C_{19}H_{16}O_8$) detected at *m/z* 371.0761 and eluted at 11.45 min. In addition, it was 14 Da (CH_2) smaller than the size of M35 and M36. The characteristic ion at *m/z* 329.0680 was yielded by dropping of acetyl. The prominent fragment ion at *m/z* 175.0389, 28 Da larger than 147.0461 of eupatorin, implying that the loss of CH_2 and acetylation reaction took place at the methoxy group at 4′ position of B ring.

Metabolites M40 and M41 ($C_{19}H_{16}O_7$) were detected as deprotonated $[M-H]^-$ ion at *m/z* 355.0813 and 355.0814, which were eluted at 11.86 min and 13.15 min, 30 Da (CH_2O) lower than that of M35 and M36. In M40, the diagnostic fragment ion at *m/z* 117.0329 produced by RDA reaction was 30 less than 147.0461 of the parent drug, while the fragment ion at *m/z* 263.0361 was 42 higher than 221.0434 of eupatorin, indicating that loss of CH_2O and acetylation reaction occurred at the position of 4′ and 5, respectively. However, in M41, the distinctive fragment ion at *m/z* 159.10931 was 12 higher than

147.0461 of eupatorin, the crucial fragment ion at *m/z* 221.0027 was similar to fragment ion at *m/z* 221.0434, suggesting that the loss of CH_2O still occurred at the methoxy group at 4′ position while acetylation reaction took place at the hydroxyl group at 3′ position. The respective Clog P values were 1.64857 and 2.87497, so M40 and M41 were ensured.

Metabolite M42 ($C_{21}H_{18}O_8$) with the $[M\text{-}H]^-$ ion of *m/z* 397.0918, which was eluted at 15.21 min, 42 Da higher than that of M40 and M41, speculating that the loss of CH_2O and di-acetylation of amines took place. The characteristic fragment ion at *m/z* 159.0462 created by RDA reaction was 12 larger than 147.0461 of the parent drug, while the product ion at *m/z* 263.0559 increased by 42 Da compared with 221.0434 of eupatorin, inferring that loss of CH_2O occurred at the position of 4′ like M40 and M41, while di-acetylation reaction happened at the hydroxyl group at 5 and 3′ position.

Metabolite M43 ($C_{20}H_{16}O_9$), eluted at 14.14 min, which was detected with the deprotonated molecular ion $[M\text{-}H]^-$ at *m/z* 399.0704, 84 Da higher than that of M4 and M5, implying that di-acetylation reaction happened on the basis of the loss of CH_2 and CH_2. The diagnostic fragment ions at *m/z* 357.0641 and 315.0523 were attributed to the loss of C_2H_2O consecutively. Moreover, according to the product ions at *m/z* 175.0034 and 147.0316 acquired by RDA reaction, it may have three possible metabolites.

Metabolite M44 ($C_{20}H_{16}O_7$) exhibited a sharp peak at an elution time of 15.24 min in the XIC with a deprotonated ion at *m/z* 367.0806 and it was 32 Da (2O) lower than M43, suggesting that the loss of CH_2O and CH_2O and di-acetylation reaction occurred. The noteworthy fragment ion at *m/z* 283.0237 was yielded by dropping of $2C_2H_2O$. M44 generated the fragment ions at *m/z* 189.0633 and 159.0348 after RDA reaction, so the possible structures were inferred according to above MS/MS information.

Metabolites M45, M46, M47 and M48, eluted at 4.57 min, 5.14 min, 5.43 min, 9.69 min, respectively, all exhibited the deprotonated ion at *m/z* 329.1028, 329.1029, 329.1029, 329.1029, implying that they were isomers with the molecular formula $C_{18}H_{18}O_6$ and were 2 Da higher than that of M9 and M10, so hydrogenation happened on the basis of the loss of O. In M45, the representative product ion at *m/z* 130.9874 obtained by RDA reaction was 16 (O) less than 147.0461 of eupatorin, while the fragment ion at *m/z* 223.0900 was twice higher than 221.0434 of eupatorin, indicating that the loss of O and hydrogenation happened at the position of 3′ and 4, respectively. Like M45, according to the crucial product ions at *m/z* 147.0535 and 207.0777, 149.0538 and 207.0618, 133.0731 and 223.0742, the structures of M46, M47 and M48 were distinguished by the analysis above. Furthermore, M45, M46, M47 and M48 were also proved by the different Clog P values of 1.71027, 1.7953, 2.28982 and 2.89116, respectively.

Metabolites M49 and M50 ($C_{18}H_{18}O_5$) were observed in the extracted chromatogram at *m/z* 313.1080 and 313.1086 with the retention time of 7.61 min and 7.88 min, 2 Da higher than that of M11, while lacked one oxygen atom compared with M45, M46, M47 and M48. In M49, the distinctive product ion at *m/z* 130.9677 obtained by the RDA reaction was 16 (O) less than 147.0461 of eupatorin, inferring that hydrogenation happened at the position of 4′. Nevertheless, the characteristic fragment ion at *m/z* 133.0654 yielded in M49 by the RDA reaction was 14 less than 147.0461, so hydrogenation happened at the position of 2 and 3. Besides, M49 and M50 were also checked by the different Clog P values of 2.5321 and 3.02662, respectively.

Metabolite M51 ($C_{17}H_{16}O_7$) was obtained with a peak at *m/z* 331.0824 in the UPLC system, which was eluted at 10.05 min, 2 Da larger than that of M1, M2 and M3. According to MS/MS spectrum, diagnostic product ions at *m/z* 316.0595, 313.1396, 223.1657, 109.0288 and 135.0450 were formed by losing CH_3, H_2O, $C_6H_4O_2$, $C_{11}H_{10}O_5$ and RDA reaction. It's worth mentioning that the fragment ion at *m/z* 135.0450 was 12 less than 147.0461 of eupatorin, so demethylation happened at the methoxy group at 4′ position and hydrogenation occurred at the position of 2 and 3.

Metabolite M52 ($C_{18}H_{20}O_7$) was eluted at the retention time of 12.78 min. Its deprotonated molecular ion $[M\text{-}H]^-$ at *m/z* 347.1140 was increased 4 Da compared with eupatorin, so di-hydrogenation occurred. According to the dominant product ion at *m/z* 149.0591 obtained by RDA reaction, 2 Da higher than 147.0461 of eupatorin, and the fragment ion at *m/z* 225.0074, 4 Da higher than 221.0434, indicating that di-hydrogenation happened at the position of 2, 3 and 4.

Metabolites M53 and M54 ($C_{18}H_{20}O_6$) were observed in the chromatogram at *m/z* 331.1186 and 331.1185 with the retention time of 3.60 min and 4.07 min, respectively, 16 Da (O) lower than that of M52, inferring that the loss of O happened on the basis of di-hydrogenation. The crucial ion at *m/z* 133.0325 was generated after the RDA cleavage, which was less than *m/z* 147.0461 one oxygen, so the loss of O occurred at the hydroxyl group at 3′ position in M53. However, according to the characteristic product ions at *m/z* 149.0680 and 209.0813, the loss of O happened at the hydroxyl group at 5 position while the di-hydrogenation was at the same position as M52. The respective Clog P values were 1.55427 and 1.6393, so M53 and M54 were verified.

Metabolite M55 ($C_{18}H_{20}O_5$) was eluted at 6.36 min possessing the deprotonated molecular ion $[M-H]^-$ at *m/z* 315.1214, which was 16 Da (O) lower than that of M53, M54 and 4 Da higher than that of M11. Based on the previous analysis of M11, M53 and M54, the structure of M55 can be inferred.

Metabolites M56, M57 and M58 ($C_{17}H_{18}O_7$): Three chromatographic peaks were eluted at 10.18 min, 10.24 min and 10.80 min with deprotonated molecular ions $[M-H]^-$ at *m/z* 333.0972, 333.0982 and 333.0979, which were 4 Da larger than the size of M1-M3 and 14 Da (CH_2) higher than that of M52. According to the prominent product ions at *m/z* 149.0642 and 135.1164, together with the information of M1–M3 and M52, the structures of M56-M58 could be identified. In addition, M56, M57 and M58 were also ensured by the different Clog P values of 0.324751, 0.371274 and 0.644751, respectively.

Metabolite M59 ($C_{16}H_{16}O_7$) was observed with a peak at *m/z* 319.0815 in the chromatogram, which was eluted at 8.47 min, 4 Da larger than that of M3 and M4. According to the MS/MS information, the typical fragment ions at *m/z* 301.0701, 211.0353, 197.0452, 149.0269 and 135.0443 were created by loss of H_2O, $C_6H_4O_2$, $C_7H_6O_2$ and RDA reaction, so there were three possible metabolites of M59.

Metabolites M60 and M61 ($C_{16}H_{16}O_5$) appeared as deprotonated molecular ions $[M-H]^-$ at *m/z* 287.0923 and 287.0927, together with the retention time of 9.97 min and 11.07 min, respectively, which were 4 Da higher than M8, indicating that M60 and M61 might undergo the loss of CH_2O and CH_2O reaction followed by di-hydrogenation. In the secondary mass spectrum of M61, it obtained the fragment ions at *m/z* 272.0695, 241.2138, 165.0166, 123.0117 and 149.0683 yielded by dropping of CH_3, CH_2O_2, $C_7H_6O_2$, $C_9H_8O_3$ and RDA cleavage, so the loss of CH_2O and CH_2O reaction happened at the positions of 6 and 7 while di-hydrogenation happened at the positions of 2, 3 and 4. However, the characteristic fragment ions at *m/z* 195.0653, 93.0325 and 119.0500 produced by the loss of C_6H_4O, $C_{10}H_{10}O_4$ and RDA cleavage. So, there were two positions (4′, 7 or 4′, 6) to have lost CH_2O. Finally, the sizes of different Clog P values were combined to determine the structure of M60.

Metabolite M62 ($C_{17}H_{18}O_5$) was eluted at 7.37 min, which displayed deprotonated molecular ion $[M-H]^-$ at *m/z* 301.1078, 4 Da larger the size of M12. In M62, the characteristic fragment ion at *m/z* 149.0605 was twice higher than 147.0461 of eupatorin, while, the prominent fragment ion at *m/z* 179.0711 was 42 times lower than fragment ion at *m/z* 221.0434, so the loss of O, CH_2O and di-hydrogenation occurred at the same positions as M12 and M52, respectively.

Metabolite M63 ($C_{15}H_{10}O_4$), the deprotonated molecular ion of *m/z* 253.0512 was observed at the retention time of 7.81 min, which was 30 Da (CH_2O) lower than that of M8. M63 comprised the typical fragment ions at *m/z* 225.0558, 209.0606, 161.0249 and 117.0351 by dropping of CO, CO_2, C_6H_4O and RDA cleavage. And the fragment ion at *m/z* 101.0246 arose by loss of O from the ion at *m/z* 117.0351, so the structure was inferred.

181.0136 tested after RDA reaction was 34 larger than 147.0461 of eupatorin, while the fragment ion at *m/z* 239.0435 was 18 times higher than that of eupatorin, so internal hydrolysis happened at C-2 and C-3 and oxidation is most likely to occur at C-5′ [29].

Metabolite M68 ($C_{19}H_{17}NO_8$) was observed with a peak at *m/z* 386.0865 in the UPLC system, which was eluted at 10.00 min, 57 Da higher than that of M1-M3. The fragment ion at *m/z* 329.0662 was acquired, corresponding to the loss of glycine. Additionally, the conspicuous fragment ions at *m/z* 264.1192 and 147.0973 were yielded through the loss of $C_7H_6O_2$ and RDA reaction, implying that the loss of CH_2 and glycine conjugation were connected to A ring. Hence, there were two possible metabolites of M68.

Metabolite M69 ($C_{19}H_{17}NO_6$) was detected at 6.09 min, which presented an accurate deprotonated ion $[M-H]^-$ at *m/z* 354.0992, 57 Da higher than that of M12, indicating that M69 might experience the loss of O and CH_2O reaction followed by glycine conjugation. A sequence of crucial fragment ions at *m/z* 324.2008, 250.9077, 174.9553 and 204.0387 were produced by the loss of $2CH_3$, $C_3H_5NO_3$, $C_9H_9NO_3$ and RDA reaction, while the characteristic fragment ion at *m/z* 204.0387 was 57 higher than 147.0461 of eupatorin, inferring that glycine conjugation was connected to B ring and the loss of O and CH_2O reaction was at the same position as M12.

Metabolite M70 ($C_{22}H_{22}N_2O_8$) exhibited a sharp peak at an elution time of 5.07 min in the XIC with a deprotonated ion at *m/z* 441.1302. The characteristic fragment ion at *m/z* 312.8496 was observed, corresponding to the loss of glutamine [24]. Furthermore, A strong ion at *m/z* 245.1021 appeared in the secondary mass spectrum of M70 after the RDA reaction, and was 98 higher than 147.0461 of eupatorin, which created the prominent fragment ion at *m/z* 117.2304 by dropping of glutamine, suggesting that the loss of CH_2O occurred at the methoxy group at 4′ position while glutamine conjugation took place at the hydroxyl group at 3′ position.

Metabolite M71 ($C_{22}H_{22}O_{12}$), displayed a peak at 5.49 min, as well as a deprotonated molecular ion $[M-H]^-$ at *m/z* 477.1036. The predominated fragment ion at *m/z* 315.6277 was attributed to the loss of glucose. In addition, the distinctive fragment ion at *m/z* 146.9654 resulted from RDA reaction was consistent with *m/z* 147.0461 of eupatorin, further noteworthy MS/MS fragment ions at *m/z* 355.0661 and 192.9548 were yielded corresponding to the consecutive loss of $C_7H_6O_2$ and glucose. Thus, the loss of CH_2 and CH_2 took place at the methoxy group at 6 and 7 position, while glucose conjugation happened at the hydroxyl group at 5 position.

The detected metabolites are listed in Table 1. Moreover, their XICs are exhibited in Figure 3.

Metabolites M64, M65 and M66 ($C_{18}H_{14}O_8$) were the isomeric metabolites with the deprotonated $[M-H]^-$ ions at *m/z* 357.0607, 357.0609 and 357.0610, 14 Da higher than that of eupatorin, which were eluted at 10.79 min, 10.81 min and 12.99 min, respectively, suggesting that ketone formation reaction occurred. Several conspicuous ions at *m/z* 342.0374, 327.0132, 313.0304, 221.0224, 235.0267, 161.0174 and 147.0012 all appeared in the secondary mass spectra after the loss of CH_3, CH_3, CO_2, $C_7H_4O_3$, $C_7H_6O_2$ and RDA reaction. Moreover, the Clog P values of M64, M65 and M66 were 2.17223, 2.27223 and 2.52223, respectively. In consequence, the structures of M64, M65 and M66 were distinguished according to the above information.

Metabolite M67 ($C_{18}H_{18}O_9$) was eluted at 12.29 min and showed the deprotonated molecular ion $[M-H]^-$ at *m/z* 377.0875, 18 Da higher than that of M14-M17, implying that M67 might undergo oxidation followed by internal hydrolysis. In the MS/MS spectrum of M67, the representative product ion at *m/z*

Table 1. Summary of metabolites of eupatorin in vivo and in vitro.

Metabolites ID	Composition	Formula	*m/z*	Error (ppm)	R.T. (min)	MS/MS Fragments	Clog P	Score (%)	Plasma	Bile	Urine	Feces	RLMs	RIF
M1	Loss of CH_2 $[M-H]^-$	$C_{17}H_{14}O_7$	329.0660	−2.0	9.93	314.0427, 313.0384, 299.0188, 285.0371, 207.7129	2.26422	83.3	-	-	+[a,b]	+[a,b]	+[I,II]	+[a,b]
M2	Loss of CH_2 $[M-H]^-$	$C_{17}H_{14}O_7$	329.0668	0.5	10.27	314.0423, 313.0344, 299.0189, 285.0393, 133.0287	2.26434	91.8	-	-	+[a,b]	+[a,b]	+[I,II]	+[a,b]
M3	Loss of CH_2 $[M-H]^-$	$C_{17}H_{14}O_7$	329.0662	−1.4	10.79	314.0436,313.0357, 299.0204, 285.0396, 207.7166	2.51422	96.3	-	-	+[a,b]	+[a,b]	+[I,II]	+[a,b]
M4a M4b	Loss of CH_2 and CH_2 $[M-H]^-$	$C_{16}H_{12}O_7$	315.0500	−3.3	7.26	300.0279, 297.1740, 269.1760, 251.1269, 133.0270	1.82034 2.07034	85.4	-	-	+[a,b]	+[a,b]	+[I]	-
M5	Loss of CH_2 and CH_2 $[M-H]^-$	$C_{16}H_{12}O_7$	315.0504	−2.0	8.50	300.0275, 297.0411, 269.1755, 251.1658, 147.0821	2.18513	93.3	-	-	+[a,b]	+[a,b]	+[I]	-
M6a M6b M6c	Loss of CH_2O $[M-H]^-$	$C_{17}H_{14}O_6$	313.0713	−1.4	13.86	298.0483, 283.0250, 221.0632, 147.0078, 117.0364	2.86048 3.08571 3.33571	90.9	-	-	-	+[a,b]	-	+[a,b]
M7	Loss of CH_2O and CH_2 $[M-H]^-$	$C_{16}H_{12}O_6$	299.0562	0.3	10.10	284.0326, 281.1787, 251.1281, 146.9687	2.74964	83.8	-	-	+[a]	-	+[I]	+[b]
M8	Loss of CH_2O and CH_2O $[M-H]^-$	$C_{16}H_{12}O_5$	283.0614	0.8	13.60	268.0379, 240.0428, 267.0306, 161.0025, 146.9655	3.30833	93.5	-	-	+[a,b]	+[b]	+[I]	+[b]
M9	Loss of O $[M-H]^-$	$C_{18}H_{16}O_6$	327.0882	2.4	4.98	308.9931, 299.1274, 281.2489, 205.0025, 146.9380	2.45814	75.3	-	-	-	-	-	+[b]
M10	Loss of O $[M-H]^-$	$C_{18}H_{16}O_6$	327.0872	−0.8	7.47	309.0800, 299.0957, 281.2493, 221.0452, 130.9716	3.44497	83.5	-	-	-	-	-	+[b]
M11	Loss of O and O $[M-H]^-$	$C_{18}H_{16}O_5$	311.0930	1.5	9.55	250.9816, 204.9868, 174.9556, 130.9658	3.19475	75.7	-	-	-	-	-	+[b]
M12a M12b	Loss of O and CH_2O $[M-H]^-$	$C_{17}H_{14}O_5$	297.0768	−0.2	7.33	267.1016, 253.0865, 175.0394, 147.0452, 145.0305	2.78433	82.1	-	-	-	-	-	+[b]
M13	Loss of CO $[M-H]^-$	$C_{17}H_{16}O_6$	315.0862	−3.8	12.74	300.0633, 285.0401, 270.0144, 193.0503, 147.0445	2.84747	87.9	-	-	-	+[a,b]	-	+[a,b]
M14	Oxidation $[M-H]^-$	$C_{18}H_{16}O_8$	359.0772	−0.2	10.01	344.0542, 329.0304, 314.0064, 220.9817, 163.0384	1.79518	82.9	-	-	+[a,b]	+[a,b]	+[I]	+[a,b]
M15	Oxidation $[M-H]^-$	$C_{18}H_{16}O_8$	359.0768	−1.3	10.50	344.0529, 329.0306, 314.0061, 221.0098, 163.0368	1.84518	82.8	-	-	+[a,b]	+[a,b]	+[I]	+[a,b]
M16	Oxidation $[M-H]^-$	$C_{18}H_{16}O_8$	359.0767	−1.6	11.47	344.0536, 329.0296, 314.0066, 221.0762, 163.0019	1.86518	81.9	-	-	+[a,b]	+[a,b]	+[I]	+[a,b]
M17	Oxidation $[M-H]^-$	$C_{18}H_{16}O_8$	359.0767	−1.4	12.23	344.0542, 329.0315, 314.0085, 237.0375, 147.0130	1.87123	85.5	-	-	+[a,b]	+[a,b]	+[I]	+[a,b]
M18	Di-Oxidation $[M-H]^-$	$C_{18}H_{16}O_9$	375.0709	−3.3	9.90	329.0669, 314.0434, 299.0191, 221.1216, 178.9947	0.9644	91.2	-	-	-	+[a,b]	-	+[a]
M19	Tri-Oxidation $[M-H]^-$	$C_{18}H_{16}O_{10}$	391.0673	0.5	12.26	345.0869, 330.0636, 315.0393, 221.0399, 195.0289	0.25226	77.1	-	-	+[b]	-	-	+[a,b]
M20a M20b	Demethylation and Oxidation $[M-H]^-$	$C_{17}H_{14}O_8$	345.0605	−3.0	9.43	330.0379, 301.1825, 221.1270, 149.0245, 125.0237	1.29734	84.9	-	-	+[a,b]	+[b]	+[I]	+[b]
M21	Demethylation and Oxidation $[M-H]^-$	$C_{17}H_{14}O_8$	345.0606	−2.8	10.29	330.0384, 301.0719, 221.0028, 149.0234, 125.0311	1.59734	87.1	-	-	+[a,b]	+[b]	+[I]	+[b]

Table 1. *Cont.*

Metabolites ID	Composition	Formula	m/z	Error (ppm)	R.T. (min)	MS/MS Fragments	Clog P	Score (%)	Plasma	Bile	Urine	Feces	RLMs	RIF
M22	Methylation $[M-H]^-$	$C_{19}H_{18}O_7$	357.0972	−2.1	10.02	342.0740, 327.0503, 312.0266, 235.0434, 147.0433	2.06632	80.6	-	-	$+^{a,b}$	$+^{a,b}$	-	$+^{a,b}$
M23	Methylation $[M-H]^-$	$C_{19}H_{18}O_7$	357.0969	−3.1	12.86	342.0737, 327.0508, 312.0266, 221.0766, 161.0269	3.18323	78.6	-	-	$+^{a,b}$	$+^{a,b}$	-	$+^{a,b}$
M24	Loss of O+Methylation $[M-H]^-$	$C_{19}H_{18}O_6$	341.1025	−1.5	7.15	326.1073, 311.0918, 235.0607, 130.9906, 107.0440	2.80306	82.8	-	-	-	-	-	$+^{b}$
M25	Loss of O+Methylation $[M-H]^-$	$C_{19}H_{18}O_6$	341.1027	−1.2	8.79	326.0798, 311.0451, 204.9196, 161.0595, 137.0553	2.9313	82.1	-	-	-	-	-	$+^{b}$
M26	Loss of CH_2O and CH_2O+Demethylation $[M-H]^-$	$C_{15}H_{10}O_5$	269.0459	1.3	9.89	253.0124, 241.0500, 225.0555, 133.0298, 117.0349	2.88784	95.6	$+^{a,b}$	-	$+^{a,b}$	-	-	$+^{a,b}$
M27	Loss of CH_2O and CH_2+Demethylation $[M-H]^-$	$C_{15}H_{10}O_6$	285.0402	−0.9	8.45	267.0130, 241.0462, 221.0063, 177.0189, 133.0307	2.31115	90.2	-	-	-	-	-	$+^{b}$
M28	Glucuronidation $[M-H]^-$	$C_{24}H_{24}O_{13}$	519.1140	−0.8	7.10	397.0442, 343.0822, 328.0587, 313.0346, 146.9662	−0.495	81.6	$+^{a}$	$+^{b}$	$+^{a,b}$	-	$+^{II}$	-
M29	Glucuronidation $[M-H]^-$	$C_{24}H_{24}O_{13}$	519.1151	1.3	8.14	343.0824, 328.0588, 323.0173, 313.0354, 221.0262	0.62193	83.1	$+^{a}$	$+^{b}$	$+^{a,b}$	-	$+^{II}$	-
M30	Demethylation and Glucuronide Conjugation $[M-H]^-$	$C_{23}H_{22}O_{13}$	505.0979	−1.8	7.88	329.0669, 309.0687, 299.0165, 285.0735	0.14693	81.3	-	-	$+^{a,b}$	-	-	-
M31	Sulfate Conjugation $[M-H]^-$	$C_{18}H_{16}O_{10}S$	423.0391	0.1	8.93	343.0830, 328.0593, 313.0355, 285.0413, 147.0037	0.27032	86.1	-	$+^{a,b}$	$+^{a,b}$	-	-	$+^{a,b}$
M32	Sulfate Conjugation $[M-H]^-$	$C_{18}H_{16}O_{10}S$	423.0387	−0.9	9.20	343.0836, 328.0606, 313.0371, 285.0457, 227.0084	1.38723	89.6	-	$+^{a,b}$	$+^{a,b}$	-	-	$+^{a,b}$
M33	Loss of CH_2+Sulfate Conjugation $[M-H]^-$	$C_{17}H_{14}O_{10}S$	409.0233	−0.4	9.01	329.0670, 314.0432, 299.0198, 212.0456, 132.0208	0.81223	91.5	$+^{a}$	-	$+^{a,b}$	-	-	-
M34	Loss of O+Sulfate Conjugation $[M-H]^-$	$C_{18}H_{16}O_9S$	407.0434	−2.1	12.65	327.0826, 301.0034, 220.9818, 131.0573	1.00706	63.3	-	-	-	-	-	$+^{b}$
M35	N-Acetylation $[M-H]^-$	$C_{20}H_{18}O_8$	385.0917	−3.1	13.36	370.0729, 355.0427, 343.0846, 263.0551, 147.0513	1.49632	74.4	-	-	-	-	-	$+^{b}$
M36	N-Acetylation $[M-H]^-$	$C_{20}H_{18}O_8$	385.0925	−0.9	13.90	370.0735, 355.0492, 343.0874, 221.0781, 189.0551	2.61323	77.9	-	-	-	-	-	$+^{b}$
M37	Loss of O+N-Acetylation $[M-H]^-$	$C_{20}H_{18}O_7$	369.0987	2.1	13.02	327.2227, 279.0680, 263.1681, 237.1098, 130.9934	2.23306	75.3	$+^{a}$	-	$+^{a,b}$	$+^{a,b}$	$+^{I}$	$+^{a,b}$
M38	Loss of O+N-Acetylation $[M-H]^-$	$C_{20}H_{18}O_7$	369.0975	−1.3	13.90	354.0755, 339.0542, 311.0594, 174.9586, 164.9289	2.3613	76.7	$+^{a}$	-	$+^{a,b}$	$+^{a,b}$	$+^{I}$	$+^{a,b}$
M39	Loss of CH_2+N-Acetylation $[M-H]^-$	$C_{19}H_{16}O_8$	371.0761	−3.2	11.45	329.0680, 314.0439, 299.0196, 220.9869, 175.0389	2.13823	76.8	-	-	-	$+^{a,b}$	-	$+^{b}$
M40	Loss of CH_2O+N-Acetylation $[M-H]^-$	$C_{19}H_{16}O_7$	355.0813	−2.9	11.86	340.0587, 325.0335, 313.1107, 263.0361, 117.0329	1.64857	78.2	-	-	$+^{a,b}$	$+^{a,b}$	$+^{I}$	$+^{a,b}$

Table 1. *Cont.*

Metabolites ID	Composition	Formula	*m/z*	Error (ppm)	R.T. (min)	MS/MS Fragments	Clog P	Score (%)	Plasma	Bile	Urine	Feces	RLMs	RIF
M41	Loss of CH_2O+N-Acetylation $[M-H]^-$	$C_{19}H_{16}O_7$	355.0814	−2.6	13.15	340.0589, 325.0356, 313.0337, 221.0027, 159.1093	2.87497	78.1	-	-	+ [a,b]	+ [a,b]	+ [I]	+ [a,b]
M42	Loss of CH_2O+Di-Acetylation of Amines $[M-H]^-$	$C_{21}H_{18}O_8$	397.0918	−2.6	15.21	382.0691, 367.0426, 313.2553, 263.0559, 159.0462	1.66306	74.8	-	-	-	+ [b]	-	+ [b]
M43a M43b M43c	Loss of CH_2 and CH_2+Di-Acetylation of Amines $[M-H]^-$	$C_{20}H_{16}O_9$	399.0704	−4.3	14.14	384.0497, 357.0641, 315.0523, 175.0034, 147.0316	1.56823	71.8	-	-	-	+ [b]	-	-
M44a M44b M44c	Loss of CH_2O and CH_2O+Di-Acetylation of Amines $[M-H]^-$	$C_{20}H_{16}O_7$	367.0806	−4.8	15.24	283.0237, 233.1253, 202.9904, 189.0633, 159.0348	2.05475 2.11133 2.40475	71.9	-	-	-	+ [a,b]	-	-
M45	Loss of O+Hydrogenation $[M-H]^-$	$C_{18}H_{18}O_6$	329.1028	−0.8	4.57	314.0861, 299.1136, 283.2624, 223.0900, 130.9874	1.71027	75.9	-	-	-	-	-	+ [a,b]
M46	Loss of O+Hydrogenation $[M-H]^-$	$C_{18}H_{18}O_6$	329.1029	−0.5	5.14	314.0905, 299.1189, 283.1288, 207.0777, 147.0535	1.7953	76.7	-	-	-	-	-	+ [a,b]
M47	Loss of O+Hydrogenation $[M-H]^-$	$C_{18}H_{18}O_6$	329.1029	−0.3	5.43	314.0384, 299.0986, 283.2612, 207.0618, 149.0538	2.28982	78.7	-	-	-	-	-	+ [a,b]
M48	Loss of O+Hydrogenation $[M-H]^-$	$C_{18}H_{18}O_6$	329.1029	−0.6	9.69	314.0226, 299.0298, 283.0982, 223.0742, 133.0731	2.89116	75.8	-	-	-	-	-	+ [a,b]
M49	Loss of O and O+Hydrogenation $[M-H]^-$	$C_{18}H_{18}O_5$	313.1080	−0.3	7.61	298.0759, 269.1191, 239.1066, 206.9936, 130.9677	2.5321	88.5	-	-	-	-	-	+ [a,b]
M50	Loss of O and O+Hydrogenation $[M-H]^-$	$C_{18}H_{18}O_5$	313.1086	1.6	7.88	298.0767, 269.1189, 239.1061, 207.0818, 133.0654	3.02662	90.7	-	-	-	-	-	+ [a,b]
M51	Demethylation and Hydrogenation $[M-H]^-$	$C_{17}H_{16}O_7$	331.0824	0.3	10.05	316.0595, 313.1396, 301.0343, 223.1657, 135.0450	1.70816	90.5	-	-	+ [a,b]	-	-	+ [a]
M52	Di-Hydrogenation $[M-H]^-$	$C_{18}H_{20}O_7$	347.1140	1.0	12.78	332.0913, 317.0676, 225.0074, 149.0591, 123.0079	0.81747	60.2	-	-	+ [a,b]	-	-	+ [b]
M53	Loss of O+Di-Hydrogenation $[M-H]^-$	$C_{18}H_{20}O_6$	331.1186	−0.2	3.60	301.1097, 299.1257, 285.1728, 225.0501, 133.0325	1.55427	53.1	-	-	+ [b]	-	-	+ [b]
M54	Loss of O+Di-Hydrogenation $[M-H]^-$	$C_{18}H_{20}O_6$	331.1185	−0.7	4.07	301.1088, 299.1304, 285.0945, 209.0813, 149.0680	1.6393	52.8	-	-	+ [b]	-	-	+ [b]
M55	Loss of O and O+Di-Hydrogenation $[M-H]^-$	$C_{18}H_{20}O_5$	315.1214	−1.6	6.36	285.1155, 271.1557, 269.1318, 241.1035, 133.0693	2.3761	83.6	-	-	+ [a]	-	-	+ [a,b]

Table 1. *Cont.*

Metabolites ID	Composition	Formula	m/z	Error (ppm)	R.T. (min)	MS/MS Fragments	Clog P	Score (%)	Plasma	Bile	Urine	Feces	RLMs	RIF
M56	Loss of CH_2+Di-Hydrogenation [M-H]$^-$	$C_{17}H_{18}O_7$	333.0972	−2.3	10.18	318.0813, 317.1125, 303.0579, 211.0624, 149.0642	0.32475	94.4	-	-	+^{a,b}	-	-	-
M57	Loss of CH_2+Di-Hydrogenation [M-H]$^-$	$C_{17}H_{18}O_7$	333.0982	0.6	10.24	315.0034, 225.2199, 179.1064, 135.1164, 109.0679	0.37127	63.2	-	-	+^{a,b}	-	-	-
M58	Loss of CH_2+Di-Hydrogenation [M-H]$^-$	$C_{17}H_{18}O_7$	333.0979	−0.3	10.80	315.0884, 300.0638, 285.0374, 211.0614, 149.0693	0.64475	90.6	-	-	+^{a,b}	-	-	-
M59a M59b M59c	Loss of CH_2 and CH_2+Di-Hydrogenation [M-H]$^-$	$C_{16}H_{16}O_7$	319.0815	−2.6	8.47	301.0701, 211.0353, 197.0452, 149.0269, 135.0443	0.19855 −0.1214 0.22768	89.0	-	-	-	+^{a,b}	-	-
M60	Loss of CH_2O and CH_2O+Di-Hydrogenation [M-H]$^-$	$C_{16}H_{16}O_5$	287.0923	−0.8	9.97	272.0689, 241.0875, 195.0653, 119.0500, 93.0325	1.35527	80.7	-	-	-	-	-	+^{b}
M61	Loss of CH_2O and CH_2O+Di-Hydrogenation [M-H]$^-$	$C_{16}H_{16}O_5$	287.0927	0.5	11.07	272.0695, 241.2138, 165.0166, 149.0683, 123.0117	1.4214	85.8	-	-	-	-	-	+^{b}
M62a M62b	Loss of O and CH_2O+Di-Hydrogenation [M-H]$^-$	$C_{17}H_{18}O_5$	301.1078	−1.2	7.37	271.1359, 255.0541, 241.0814, 179.0711, 149.0605	1.9972	87.9	+^{a}	-	-	+^{a,b}	-	-
M63	Loss of CH_2O and CH_2O+Loss of Hydroxymethylene [M-H]$^-$	$C_{15}H_{10}O_4$	253.0512	2.1	7.81	225.0558, 209.0606, 161.0249, 117.0351, 101.0246	3.4753	70.9	+^{b}	-	+^{a,b}	-	-	+^{b}
M64	Ketone Formation [M-H]$^-$	$C_{18}H_{14}O_8$	357.0607	−2.5	10.79	342.0374, 327.0132, 313.0304, 221.0224, 161.0174	2.17223	79.1	-	-	-	+^{b}	+^{I}	-
M65	Ketone Formation [M-H]$^-$	$C_{18}H_{14}O_8$	357.0609	−1.8	10.81	342.0379, 327.0146, 313.0349, 235.0241, 147.0012	2.27223	76.1	-	-	-	+^{b}	+^{I}	-
M66	Ketone Formation [M-H]$^-$	$C_{18}H_{14}O_8$	357.0610	−1.6	12.99	342.0385, 327.0198, 313.0421, 235.0267, 147.0503	2.52223	77.4	-	-	-	+^{b}	+^{I}	-
M67	Oxidation and Internal Hydrolysis [M-H]$^-$	$C_{18}H_{18}O_9$	377.0875	−0.8	12.29	362.0508, 349.0873, 239.0435, 181.0136, 139.0054	0.21452	83.4	-	-	-	+^{a,b}	-	+^{a,b}
M68a M68b	Loss of CH_2+Glycine Conjugation [M-H]$^-$	$C_{19}H_{17}NO_8$	386.0865	−4.2	10.00	329.0662, 314.0421, 299.0189, 264.1192, 147.0973	2.15391 2.40391	80.8	-	-	-	+^{b}	-	-
M69a M69b	Loss of O and CH_2O+Glycine Conjugation [M-H]$^-$	$C_{19}H_{17}NO_6$	354.0992	2.5	6.09	324.2008, 250.9077, 221.0751, 204.0387, 174.9553	2.66894	73.9	-	+^{a}	-	-	-	-
M70	Loss of CH_2O+Glutamine Conjugation [M-H]$^-$	$C_{22}H_{22}N_2O_8$	441.1302	−0.2	5.07	312.8496, 245.1021, 221.5883, 117.2304	0.99254	50.9	-	-	-	-	-	+^{b}
M71	Loss of CH_2 and CH_2+Glucose Conjugation [M-H]$^-$	$C_{22}H_{22}O_{12}$	477.1036	−0.6	5.49	355.0661, 315.6277, 192.9548, 146.9654, 123.0795	−0.3982	52.9	-	-	-	-	-	+^{b}

+, Detected; -, Undetected. RLMs, rat liver microsomes; RIF, rat intestinal flora. (a) Metabolites obtained by methanol precipitation protein; (b) metabolites obtained by ethyl acetate extraction; a+b metabolites obtained by methanol precipitation protein and ethyl acetate extraction. (I) Phase I metabolites obtained from liver microsomes; (II) phase II metabolites obtained from liver microsomes.

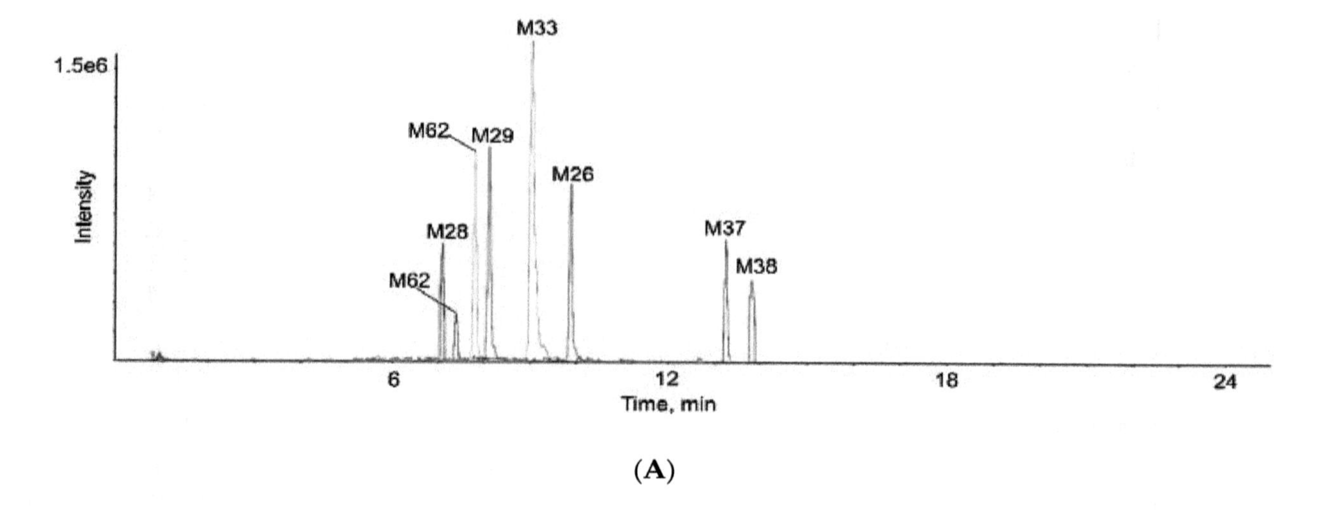

(A)

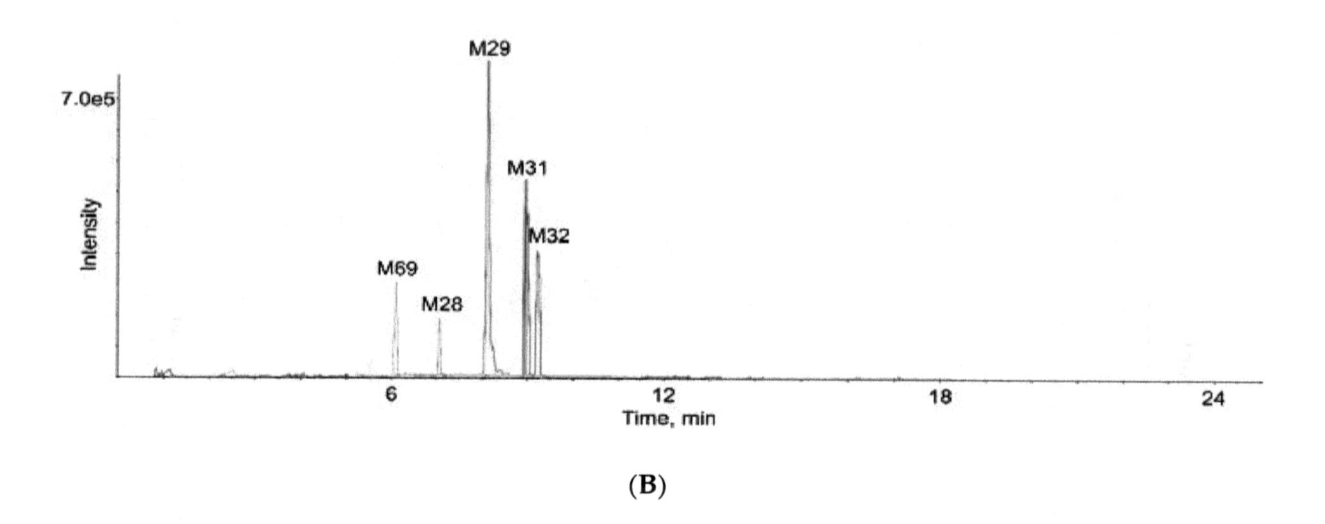

(B)

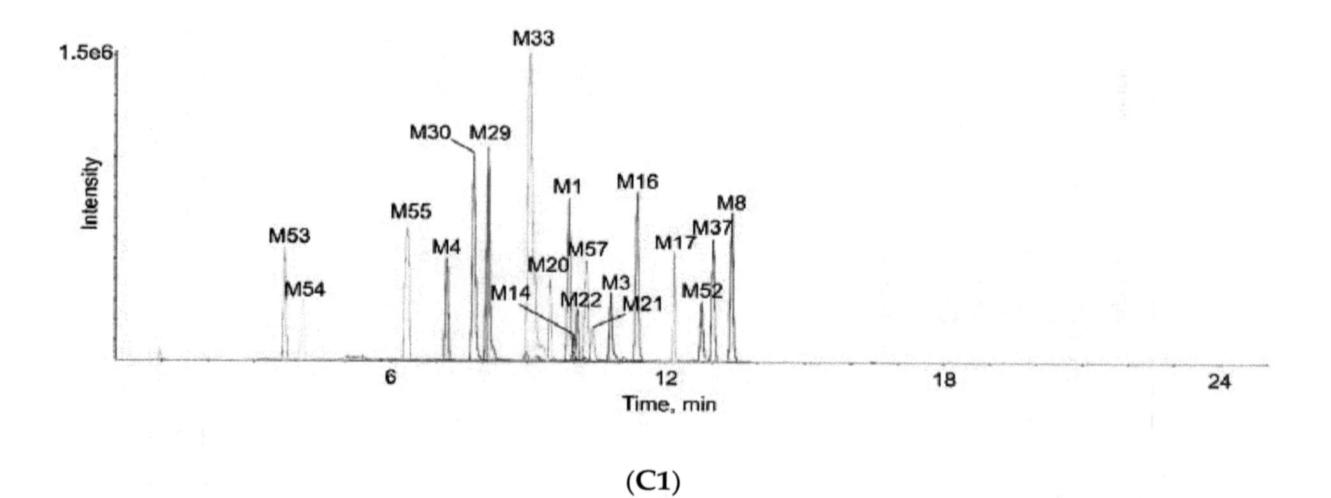

(C1)

Figure 3. *Cont.*

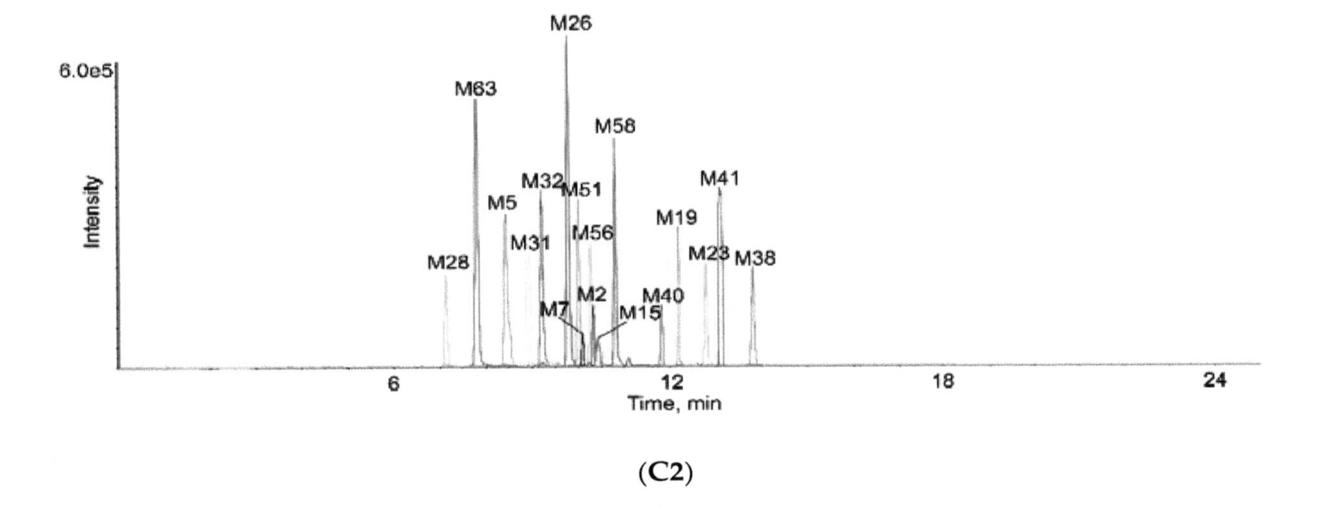

(**C2**)

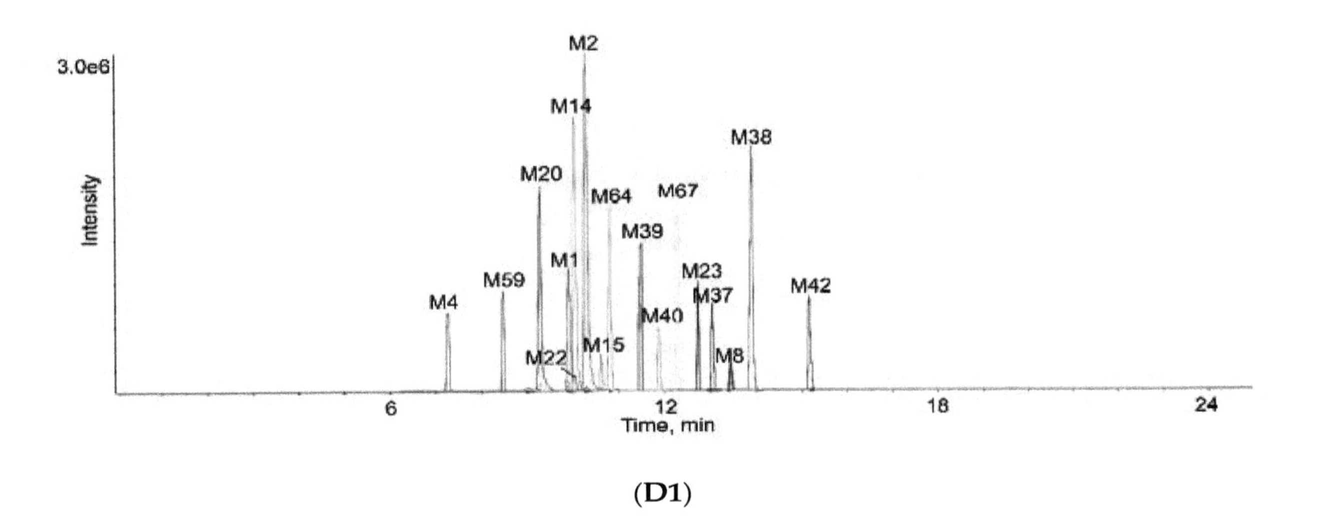

(**D1**)

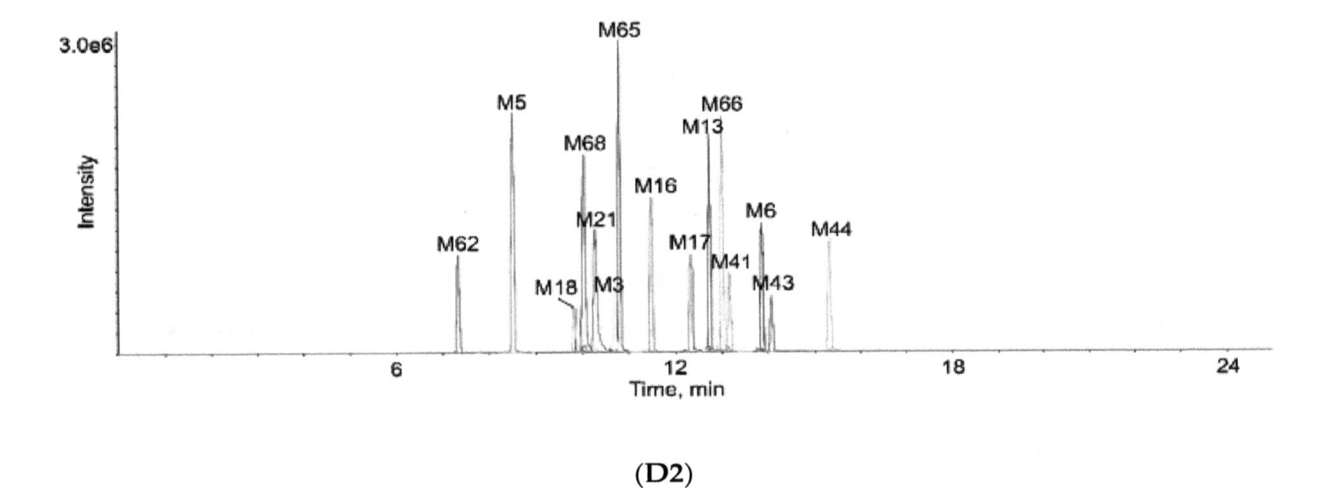

(**D2**)

Figure 3. *Cont.*

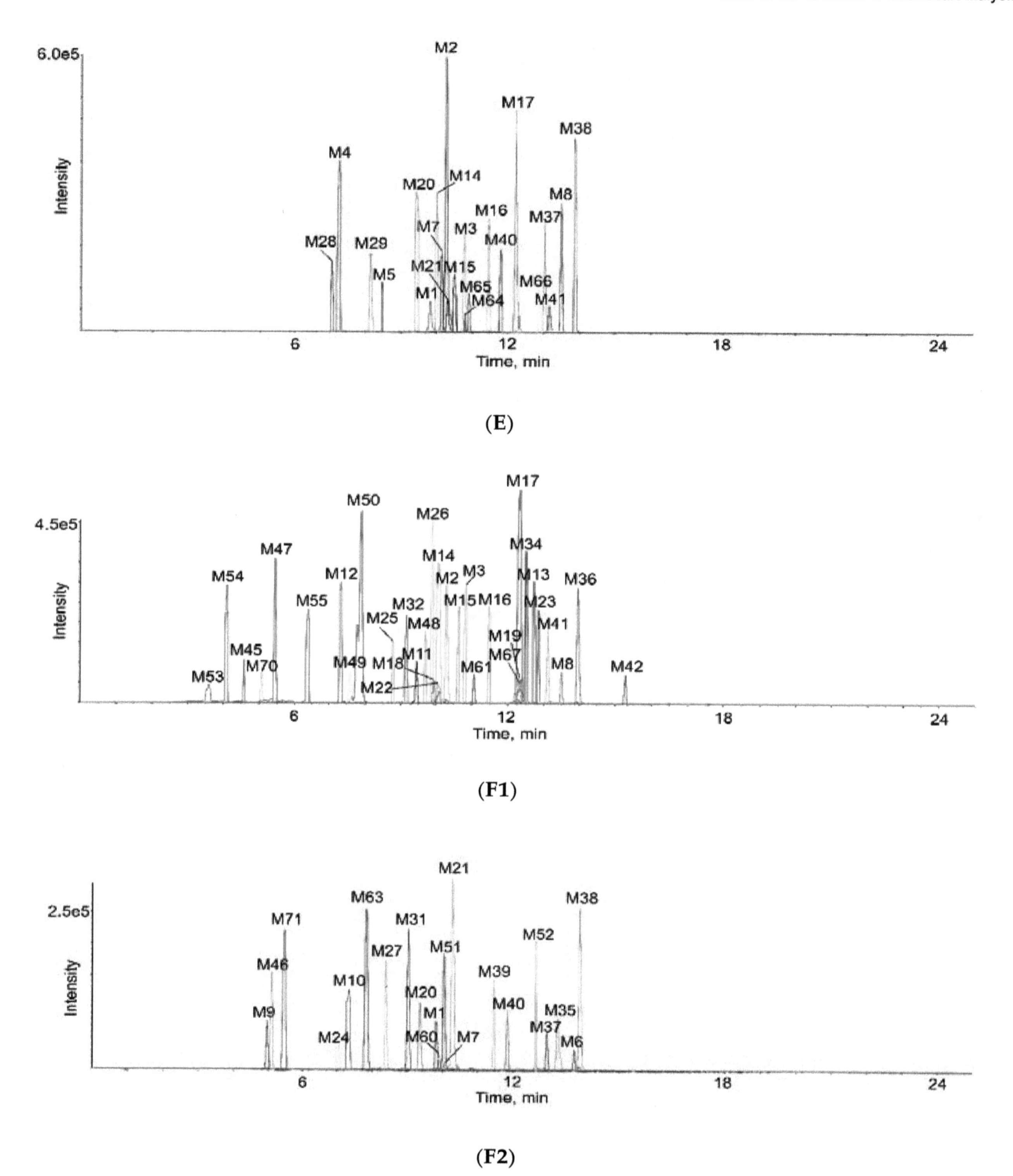

Figure 3. Extracted ion chromatograms of all metabolites of eupatorin in vivo and in vitro (**A**—in rat plasma sample, **B**—in rat bile sample, **C1**,**C2** in rat urine sample, **D1**,**D2** in rat feces sample, **E**—in rat liver microsomes, **F1**,**F2** in rat intestinal flora).

2.4. Metabolic Pathways of Eupatorin

The metabolites of eupatorin in rats after oral administration, in liver microsomes and intestinal flora through incubation was identified in this study. As a result, a total of 51 metabolites in vivo were detected, including 8 metabolites in plasma, 5 metabolites in bile, 36 metabolites in urine and 32 metabolites in feces. Meanwhile, 60 metabolites in vitro were observed, including 22 metabolites in liver microsomes and 53 metabolites in intestinal flora. The proposed metabolic pathways of eupatorin in vivo, in rat liver microsomes and in rat intestinal flora were shown in Figure 4. It is worth mentioning that the loss of CH_2, CH_2O, O, oxidation, glucuronidation and ketone formation was the primary metabolic step that produced further reactions such as sulfate conjugation, hydrogenation, N-acetylation, methylation, demethylation, internal hydrolysis, glycine conjugation, glutamine conjugation and glucose conjugation. Moreover, all metabolic changes above had taken place in vivo and in vitro. However, glycine conjugation was just present in vivo, while glutamine conjugation and glucose conjugation merely existed in vitro.

2.5. Comparison of Metabolites in Vivo and in Vitro

Drug metabolism plays a significant impact on various fields of pharmaceutical mechanisms as well as drug development and clinical use. In this work, the metabolism of eupatorin in vivo (plasma, bile, urine and feces) and in vitro (rat liver microsomes and intestinal flora) was investigated. In vivo; rat urine and feces possessed high activity for eupatorin metabolism, which were identified as having 36 and 32 metabolites, respectively. Nevertheless, only 8 metabolites were observed in rat plasma and 5 metabolites were detected in rat bile, suggesting that the rat plasma and bile might hold low biotransformation activity [30]. In vitro, 53 metabolites were obtained in rat intestinal flora while 22 metabolites were identified in rat liver microsomes, which implied that most metabolites could be excreted in intestinal flora samples and intestinal tract was more suitable for rapid identification of metabolites of eupatorin in vitro, with enormous catalytic and metabolic capacity which exceeds that of the liver microsomes [24]. Thus, the intestinal tract is considered as an extremely vital organ in the biotransformation of eupatorin.

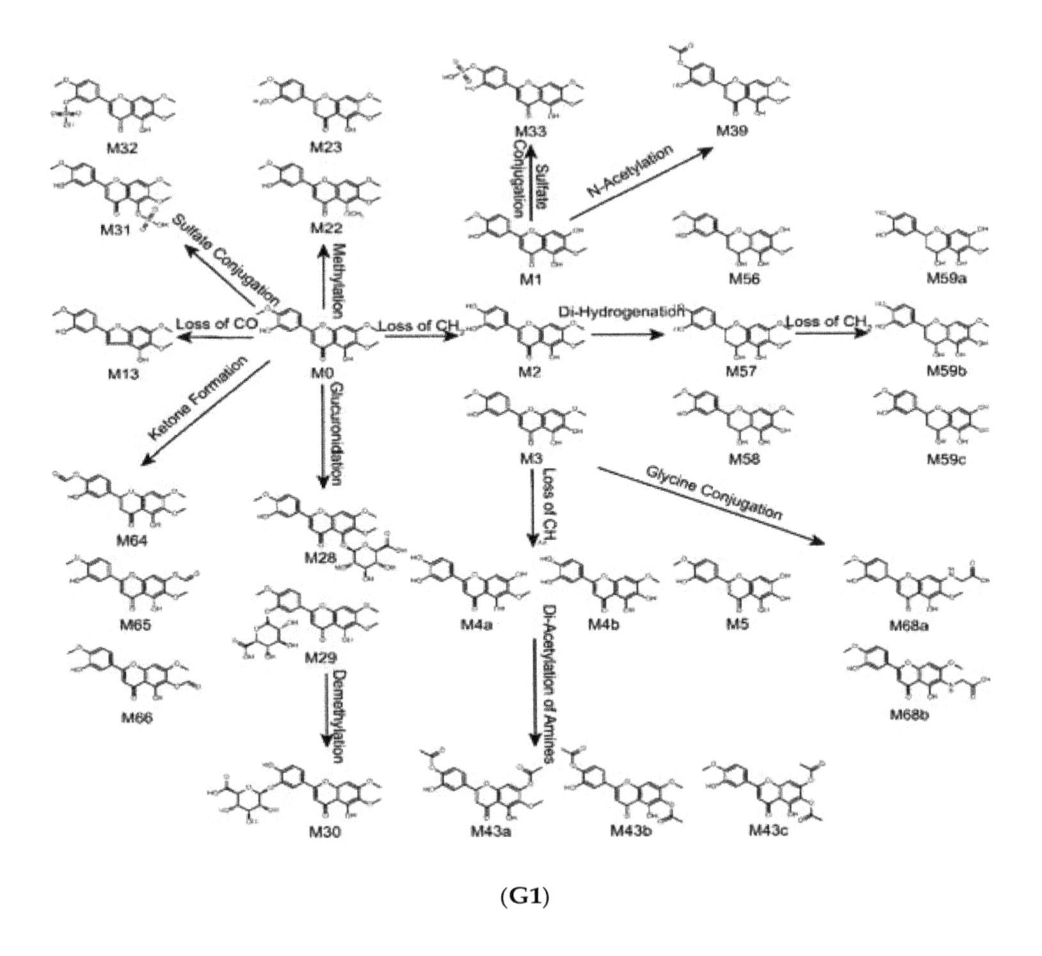

(G1)

Figure 4. *Cont.*

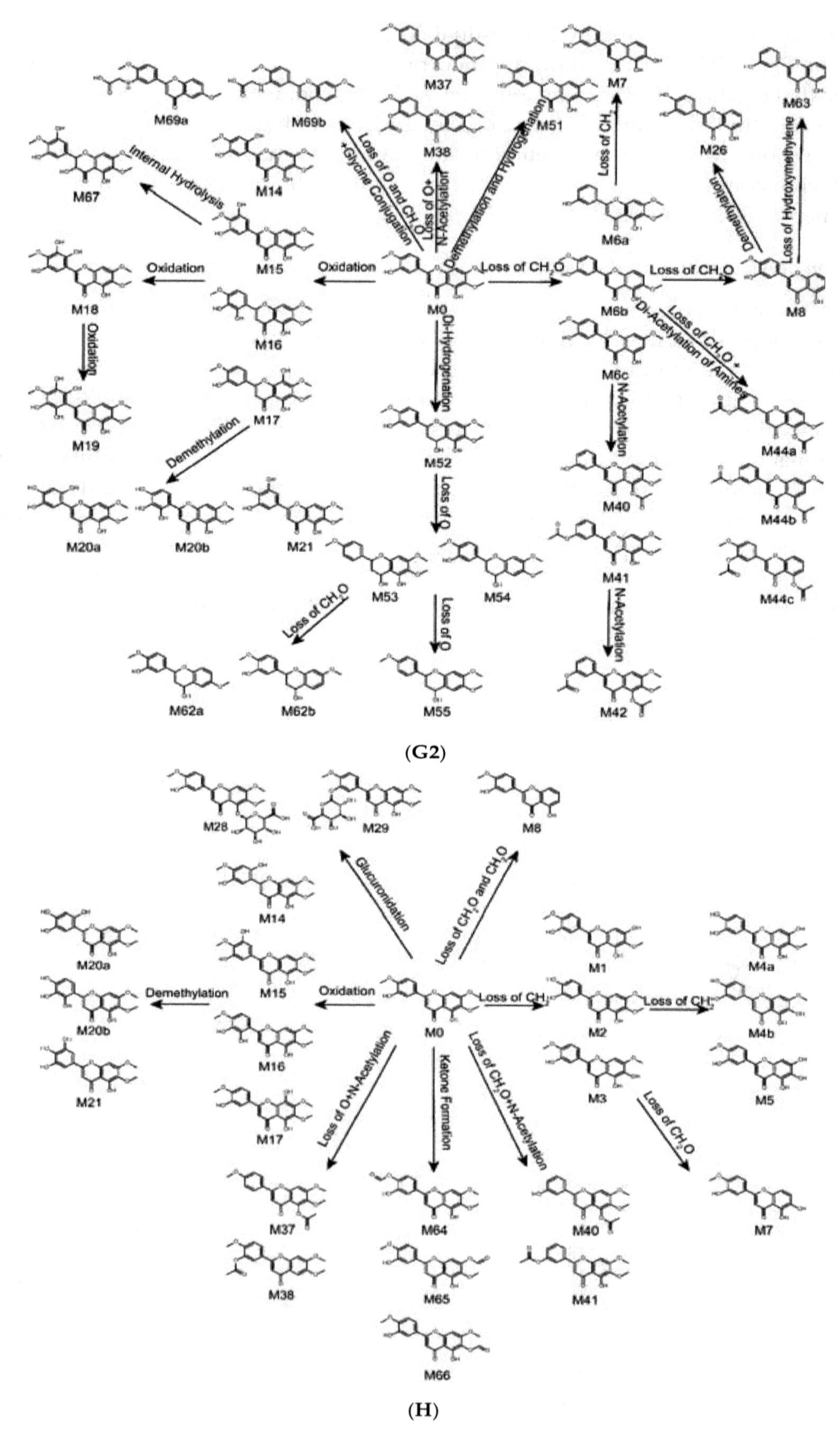

(G2)

(H)

Figure 4. *Cont.*

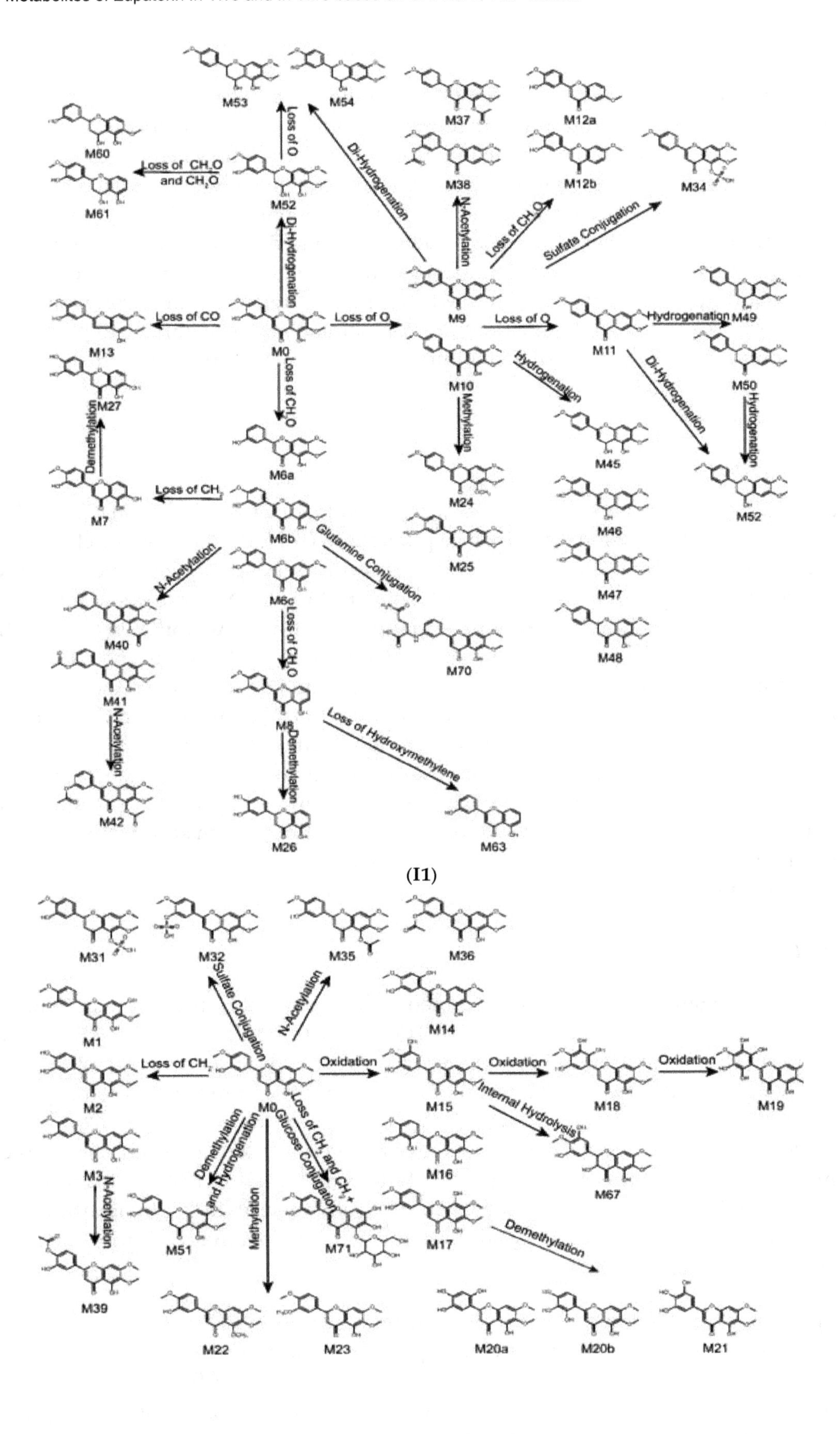

Figure 4. Metabolic profile and proposed metabolic pathways of eupatorin in vivo and in vitro (**G1,G2** in vivo, **H** in rat liver microsomes, **I1,I2** in rat intestinal flora).

2.6. *Metabolite Activity of Eupatorin*

It has been reported in the literature that OS was taken as a beverage to improve health and for treatment of kidney disease, bladder inflammation and urethritis [1,2]. As its major active ingredient, eupatorin has also been reported to have meaningful anti-inflammatory activity [15,16]. In this study, the metabolites of eupatorin in urine samples were the largest, which may be related to the therapeutic effects of cystitis, nephritis and urethritis. In addition, many of the metabolites of eupatorin have been studied. For example, M4a namely nepetin, is a natural flavonoid present in different plants. In recent years, accumulating evidence has shown that nepetin exhibits various pharmacological activities, especially potent anti-inflammatory properties, which might be related to the strong anti-inflammatory activity of eupatorin [30–32]. Overall, the identification of metabolites of eupatorin provides a basis for new pharmacological studies and these metabolites will be further explored in the future.

3. Material and Methods

3.1. *Chemicals and Materials*

Eupatorin (855-96-9, purity > 98.94%) was purchased from Chengdu Desite Co., Ltd. (Chengdu, China). Beta-nicotinamide adenine dinucleotide phosphate (β-NADPH) was purchased from Sigma Chemical (St. Louis, MO, USA). Alamethicin and uridine 5′-diphosphoglucuronic acid trisodium salt (UDPGA) were purchased from BD Biosciences (Woburn, MA, USA). Phosphate buffer saline (PBS) was purchased from Sangon Biotech Co., Ltd. (Shanghai, China). Acetonitrile and methanol were all HPLC grade and were purchased from J.T.-Baker Company (Phillipsburg, NJ, USA). Formic acid (HPLC grade) was provided by Diamond Technology (Dikma Technologies Inc., Lake Forest, CA, USA). Purified water was purchased from Wahaha (Hangzhou Wahaha Group Co., Ltd., Hangzhou, China). L-ascorbic acid, L-cysteine, eurythrol, tryptone and nutrient agar were purchased from Beijing AoBoXing Bio-tech Co., Ltd. (Beijing, China). Sodium carboxymethyl cellulose (CMC-Na), sodium carbonate (Na_2CO_3), magnesium chloride ($MgCl_2$), potassium dihydrogen phosphate (KH_2PO_4), dipotassium phosphate (K_2HPO_4), calcium chloride ($CaCl_2$), ammonium sulfate ($(NH_4)_2SO_4$), sodium chloride (NaCl) and magnesium sulfate ($MgSO_4$) were obtained from Tianjin Guangfu Technology Development Co., Ltd. (Tianjin, China).

3.2. *Instruments and Conditions*

UHPLC-Q-TOF-MS/MS analysis was performed on a Nexera-X2 UHPLC system (Shimadzu Corp., Kyoto, Japan), which was combined with a triple TOF^{TM} 5600$^+$ MS/MS system (AB SCIEX, Concord, Ontario, Canada). The chromatographic separation was achieved on Poroshell 120 EC-C_{18} column (2.1 × 100 mm, 2.7 μm) with a SecurityGuard® UHPLC C18 pre-column (Poroshell).

The mobile phase was composed of 0.1% aqueous formic acid (eluent A) and acetonitrile (eluent B). The gradient elution program was as follows: 10–55% B from 0 to 15 min, 55–95% B from 15 to 20 min, 95–95% B from 20 to 25 min. The column temperature remained at 40 °C. In addition, the injection flow rate and the volume were set at 0.3 mL/min and 3 μL, respectively. Before the next injection, equilibration was performed for 3 min.

Mass spectrometric detection was carried out by a Triple TOF^{TM} 5600 system equipped with Duo-Spray^{TM} ion sources in the negative electrospray ionization (ESI) mode. The following mass spectrometry parameter settings were applied: ion spray voltage (IS), −4.5 kV; the turbo spray temperature, 550 °C; the optimized delustering potential (DP), −60 V; collision energy (CE), −10 eV; and the collision energy spread (CES), 15 eV. Moreover, the nebulizer gas (gas 1), the heater gas (gas 2) and the curtain gas were set to 55, 55 and 35 L/min, respectively.

3.3. *Metabolism in Vivo*

3.3.1. Animals and Drug Administration

Eighteen male Sprague-Dawley (SD) rats (220–220 g, 12–14 weeks old) were purchased from the Experimental Animal Research Center of Hebei Medical University (Certificate No.1811164). The conditions of temperature (22–25 °C), humidity (55–60%) and light (12 h light/dark cycle) were standard for 7 days before being used. All rats were fasted for 12 h but allowed water before the experiments. These rats were divided into six groups randomly with three rats per group. Groups 1, 3 and 5 were the control groups for blank blood, blank bile, blank urine and feces, respectively. Groups 2, 4 and 6 were the drug groups for blood, bile, urine and feces, respectively. Rats in groups 2, 4, 6 were given eupatorin by gavage, which dissolved in a 0.5% CMC-Na solution at a dose of 50 mg/kg. Nevertheless, an equal 0.5% CMC-Na solution without eupatorin was orally given to groups 1, 3, 5. All rat experiments were conducted in accordance with the committee's guidelines on the Care and Use of Laboratory Animals.

3.3.2. Bio-Sample Collection

The plasma samples collection: About 300 μL–500 μL for each blood sample was gathered from the eye canthus of rat into 1.5 mL heparinized tubes at 0.083, 0.167, 0.25, 0.5, 1, 2, 3, 6, 9, 12 and 24 h after gavage. Every blood sample were centrifuged immediately at 1920 g for 10 min at 4 °C to collect the supernatant. After that all collected plasma samples were combined and stored at −80 °C.

The bile collection: Each rat was injected 20% urethane solution intraperitoneally with 1–2 mL to anesthetize the rats after gavage. Then the rats were performed with bile duct cannulation operation and the bile samples were gathered during 0–1 h, 1–3 h, 3–5 h, 5–8 h, 8–12 h, 12–20 h and 20–24 h with PE-10 tubes (ID = 0.07 cm) [33,34]. Lastly, all bile samples were consolidated and frozen at −80 °C.

The urine and feces collection: The rats were separately housed in metabolic cages with free access to deionized water to collect the urine and feces samples over a 0–72 h period after gavage [35,36]. Finally, all the urine and feces samples were separately mixed, and they were placed at −80 °C before pretreatment was conducted.

3.3.3. Bio-Sample Pretreatment

All biological samples were disposed with two methods: Protein precipitation and liquid-liquid extraction were performed on the combined plasma, bile and urine with three times of methanol and ethyl acetate, respectively. Next, the mixture was vortexed for 5 min and centrifuged at 21,380× *g* for 10 min at 4 °C to obtain the supernatant, which was then collected and dried under nitrogen flow.

Dried and powdered feces samples were severally added to 3-fold methanol and ethyl acetate and then were ultrasonically extracted for 45 min. After centrifugation for 10 min at 21,380× *g*, they were dried under nitrogen gas like the supernatant in plasma, bile and urine samples.

150 μL methanol was added to the residua above with an ultrasonic operation for 15 min, centrifugation at 21,380× *g* for 10 min to gain the supernatant which were ultimately passed through the 0.22 μm millipore filters before injecting into the chromatographic system for further analysis. The control group was handled the same as the drug group.

3.4. *Metabolism in Vitro by Rat Liver Microsomes*

3.4.1. Phase I Metabolism

The typical incubation mixture was carried out in a PBS buffer (pH 7.4) with a final volume of 200 μL, which consisted of liver microsomal protein (1.0 mg/mL), eupatorin (100 μmol/L), $MgCl_2$ (3.3 mmol/L), and β-NADPH (1.3 mmol/L) [37]. Preincubation was conducted at 37 °C for 5 min, subsequently NADPH was added to start the reaction. After incubation at 37 °C for 90 min, the reaction was terminated by adding 1 mL of ethyl acetate. Next, vortex and centrifugation for 5 and 10 min, respectively, and then the organic phase was gathered and evaporated under nitrogen gas. 100 μL of acetonitrile was put in the residua and they were eventually passed through the 0.22 μm millipore filters and placed at −20 °C before analysis. Groups contained blank groups incubated without the addition of eupatorin, the control groups incubated without the addition of NADPH and the sample groups, which were implemented in triplicate with the same treatment [38,39].

3.4.2. Phase II Metabolism

The representative incubation mixture was performed in a PBS buffer (pH 7.4) with a final volume of 200 μL, which including liver microsomal protein (1.0 mg/mL), eupatorin (100 μmol/L), $MgCl_2$ (3.3 mmol/L), and UDPGA (2 mmol/L). Preincubation was implemented at 37 °C for 20 min, subsequently UDPGA was added to begin the reaction. After incubation at 37 °C for 1 h, the reaction was ceased by adding 200 μL of ice-acetonitrile. Next, vortex and centrifugation for 5 and 10 min, respectively. In addition, the supernatant was passed through the 0.22 μm millipore filter before injecting into the UHPLC-Q-TOF-MS/MS system for analysis. Groups contained blank groups incubated without the addition of eupatorin, the control groups incubated without the addition of UDPGA and the sample groups, which were carried out in triplicate with the same treatment.

3.5. Metabolism in Vitro by Rat Intestinal Flora

3.5.1. Preparation of Anaerobic Culture Medium

Solution A: K_2HPO_4 (0.78%) 37.5 mL; Solution B: KH_2PO_4 (0.47%), NaCl (1.18%), $(NH_4)_2SO_4$ (1.2%), $CaCl_2$ (0.12%) and $MgSO_4$ (0.25%) 37.5 mL; Solution C: Na_2CO_3 (8%) 50 mL; Solution D: L-ascorbic acid (25%) 2 mL together with L-cysteine 0.5 g, eurythrol 1 g, tryptone 1 g and nutrient agar 1 g, which were all mixed up. Ultrapure water was added to 1 L and then HCl (1 mol/L) was put to adjust the pH of the solution to 7.5–8.0.

3.5.2. Preparation of Intestinal Flora Culture Solution

Fresh intestinal contents (3 g) taken from SD rats were combined with anaerobic culture medium (30 mL) instantly. After stirring with a glass rod, filtered with gauze to obtain the intestinal bacterial liquid.

3.5.3. Sample Preparation

Eupatorin (1 mg/ mL,100 μL) was added to intestinal flora culture medium (1 mL), which was then filled with nitrogen without oxygen. The reactions were terminated by adding 3 volumes of methanol after incubation for 12 h. Next, the mixtures were vortexed for 5 min and centrifuged for 10 min at 21,380 g. Subsequently, the organic phases were collected and evaporated under nitrogen gas, and 100 μL of methanol was added to the residua, vortexed and centrifuged again for 5 and 10 min, respectively. Before analysis, the supernatant was passed through the 0.22 μm millipore filter. Blank groups were incubated without eupatorin, meanwhile the control groups were incubated not in intestinal flora culture solution but in anaerobic culture medium, but others were the same.

4. Conclusions

In conclusion, the identification of metabolites of eupatorin in vivo and in vitro had achieved great success firstly by means of UHPLC-Q-TOF-MS/MS combined with a powerful and efficient data acquisition and processing method. The results displayed that a total of 71 metabolites were characterized: 51 metabolites were identified in vivo (8 metabolites in the plasma, 5 metabolites in the bile, 36 metabolites in the urine and 32 metabolites in the feces), while 60 metabolites were detected in vitro (22 metabolites in the rat liver microsomes and 53 metabolites in rat intestinal flora). This study was expected to benefit future efficacy and safety studies on eupatorin and provide guidelines for intake of OS. There is no doubt that further studies are needed to confirm the impact of these metabolites on human health and safety, thus providing reasonable recommendations for the consumption of foods and drugs containing eupatorin.

Author Contributions: Conceptualization, L.-T.Z. and L.-Y.L.; investigation, L.-Y.L., X.F., J.-T.Y., S.-H.L. and Y.-P.S.; formal analysis, L.-Y.L., Y.-T.C. and X.F.; writing—original draft preparation, review, and editing, L.-Y.L. and Y.-T.C. All authors read and approved the final manuscript.

References

1. Pariyani, R.; Ismail, I.S.; Azam, A.; Khatib, A.; Abas, F.; Shaari, K.; Hamza, H. Urinary metabolic profiling of cisplatin nephrotoxicity and nephroprotective effects of *Orthosiphon stamineus* leaves elucidated by ^{1}H NMR spectroscopy. *J. Pharm. Biomed. Anal.* **2017**, *135*, 20–30. [CrossRef] [PubMed]
2. Hossain, M.A.; Rahman, S.M. Isolation and characterisation of flavonoids from the leaves of medicinal plant *Orthosiphon stamineus*. *Arab. J. Chem.* **2015**, *8*, 218–221. [CrossRef]
3. Yuliana, N.D.; Khatib, A.; Link-Struensee, A.M.; Ijzerman, A.P.; Rungkat-Zakaria, F.; Choi, Y.H.; Verpoorte, R. Adenosine A1 receptor binding activity of methoxy flavonoids from *Orthosiphon stamineus*. *Planta Med.* **2009**, *75*, 132–136. [CrossRef] [PubMed]
4. Ho, C.-H.; Noryati, I.; Sulaiman, S.-F.; Rosma, A. In vitro antibacterial and antioxidant activities of *Orthosiphon stamineus* Benth. extracts against food-borne bacteria. *Food Chem.* **2010**, *122*, 1168–1172. [CrossRef]
5. Zhang, J.; Wen, Q.; Qian, K.; Feng, Y.; Luo, Y.; Tan, T. Metabolic profile of rosmarinic acid from Java tea (*Orthosiphon stamineus*) by ultra-high-performance liquid chromatography coupled to quadrupole-time-of-flight tandem mass spectrometry with a three-step data mining strategy. *Biomed. Chromatogr.* **2019**. [CrossRef] [PubMed]
6. Guo, Z.; Liang, X.; Xie, Y. Qualitative and quantitative analysis on the chemical constituents in *Orthosiphon stamineus* Benth. using ultra high-performance liquid chromatography coupled with electrospray ionization tandem mass spectrometry. *J. Pharm. Biomed. Anal.* **2019**, *164*, 135–147. [CrossRef] [PubMed]
7. Saidan, N.H.; Aisha, A.F.A.; Hamil, M.S.R.; Majid, A.M.S.A.; Ismail, Z. A novel reverse phase high-performance liquid chromatography method for standardization of *Orthosiphon stamineus* leaf extracts. *Pharmacogn. Res.* **2015**, *7*, 23–31.
8. Yam, M.F.; Mohamed, E.A.H.; Ang, L.F.; Pei, L.; Darwis, Y.; Mahmud, R.; Asmawi, M.Z.; Basir, R.; Ahmad, M. A simple isocratic HPLC method for the simultaneous determination of sinensetin, eupatorin, and 3′-hydroxy-5,6,7,4′-tetramethoxyflavone in *Orthosiphon stamineus* extracts. *J. Acupunct. Meridian Stud.* **2012**, *5*, 176–182. [CrossRef] [PubMed]
9. Akowuah, G.; Zhari, I.; Norhayati, I.; Sadikun, A.; Khamsah, S. Sinensetin, eupatorin, 3′-hydroxy-5,6,7,4′-tetramethoxyflavone and rosmarinic acid contents and antioxidative effect of *Orthosiphon stamineus* from Malaysia. *Food Chem.* **2004**, *87*, 559–566. [CrossRef]
10. Razak, N.A.; Abu, N.; Ho, W.Y.; Zamberi, N.R.; Tan, S.W.; Alitheen, N.B.; Long, K.; Yeap, S.K. Cytotoxicity of eupatorin in MCF-7 and MDA-MB-231 human breast cancer cells via cell cycle arrest, anti-angiogenesis and induction of apoptosis. *Sci. Rep.* **2019**, *9*, 1514. [CrossRef]
11. Lee, K.; Hyun Lee, D.; Jung, Y.J.; Shin, S.Y.; Lee, Y.H. The natural flavone eupatorin induces cell cycle arrest at the G2/M phase and apoptosis in HeLa cells. *Appl. Biol. Chem.* **2016**, *59*, 193–199. [CrossRef]
12. Estevez, S.; Marrero, M.T.; Quintana, J.; Estevez, F. Eupatorin-induced cell death in human leukemia cells is dependent on caspases and activates the mitogen-activated protein kinase pathway. *PLoS ONE* **2014**, *9*, e112536. [CrossRef] [PubMed]
13. Androutsopoulos, V.; Arroo, R.R.J.; Hall, J.F.; Surichan, S.; Potter, G.A. Antiproliferative and cytostatic effects of the natural product eupatorin on MDA-MB-468 human breast cancer cells due to CYP1-mediated metabolism. *Breast Cancer Res.* **2008**, *10*. [CrossRef] [PubMed]
14. Doleckova, I.; Rarova, L.; Gruz, J.; Vondrusova, M.; Strnad, M.; Krystof, V. Antiproliferative and antiangiogenic effects of flavone eupatorin, an active constituent of chloroform extract of *Orthosiphon stamineus* leaves. *Fitoterapia* **2012**, *83*, 1000–1007. [CrossRef] [PubMed]
15. Laavola, M.; Nieminen, R.; Yam, M.F.; Sadikun, A.; Asmawi, M.Z.; Basir, R.; Welling, J.; Vapaatalo, H.; Korhonen, R.; Moilanen, E. Flavonoids eupatorin and sinensetin present in *Orthosiphon stamineus* leaves inhibit inflammatory gene expression and STAT1 activation. *Planta Med.* **2012**, *78*, 779–786. [CrossRef]

16. Yam, M.F.; Lim, V.; Salman, I.M.; Ameer, O.Z.; Ang, L.F.; Rosidah, N.; Abdulkarim, M.F.; Abdullah, G.Z.; Basir, R.; Sadikun, A.; et al. HPLC and anti-inflammatory studies of the flavonoid rich chloroform extract fraction of *Orthosiphon stamineus* leaves. *Molecules* **2010**, *15*, 4452–4466. [CrossRef]
17. Yam, M.F.; Tan, C.S.; Ahmad, M.; Shibao, R. Mechanism of vasorelaxation induced by eupatorin in the rats aortic ring. *Eur. J. Pharmacol.* **2016**, *789*, 27–36. [CrossRef]
18. Liao, M.; Cheng, X.; Diao, X.; Sun, Y.; Zhang, L. Metabolites identificaion of two bioactive constituents in Trollius ledebourii in rats using ultra-high-performance liquid chromatography coupled to quadrupole time-of-flight mass spectrometry. *J. Chromatogr. B* **2017**, *1068*, 297–312. [CrossRef]
19. Almazroo, O.A.; Miah, M.K.; Venkataramanan, R. Drug Metabolism in the Liver. *Clin. Liver Dis.* **2017**, *21*, 1–20. [CrossRef]
20. Li, H.; He, J.; Jia, W. The influence of gut microbiota on drug metabolism and toxicity. *Expert Opin. Drug Metab. Toxicol.* **2016**, *12*, 31–40. [CrossRef]
21. Noh, K.; Kang, Y.R.; Nepal, M.R.; Shakya, R.; Kang, M.J.; Kang, W.; Lee, S.; Jeong, H.G.; Jeong, T.C. Impact of gut microbiota on drug metabolism: An update for safe and effective use of drugs. *Arch. Pharm. Res.* **2017**, *40*, 1345–1355. [CrossRef] [PubMed]
22. Zhang, J.Y.; Wang, Z.J.; Li, Y.; Liu, Y.; Cai, W.; Li, C.; Lu, J.-Q.; Qiao, Y.-J. A strategy for comprehensive identification of sequential constituents using ultra-high-performance liquid chromatography coupled with linear ion trap-Orbitrap mass spectrometer, application study on chlorogenic acids in Flos Lonicerae Japonicae. *Talanta* **2016**, *147*, 16–27. [CrossRef] [PubMed]
23. Yao, D.; Wang, Y.; Huo, C.; Wu, Y.; Zhang, M.; Li, L.; Shi, Q.; Kiyota, H.; Shi, X. Study on the metabolites of isoalantolactone in vivo and in vitro by ultra performance liquid chromatography combined with Triple TOF mass spectrometry. *Food Chem.* **2017**, *214*, 328–338. [CrossRef] [PubMed]
24. Chen, Y.; Feng, X.; Li, L.; Zhang, X.; Song, K.; Diao, X.; Sun, Y.; Zhang, L. UHPLC-Q-TOF-MS/MS method based on four-step strategy for metabolites of hinokiflavone in vivo and in vitro. *J. Pharm. Biomed. Anal.* **2019**, *169*, 19–29. [CrossRef] [PubMed]
25. Yuan, L.; Liang, C.; Diao, X.; Cheng, X.; Liao, M.; Zhang, L. Metabolism studies on hydroxygenkwanin and genkwanin in human liver microsomes by UHPLC-Q-TOF-MS. *Xenobiotica* **2018**, *48*, 332–341. [CrossRef] [PubMed]
26. Zhang, X.; Yin, J.; Liang, C.; Sun, Y.; Zhang, L. A simple and sensitive UHPLC-Q-TOF-MS/MS method for sophoricoside metabolism study in vitro and *in vivo*. *J. Chromatogr. B* **2017**, *1061*, 193–208. [CrossRef]
27. Ma, Y.; Xie, W.; Tian, T.; Jin, Y.; Xu, H.; Zhang, K.; Du, Y. Identification and comparative oridonin metabolism in different species liver microsomes by using UPLC-Triple-TOF-MS/MS and PCA. *Anal. Biochem.* **2016**, *511*, 61–73. [CrossRef] [PubMed]
28. Liang, C.; Zhang, X.; Diao, X.; Liao, M.; Sun, Y.; Zhang, L. Metabolism profiling of nevadensin in vitro and in vivo by UHPLC-Q-TOF-MS/MS. *J. Chromatogr. B* **2018**, *1084*, 69–79. [CrossRef]
29. Zhang, X.; Yin, J.; Liang, C.; Sun, Y.; Zhang, L. UHPLC-Q-TOF-MS/MS Method Based on Four-Step Strategy for Metabolism Study of Fisetin in vitro and *in vivo*. *J. Agric. Food Chem.* **2017**, *65*, 10959–10972. [CrossRef]
30. Chen, X.; Han, R.; Hao, P.; Wang, L.; Liu, M.; Jin, M.; Kong, D.; Li, X. Nepetin inhibits IL-1beta induced inflammation via NF-kappaB and MAPKs signaling pathways in ARPE-19 cells. *Biomed. Pharmacother.* **2018**, *101*, 87–93. [CrossRef]
31. Clavin, M.; Gorzalczany, S.; Macho, A.; Muñoz, E.; Ferraro, G.; Acevedo, C.; Martino, V. Anti-inflammatory activity of flavonoids from Eupatorium arnottianum. *J. Ethnopharmacol.* **2007**, *112*, 585–589. [CrossRef] [PubMed]
32. Lee, Y.S.; Yang, W.K.; Yee, S.M.; Kim, S.M.; Park, Y.C.; Shin, H.J.; Han, C.K.; Lee, Y.C.; Kang, H.S.; Kim, S.H. KGC3P attenuates ovalbumin-induced airway inflammation through downregulation of p-PTEN in asthmatic mice. *Phytomedicine* **2019**, *62*, 152942. [CrossRef] [PubMed]
33. Liao, M.; Diao, X.; Cheng, X.; Sun, Y.; Zhang, L. Nontargeted SWATH acquisition mode for metabolites identification of osthole in rats using ultra-high-performance liquid chromatography coupled to quadrupole time-of-flight mass spectrometry. *RSC Adv.* **2018**, *8*, 14925–14935. [CrossRef]
34. Diao, X.; Liao, M.; Cheng, X.; Liang, C.; Sun, Y.; Zhang, X.; Zhang, L. Identification of metabolites of Helicid in vivo using ultra-high performance liquid chromatography-quadrupole time-of-flight mass spectrometry. *Biomed. Chromatogr.* **2018**, *32*, e4263. [CrossRef] [PubMed]
35. Xie, W.; Jin, Y.; Hou, L.; Ma, Y.; Xu, H.; Zhang, K.; Zhang, L.; Du, Y. A practical strategy for the characterization

of ponicidin metabolites in vivo and in vitro by UHPLC-Q-TOF-MS based on nontargeted SWATH data acquisition. *J. Pharm. Biomed. Anal.* **2017**, *145*, 865–878. [CrossRef] [PubMed]

36. Tian, T.; Jin, Y.; Ma, Y.; Xie, W.; Xu, H.; Zhang, K.; Zhang, L.; Du, Y. Identification of metabolites of oridonin in rats with a single run on UPLC-Triple-TOF-MS/MS system based on multiple mass defect filter data acquisition and multiple data processing techniques. *J. Chromatogr. B* **2015**, *1006*, 80–92. [CrossRef] [PubMed]
37. Zhang, X.; Liang, C.; Yin, J.; Sun, Y.; Zhang, L. Identification of metabolites of liquiritin in rats by UHPLC-Q-TOF-MS/MS: Metabolic profiling and pathway comparison in vitro and in vivo. *RSC Adv.* **2018**, *8*, 11813–11827. [CrossRef]
38. Jia, P.; Zhang, Y.; Zhang, Q.; Sun, Y.; Yang, H.; Shi, H.; Zhang, X.; Zhang, L. Metabolism studies on prim-O-glucosylcimifugin and cimifugin in human liver microsomes by ultra-performance liquid chromatography quadrupole time-of-flight mass spectrometry. *Biomed. Chromatogr.* **2016**, *30*, 1498–1505. [CrossRef]
39. Yisimayili, Z.; Guo, X.; Liu, H.; Xu, Z.; Abdulla, R.; Aisa, H.A.; Huang, C. Metabolic profiling analysis of corilagin in vivo and in vitro using high-performance liquid chromatography quadrupole time-of-flight mass spectrometry. *J. Pharm. Biomed. Anal.* **2019**, *165*, 251–260. [CrossRef]

10

Determination of Kanamycin by High Performance Liquid Chromatography

Xingping Zhang [1,2], Jiujun Wang [1], Qinghua Wu [1], Li Li [1], Yun Wang [1] and Hualin Yang [1,2,*]

1 College of Life Science, Yangtze University, Jingzhou 434025, China; xpzhang1987@163.com (X.Z.); wangjiujun2018@126.com (J.W.); wqh212@hotmail.com (Q.W.); lily2012@yangtzeu.edu.cn (L.L.); 1wangyun@yangtzeu.edu.cn (Y.W.)

2 Research and Development Sharing Platform of Hubei Province for Freshwater Product Quality and Safety, Yangtze University, Jingzhou 434025, China

* Correspondence: yanghualin2005@126.com

Academic Editors: Marcello Locatelli, Angela Tartaglia, Dora Melucci, Abuzar Kabir, Halil Ibrahim Ulusoy and Victoria Samanidou

Abstract: Kanamycin is an aminoglycoside antibiotic widely used in treating animal diseases caused by Gram-negative and Gram-positive infections. Kanamycin has a relatively narrow therapeutic index, and can accumulate in the human body through the food chain. The abuse of kanamycin can have serious side-effects. Therefore, it was necessary to develop a sensitive and selective analysis method to detect kanamycin residue in food to ensure public health. There are many analytical methods to determine kanamycin concentration, among which high performance liquid chromatography (HPLC) is a common and practical tool. This paper presents a review of the application of HPLC analysis of kanamycin in different sample matrices. The different detectors coupled with HPLC, including Ultraviolet (UV)/Fluorescence, Evaporative Light Scattering Detector (ELSD)/Pulsed Electrochemical Detection (PED), and Mass Spectrometry, are discussed. Meanwhile, the strengths and weaknesses of each method are compared. The pre-treatment methods of food samples, including protein precipitation, liquid-liquid extraction (LLE), and solid-phase extraction (SPE) are also summarized in this paper.

Keywords: Kanamycin; HPLC; sample pre-treatment; different detectors; food contamination

---

1. Introduction

Kanamycin is widely used in the treatment of animal infections, added as growth promoters or feed additives for preventive therapy [1]. The antibacterial mechanism of kanamycin is that it can irreversibly bind to the bacterial ribosomal 30S subunit and inhibit its protein synthesis [2]. Because of its potential ototoxicity and nephrotoxicity [3–6], the indiscriminate use of kanamycin will enhance bacterial resistance and cause kanamycin-residue accumulation in animal-derived food, which threatens human health. Therefore, the European Union has promulgated regulations on the maximum residue limits (MRLs) of kanamycin in different food matrices (100 μg/kg for muscle, 100 μg/kg for egg, 600 μg/kg for liver, 2500 μg/kg for kidney, 150 μg/kg for milk and 100 μg/kg for chicken meat) [7].

Kanamycin was isolated in 1957 [8]. It is a mixture of several closely related compounds, such as main constituent kanamycin A (>95%), as well as minor constituents kanamycin B, C, and D (<5%). The major components are shown in Figure 1 [9]. In addition, degradation products such as 2-deoxystreptamine and paromamine can also be present [10]. Kanamycin A and C are isomers, whereas kanamycin B and D have different functional groups [9].

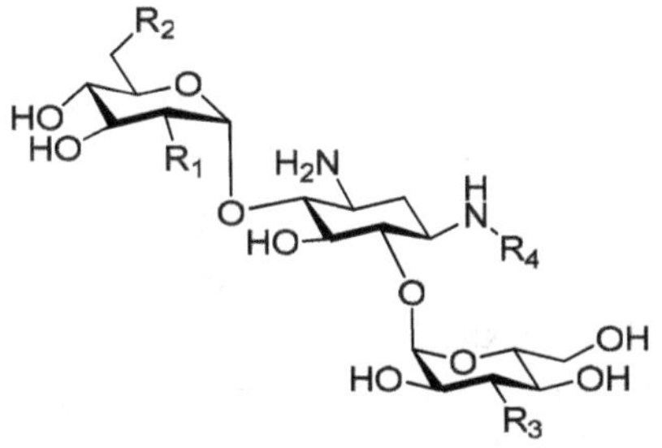

	R1	R2	R3	R4
kanamycin A	OH	NH_2	NH_2	H
kanamycin B	NH_2	NH_2	NH_2	H
kanamycin C	NH_2	OH	NH_2	H
kanamycin D	OH	NH_2	OH	H
amikacin	OH	NH_2	NH_2	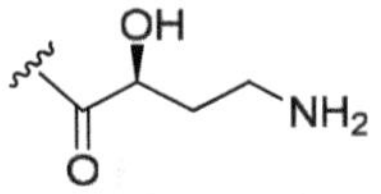

Figure 1. Structure of kanamycin A, B, C, and D and amikacin.

2. The Pre-Treatment Methods of Food Sample

The key point of detecting kanamycin is to remove the impurities or extract kanamycin from matrices. The usual techniques for extraction and cleanup of kanamycin from matrices include protein precipitation, liquid-liquid extraction (LLE), and solid-phase extraction (SPE) [11]. Based on these techniques, pre-treatment methods for kanamycin detection in food samples are summarized as follows.

2.1. Protein Precipitation

Deproteinization was commonly used in the extraction of kanamycin from biological matrices because removal of interferences is necessary to retain good recoveries. Acetonitrile, acidified methanol, and trichloroacetic acid were commonly used precipitation reagents.

In human plasma sample, the simple organic solvent of acetonitrile was used for deproteinization with kanamycin recovery range from 92.3% to 100.8% [12,13]. The acidified methanol with a final concentration of 0.13 mol/L hydrochloric acid (HCl) can also be used for deproteinization of human plasma, and kanamycin recovery ranges from 91.2% to 93.4% [14].

In rat plasma samples, trichloroacetic acid (TCA) with a final concentration of 25–30% was a good precipitation reagent and offers best recovery [15].

In human serum sample, the acidified methonal with a final concentration of 0.14 mol/L HCl can be used to extract kanamycin [16]. Meanwhile, TCA with a final concentration of 40% can be applied in human serum deproteinization, and recovery of kanamycin ranges from 93.9% to 98.4% [17].

Dried blood spots (DBSs) were more convenient than traditional venous blood sampling. In one anti-TB drug analysis, 0.1 mol/L HCl in mixed methonal solution was used for deproteinization of DBS samples [18].

Pig feeds samples were extracted with 0.1 mol/L HCl and kanamycin recovery ranged from 89.4% to 92.8% [19].

In bovine milk, swine and poultry muscle, samples were first precipitated by 15% TCA and then purified with bulk C18 resin. The recoveries of kanamycin were 92% in milk and 36.8–67% in muscle [20].

The chicken meat samples were extracted and precipitated with a mixture of acetonitrile (ACN)-2% TCA (45:55, *v*/*v*), followed by on-line clean-up using turbulent flow chromatography [21]. This automated on-line technique enabled a larger number of samples to be analyzed per day than the traditional clean-up technique. Kanamycin recovery ranged from 109% to 120% in chicken meat [21].

2.2. Liquid-Liquid Extraction

Liquid-liquid extraction (LLE) has been exploited as an extraction procedure for kanamycin from complex matrices. In a published method, veal muscle samples were extracted using CH_3CN-H_2O (86:14 *v*/*v*), followed by a defatting step using hexane liquid-liquid extraction [22].

A new pre-preparation technique of dispersive liquid-liquid microextraction based on solidification of floating organic droplet (DLLME-SFO) was developed, which is a new kind of LLE method that could be applied in the analysis of volatile and polar compounds, like kanamycin. In wastewater and soil, kanamycin is extracted with dodecanol (extraction solvent) and ethanol (dispersive solvent) [23]. Compared with conventional sample preparation methods, the proposed derivatization followed by DLLME-SFO procedure significantly reduced the consumption of organic solvent with high enrichment factor. The DLLME-SFO method facilitated high extraction efficiency and further wide linear range, with good precision, and lower detection limit. The recovery was found to be between 91.3–102.7% for wastewater and 90.3–107.7% for soil. The linearity range was 0.5–500 ng/mL. The LOD was 0.012 ng/mL and LOQ was 0.05 ng/mL [23].

2.3. Solid-Phase Extraction

In many cases, solid-phase extraction (SPE) have been extensively used to extract and concentrate trace organic materials from samples [24–26]. According to packing materials, the solid phase extraction can be classified into four types: Bonded silica gel particle, high polymer material, adsorptive packing material, and mix-mode and specialized column. In this review, the sorbents used for kanamycin analysis mostly belong to the bonded silica gel type, except for molecularly imprinted polymers (MIPs) [27,28] and Chromabond HR-XC [29], which are a high polymer type sorbent.

According to different retention mechanisms, the SPE sorbents used in this review could be further classified into reversed phases sorbents, ion exchange phases sorbents (cation exchanger and anion exchanger), and normal phases sorbents, as shown in Figure 2. The SPE sorbents included in this review are reversed phase sorbents (ODS-C18, Sep-pak tC18, Oasis HLB), strong cation exchanger (Oasis MCX, Chromabond HR-XC), and weak cation exchanger (WCX, CBA, CBX).

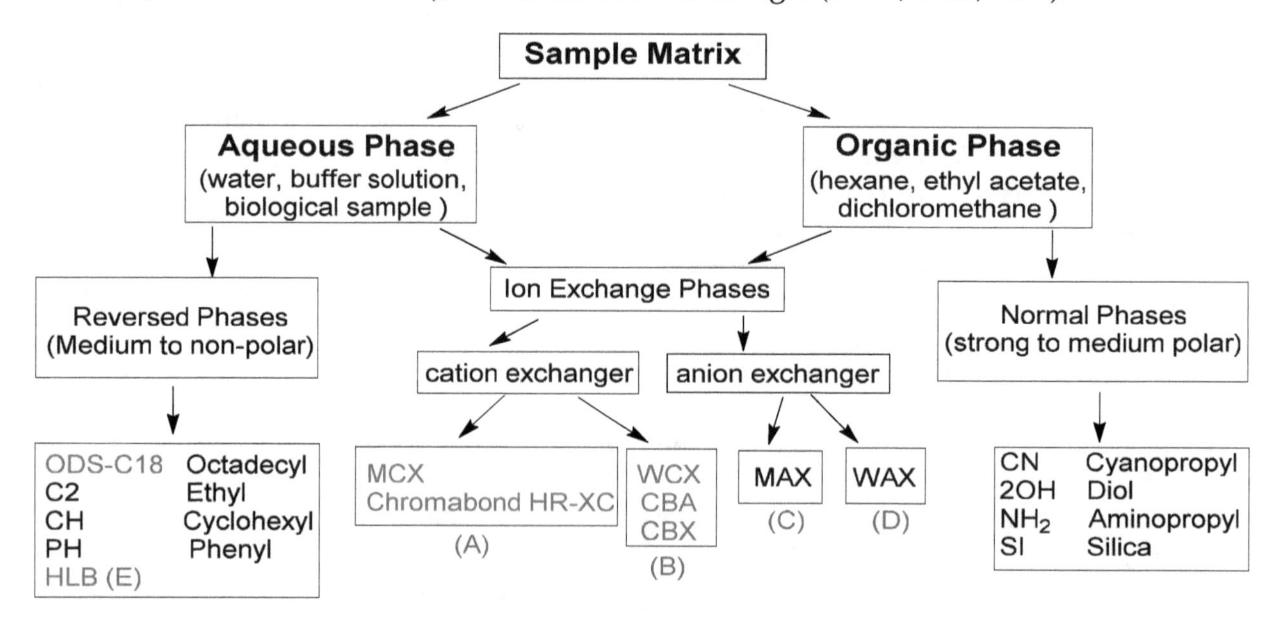

Figure 2. The classification and choice of solid-phase sorbents. (A) Strong cation exchanger; (B) Weak cation exchanger; (C) Strong anion exchanger; (D) Weak anion exchanger; (E) Hydrophilic-lipophilic balanced co-polymer-reversed phased retention.

The reversed phase sorbent Sep-pak tC18 [30] and ODS-C18 [31] was able to extract the non-polar compound from the aqueous sample. The porous silica particles surface bonded with C18 or other hydrophobic alkyl groups. Because of its hydrophobic distribution mechanism, it has strong retention with hydrophobic compounds, but weak retention with hydrophilic compounds. Before use, the cartridge must be preconditioned with a water-soluble organic solvent to solvate the alkyl chains, and then equilibrated with water. It must then be loaded with aqueous samples, followed by eluting with water. A drawback is that before loading the sample, the sorbent must be kept wet, otherwise it will result in low analyte recovery or poor reproducibility. The AccuBOND ODS-C18 cartridge was used for cleanup in soil samples with a kanamycin recovery range from 72.3% to 92.5% [31].

The HLB cartridge has both hydrophilic and lipophilic functional groups, which is a new hydrophilic-lipophilic balanced wettable reversed-phase sorbent [32]. It can overcome the limitations of traditional reversed phase sorbents. Firstly, it is wettable with water, so it has good recovery and reproducibility even the cartridge runs dry during processing. Secondly, it is available for a wide range of compounds including both polar and non-polar chemicals. In muscle, kidney, liver, honey and milk samples, kanamycin was extracted through two consecutive Oasis HLB cartridges (3 mL/60 mg) with a recovery range from 71% to 104% [33].

Ion exchange sorbents (MCX, WCX, MAX, WAX) were found to extract ionizable compounds from the aqueous sample. Because of the ion exchange and hydrophobic distribution mechanisms, the ion exchange sorbents have a strong retention to ionic compounds that have the opposite electric charge of the sorbent carrier, but very weak retention to other compounds [34].

The MCX cartridge is a mixed-mode reversed-phase strong cation exchanger with a pKa of less than 1.0; its sulfonic acid groups have high selectivity to alkaline compounds. Prior to use, it was preconditioned with MeOH, followed by water, then loaded with the extracted sample. Kanamycin is a weak alkaline compound with a pKa of 7.2. At pH lower than 5, the kanamycin was essentially charged and absorbed in the cation cartridge; thus, the sample was extracted with strong acid of 0.1 mol/L HCl [35], 10% TCA [36] or 9% FA [37] aqueous solution prior to loading. At pH higher than 9.0, the kanamycin was neutralised, and the elution could take place. Thus, ammonium methanol solution (1–25%, pH 9.2) was applied to elute kanamycin from the sorbent.

The MCX cartridge was used to extract samples in animal feeds [35], swine tissue [36] and human serum [37] with a kanamycin recovery of 98.4–106% [35] and 80.7% to 91.3% [36], respectively.

The Chromabond HR-X cartridge was styrene-divinylbenzene copolymer based strong cation exchanger. Its surface bonded to benzenesulfonic acid groups [38]. Thus, its retention mechanism was similar to the Oasis MCX sorbent. It was used for cleanup in muscle, kidney and milk samples, with kanamycin recovery ranging from 95% to 102% [29].

The WCX cartridge is a mixed-mode reversed-phase weak cation exchanger with pKa of about 5.0. Its carboxyl groups have high selectivity to strong alkaline compounds. Prior to use, it was preconditioned with MeOH, followed by water, then loaded with the extracted sample. At pH over 6.5, this sorbent was essentially charged to retain kanamycin, so the PH of extracted sample was adjusted to 6.5~7.5 with NaOH and HCl prior to loading. At pH lower than 3.0, the charge on the sorbent was neutralised, and the elution could take place. So, ammonium formate buffer solution (pH 3) [39–41] orformic acid 10% [42], 40% methanol solution [30] were applied to elute kanamycin from the sorbent.

The WCX cartridge (Accell plus CM) was used for cleanup in honey and kidney samples, with kanamycin recovery range from 82% to 105% [40]. The Taurus WCX cartridge was used in honey, milk and liver samples, with a kanamycin recovery range from 58% to 96% [41]. Consecutive SPE cleanup using Sep-pak tC18 and Oasis WCX were applied in milk sample, with a reduced matrix effect and improved absolute kanamycin recoveries from 69.9% to 77.9% [30]. Lehotay et al. used DPX SPE (conducted in a pipet tip rather than a cartridge or centrifuge tube) with 5 mL tips (10 per row) containing 50 mg WCX sorbent for the cleanup of bovine kidney, liver, and muscle samples. The recovery of kanamycin was from 82% to 94% at a spiking level of 0.1 μg/g [42].

The carboxylic acid (CBA) cartridge was a weak cation exchanger with pKa of about 4.8, similar to the Oasis WCX cartridge. Ammonium acetate (pH 7.0) was chosen as the conditioning solution.

The pH of the extracts was adjusted to pH 7.5. The 2% FA in methanol was applied to elute kanamycin from the sorbent. It was used to purify the extracts in muscle, liver, kidney, milk and egg samples [43].

The carboxypropyl (CBX) cartridge was a weak cation exchanger, similar to the Oasis WCX cartridge. The pH of the tissue extract was adjusted to pH 7.0, and then passed slowly through the CBX column. The column was washed with water and then eluted with 5 mL of acetic acid-water-methanolmixture (1:1:8) to get kanamycin; final recoveries range from 81.1% to 104% [44].

Recently, novel sorbents such as molecularly imprinted polymers (MIPs) have emerged [28,45,46]; they are synthetic materials that provide complementary binding sites to specifically capture the target analyte kanamycin. Thus, they are ideal for selective extraction and to reduce the matrix effect. MISPE-Aminoglycoside cartridges (50 mg, 3 mL) were used for extraction and clean-up processes for honey, pork and milk samples, achieving kanamycin recoveries within 78.2–97% and 70–106%, respectively [27,28]. The matrix effect results were both lower than 15%, showing that this method provided very clean extracts [27,28].

3. Liquid Chromatography Methods

HPLC is a conventional analytical method because of its low demand for instruments, and has been widely used in the analysis of kanamycin in different samples [36]. Depending on the retention mechanisms, the chromatographic columns used in this review were mainly divided into three types: Reversed-phase (RP) column, mixed-mode column, and hydrophilic interaction chromatography (HILIC) column. Each column type is marked in Tables 1–3. The different detectors coupled with HPLC mainly include UV/Fluorescence, Evaporative Light Scattering Detector (ELSD)/Pulsed Electrochemical Detection (PED), and Mass Spectrometry. The following content will be unfolded mainly on the basis of the classification above.

3.1. UV and Fluorescence-Reserved Phase Liquid Chromatography after Derivatization

Kanamycin is very polar and lacks chromophore or fluorophore, which makes it difficult to separate using traditionally reverse phase liquid chromatography (RPLC) recruiting UV or fluorescence monitoring. To overcome this problem, researchers have employed many different pre-column or post-column derivatization agents [47].

Derivatization of kanamycin is mainly focused on modifying its primary amine functions. The commonly used pre-column derivatization reagents include Phenylisocyanate (PIC) [48], 4-chloro-3,5-dinitrobenzotrifluoride(CNBF) [31], 1-naphthyl isothiocyanate (NITC) [13] and 9-fluorenylmethyl chloroformate (FMOC-Cl) [23]. Another reagent *o*-phthaldialdehyde (OPA) [35] can also be employed both in pre-column and post-column derivatization. Table 1 shows HPLC applications in the analysis of kanamycin with UV and fluorescence detection.

3.1.1. Pre-Column Derivatization

Pre-column derivatization of kanamycin changes its polarity, which optimizes its applicability for being analyzed through conventional RPLC. For example, CNBF was used as a pre-column derivatization reagent in kanamycin analysis in different kinds of soil samples with a UV detector at 245 nm with the reaction scheme as presented in Table 4 [31]. CNBF was able to react with primary and secondary amines in alkali condition, producing stable N-substituted-2, 6-dinitro-4-(trifluoromethyl)-benzamine derivative [49]. Unlike FOMC-Cl, CNBF does not need to be removed after derivatization. The analytical column was a kromasil C18 ODS column (250 × 4.6 mm, 5 μm). The SPE column was an AccuBOND ODS-C18 (3 mL/200 mg). Linearity range was 0.01–10.0 mg/kg, and LOD was 0.006 mg/kg. The HPLC-UV Chromatogram of CNBF-kanamycin A derivative is shown in Figure 3 [31].

Table 1. HPLC applications in the analysis of kanamycin A with UV and fluorescence detection.

Detection	Matrix	Compound of Interest	Derivatization Agent and Condition	Extraction and Cleanup Methods	HPLC Type	Column Type and Temperature	Mobile Phases	Detector Wave Length	Recovery (%)	LOD	LOQ	Refs
UV	solvent	KANA A	disodium tetraborate, added in mobile phase	-	IPLC	Reversed-phase column, XBridge C18 (250 × 4.6 mm, 5 μm), 50 °C	0.1 M disodium tetraborate (pH 9.0)-water (20:80, *v*/*v*) with 0.5 g/L sodium octanesulphonate, isocratic.	205 nm	-	38 mg/L	128 mg/L	[9]
UV	soil	KANA A	CNBF, PH 9.0, 10 min, 70 °C	**SPE**, AccuBOND ODS-C18	RPLC	Reversed-phase column, kromasil C18 (250 × 4.6 mm, 5 μm).	methanol-0.1% TFA in water, gradient	245 nm	72.3–92.5	0.006 mg/kg	0.01 mg/kg	[31]
UV	solvent	KANA A	PIC, 10 min, 70 °C	-	RPLC	Reversed-phase column, Phenomenex C18 (250 × 4.6 mm, 5 μm)	ACN-1% tris buffer (40:60, *v*/*v*) pH adjusted to 6.5 with 1 N sulfuric acid, isocratic	242 nm	92–98	0.597 μg/mL	1.021 μg/mL	[48]
UV	human plasma	KANA A	NITC in pyridine, 70 °C	**Protein Precipitation** with can	RPLC	Reversed-phase column, LichrocartPurospher STAR C18 (55 × 4 mm, 3 μm)	water-methanol (33:67, *v*/*v*), isocratic	230 nm	95.9–100.8	0.3 μg/mL	1.2 μg/mL	[13]
FL	animal feeds	KANA A	OPA-ME	**SPE**, Oasis MCX cartridge (3 cc, 60 mg)	RPLC	Reversed-phase column, XTerra™C18 (250 × 4.6 mm, 5 μm)	ammonium acetate solution-ACN (50:50, *v*/*v*), isocratic	Ex:230 nm; Em: 389 nm	98.4–106	5 g/ton	10 g/ton	[35]
FL	swine tissue	KANA A	FMOC-Cl, 15 min, RT	**SPE**, Oasis MCX cartridge (3 cc, 60 mg)	RPLC	Reversed-phase column, Waters symmetry C18 (150 × 3.9 mm, 5 μm)	ACN-water, gradient	Ex: 260 nm; Em: 315 nm	80.7–91.3	muscle: 0.03 mg/kg; liver: 0.06 mg/kg; kidney: 0.18 mg/kg	muscle: 0.1 mg/kg; liver: 0.2 mg/kg; kidney: 0.6 mg/kg.	[36]
FL	human plasma	KANA A	FMOC-Cl, 30 min, 25 °C	**Protein Precipitation** with ACN	RPLC	Reversed-phase column, Eclipse XDB C8 (150 × 4.6 mm, 5 μm), 25 °C	70% ACN, isocratic	Ex: 268 nm; Em: 318 nm	92.3–100.8	0.01 μg/mL	0.05 μg/mL	[12]
FL	wastewater and soil	KANA A	FMOC-Cl, 15 min, RT	**DLLME-SFO**, extraction solvent: dodecanol. dispersive solvent: ethanol	RPLC	Reversed-phase column, Diamonsil C18 (250 × 4.6 mm, 5 μm), 40 °C	ACN-water (84:16 *v*/*v*), isocratic	Ex: 265 nm; Em: 315 nm	Wastewater: 91.3–102.7; soil: 90.3–107.7	0.012 ng/mL	0.05 ng/mL	[23]
FL	pig feeds	KANA A	OPA-ME	**Protein Precipitation** with 0.1 M HCl	RPLC	Reversed-phase column, Eluent B: TSK ODS 120T (150 × 4.6 mm, 5 μm). Eluent C: Hypersil ODS for (150 × 3.2 mm, 5 μm), RT	Eluent B: THF-15 mM sodium sulphate, 3:97(*v*/*v*). Eluent C: 10 mM acetic acid-10 mM pentane sulphonate, 1.5:98.5 (*v*/*v*)	Ex: 355 nm; Em: 415 nm	89.4–92.8	0.2 mg/L	0.4 mg/L	[19]

FL: Fluorescence, KANA: Kanamycin, ACN: Acetonitrile.

Table 2. HPLC applications in the analysis of kanamycin with ELSD and PED.

Detection	Matrix	Compound of Interest	Extraction and Cleanup Methods	HPLC Type	Column Type and Temperature	Mobile Phases & IPR	Recovery (%)	LOD	LOQ	Refs
ELSD	solution	Kanamycin A, B and Sulfates	-	reversed-phase IPLC	Reversed-phase column, Spherisorb ODS-2 C18 (250 × 4.6 mm, 5 µm), RT	water-ACN (60:40, *v/v*), 11.6 mM HFBA, isocratic	Kanamycin A: 95–103, Kanamycin B: 95–105	Kanamycin A: 0.2 µg/mL, Kanamycin B: 1.4 µg/mL.	Kanamycin A: 0.6 µg/mL, Kanamycin B: 4 µg/mL	[50]
ELSD	solution	Kanamycin sulfate	-	HILIC	Mixed-mode column, Click TE-Cys (150 × 4.6 mm, 5 µm), 30 °C	ammonium formate aqueous solution (A)-ACN (B)-water(C), gradient	-	-	-	[51]
ELSD	fermentation broth	Kanamycin B	CD180 resin column	reversed-phase IPLC	Reversed-phase column, Agilent SB-Aq C18, RT	water-ACN (65:35, *v/v*), 11.6 mM HFBA, isocratic	93–96	-	0.05 mg/mL	[52]
PED	solution	Kanamycin	-	IPLC	Reversed-phase column, Platinum EPS (150 × 4.6 mm, 3 µm), 45 °C	**MPA**: sodium sulphate (5.0 g/L), sodium octanesulphonate (0.5 g/L) and 0.2M phosphate buffer pH 3.0 (50.0 mL/L) **MPB**: sodium sulphate (15 g/L), sodium octanesulphonate (0.5 g/L) and 0.2M phosphate buffer pH 3.0 (50.0 mL/L) Octanesulfonate as IPR, gradient	-	1.7 ng	5 ng	[10]
PED	solution	Kanamycin B,D	-	IPLC	Reversed-phase column, PLRP-S column packed with polystyrene-divinylbenzene, (250 × 4.6 mm, 8 µm), 45 °C	**MPA**: sodium sulphate (20 g/L), sodium octanesulphonate (1.3 g/L) and 0.2M phosphate buffer pH 3.0 (50.0 mL/L) **MPB**: sodium sulphate (60 g/L), sodium octanesulphonate (1.3 g/L) and 0.2M phosphate buffer pH 3.0 (50.0 mL/L). Octanesulfonate as IPR, gradient	-	Kanamycin D: 3 ng, Kanamycin B: 5 ng	Kanamycin D: 10 ng, Kanamycin B: 15 ng	[53]

ACN: Acetonitrile.

Table 3. HPLC applications in the analysis of kanamycin with MS detection.

Detection	Matrix	Compound of Interest	Extraction and Clean-up Methods	HPLC Type	Column Type and Temperature	Mobile Phases	Recovery (%)	LOD	LOQ	Refs
LC-MS/MS	rat plasma	kanamycin	**Protein Precipitation** with TCA added in sample	IPLC	Reversed-phase column, PhenomenexSynergi C12 Max-RP (50 × 2.0 mm, 4 µm)	0.1% FA in water-0.1% FA in ACN, gradient	-	-	20 ng/mL	[15]
UPLC-MS/MS TQD	bovine kidney, liver, muscle	kanamycin sulfate	**SPE**, Disposable pipet extraction (DPX), 5 mL tips containing 50 mg WCX sorbent	IPLC	Reversed-phase column, Waters BEH C18 (50 × 2.1 mm, 1.7 µm)	20 mM HFBA in 5% ACN in H_2O-20 mM HFBA in ACN, gradient, 3min. Tobramycin as IS	84–92	-	0.005 µg/g	[42]

Table 3. *Cont.*

Detection	Matrix	Compound of Interest	Extraction and Clean-up Methods	HPLC Type	Column Type and Temperature	Mobile Phases	Recovery (%)	LOD	LOQ	Refs
LC-MS/MS	muscle, kidney, liver, honey and milk	kanamycin	**SPE**, two-coupled Oasis HLB columns (3 mL/60 mg)	IPLC	Reversed-phase column, Capcell Pak C18 UG120 (150 × 2.0 mm, 5 μm), 30 °C	20 mM HFBA in 5% ACN-20 mM HFBA in 50% ACN, gradient,10 min. 500 ng/mL Tobramycin as IS	71–104	CCα (μg/kg) 49.5 for muscle, 48.9 for liver, 49.1 for kidney, 9.8 for honey, 121.5 for milk	CCβ (μg/kg) 60.2 for muscle, 58.4 for liver, 59.4 for kidney, 12.4 for honey, 146.4 for milk.	[33]
LC-MS/MS TQD	chicken meat	kanamycin	**Protein Precipitation** with ACN: 2% TCA (45:55, *v/v*). On-line clean-up using column: Thermo Cyclone P (50 × 0.5 mm, 60 μm)	IPLC	Mixed-mode column, Thermo Betasil phenyl hexyl (50 × 2.1 mm,3 μm), RT	1 mM HFBA and 0.5% FA in water-0.5% FA in ACN/methanol (1:1, *v/v*), gradient, 19 min	109–120	LOD: 10.0 μg/kg CCα: 121.3 μg/kg	LOQ: 25.0 μg/kg CCβ: 142.5 μg/kg	[21]
LC–MS/MS TQD	human serum	kanamycin	**Protein Precipitation** with TCA added in sample	IPLC	Reversed-phase column, Thermo Scientific™HyPURITY™C18 (5.0 × 2.1 mm, 3 μm)	water/methanol, containing 0.05% HFBA, gradient. Apramycin as IS	93.9–98.4	-	100 ng/mL	[17]
LC-MS-TOF	bovine milk & swine, poultry muscle	kanamycin	**Protein Precipitation** with TCA added in sample, then supernatant go through a tube containing bulk C18 resin	IPLC	Reversed-phase column, Waters X-Terras C18 (100 × 2.1 mm, 3.5 μm)	10 mM NFPA in H_2O-10 mM NFPA in ACN, gradient	milk: 92, muscle: 36.8–67	15 ng/g for milk and muscle	37.5 ng/g for milk, 25 ng/g for muscle	[20]
LC-MS/MS	muscle, liver, kidney, milk, egg	kanamycin	**SPE**, CBA cartridge (10 CC, 500 mg)	HILIC	HILIC column, CAPCELL PAK ST (150 × 2.0 mm, 4 μm), 30 °C	ACN-0.1%TFA, gradient	75–98	CCα (μg/kg) 8.1 for muscle, 8.5 for egg, 10.0 for liver, 11.2 for kidney, 11.5 for milk	CCβ (μg/kg) 17.6 for muscle, 21.9 for egg, 19.1 for liver, 17.4 for kidney, 18.5 for milk	[43]
LC-MS/MS	milk	kanamycin	Consecutive **SPE** of Sep-pak tC18 (6 mL/500 mg) and Oasis WCX (6 mL/500 mg), extraction with 3% TCA	HILIC	HILIC column, Click TE-Cys HILIC (150 × 3 mm, 3 μm)	1% FA in H_2O-1% FA in 80% ACN, both containing 30 mM ammonium formate, gradient, 10 min	69.9–77.9	6.1μg/kg	19.4μg/kg	[30]
UPLC-MS/MS	milk sample	Kanamycin acid salt	**SPE**, Supel, MISPE-Aminoglycoside cartridge (3 mL/50 mg)	HILIC	HILIC column, PhenomenexKinetex HILIC (100 × 2.1 mm, 1.7 μm), 35 °C	150 mM ammonium acetate in 1% FA(A)-ACN(B), gradient, 12 min	70–106	13.6 μg/kg	45.5 μg/kg	[28]
LC-MS/MS	human plasma	kanamycin	**Protein Precipitation** with acidified methanol using HCl	HILIC	HILIC column, Atlantis HILIC (150 × 2.1 mm, 3 μm), 35 °C	0.1% FA in water-0.1% FA in ACN, gradient, 9.0 min	91.2–93.4	-	1 μg/mL	[14]

Table 3. *Cont.*

Detection	Matrix	Compound of Interest	Extraction and Clean-up Methods	HPLC Type	Column Type and Temperature	Mobile Phases	Recovery (%)	LOD	LOQ	Refs
UPLC-MS/MS	human serum	kanamycin	**Protein Precipitation** with acidified methanol using HCl	HILIC	Reversed-phase column, Waters HSS T3 (50 × 2.1 mm, 1.8 µm), RT	10 mM ammonium formate in 0.1% FA-ACN in 0.1% FA, gradient, 3 min	-	0.5 µg/mL	2.5 µg/mL	[16]
UPLC-MS/MS	dried blood spots	kanamycin	**Protein Precipitation** with acidified methanol using HCl	HILIC	Reversed-phase column, Waters Acquity HSS T3 (50 × 2.1 mm, 1.8 µm)	10 mM ammonium formate in 0.1% FA-ACN in 0.1% FA, gradient, 4 min	-	0.3 µg/mL	5.0 µg/mL	[18]
Quattro Ultima LC-MS/MS	muscle and kidney	kanamycin	**SPE**, CBX cartridge (500 mg)	ZIC-HILIC	HILIC column, SeQuant ZIC-HILIC (100 × 2.1 mm, 5 µm), 32 °C	1% FA in 150 mM ammonium acetate-ACN, gradient, 19 min	81.1–104	18 ng/g	58 ng/g	[44]
LC-MS/MS	muscle, kidney and milk	kanamycin	**SPE**, Chromabond HR-X cartridge (6 mL/500 mg)	ZIC-HILIC	HILIC column, SeQuant ZIC-HILIC (100 × 2.1 mm, 5 µm), 30 °C	1% FA with 200 mM ammonium acetate in 5% ACN-ACN, gradient, 16 min	95–102	CCα (µg/kg) 118 for muscle, 2829 for kidney, 172 for milk	CCβ (µg/kg) 153 for muscle, 3401 for kidney, 215 for milk	[29]
Quattro Ultima UPLC-MS/MS	honey samples	Kanamycin A disulphatedihydrate	**SPE**, WCX cartridge, Accell Plus CM (6 mL/500 mg)	ZIC-HILIC	HILIC column, SeQuant ZIC-HILIC (150 × 2.1 mm, 3.5 µm), 40 °C	pH 4.5 with 125 mM ammonium formate-0.2% FA in ACN, gradient, 6 min. Amikacin as IS	68–112	8 µg/L	27 µg/L	[39]
Quattro Premier	kidney and honey	Kanamycin A disulphatedihydrate	**SPE**, WCX cartridge, Accell Plus CM (6 mL/500 mg)	ZIC-HILIC	HILIC column, SeQuant ZIC-HILIC (150 × 2.1 mm,3.5 µm), 40 °C	PH 4.5 with 175 mM ammonium formate-0.2% FA in ACN, gradient, 6 min. Amikacin as IS	82–105	CCα (µg/kg): 50 for honey, 2733 for kidney	CCβ (µg/kg) 67 for honey, 2965 for kidney. LOQ (µg/kg): 41 for honey, 85 for kidney.	[40]
UPLC-MS/MS	honey, milk and liver	kanamycin	**SPE**, Taurus WCX cartridge	ZIC-HILIC	Mixed-mode column, Obelisc R (100 × 2.1 mm, 5 µm), 40 °C	0.1% FA in water-0.1% FA in ACN, gradient, 8.0 min	58–96	LOD (µg/kg): 1 for honey, 1 for milk. CCα: 3 for honey, 172 for milk, 793 for liver	LOQ (µg/kg): 3 for honey, 5 for milk. CCβ: 5 for honey, 175 for milk 881 for liver	[41]
HILIC-MS/MS	honey, milk and pork	kanamycin disulfate salt	**SPE**, Supel MISPE-Aminoglycoside cartridge (3 mL/50 mg)	ZIC-HILIC	HILIC column, Zwitterionic HILIC (50 × 2.1 mm, 3.5 µm), 40 °C	175 mmol/L ammonium formate and 0.3% FA-methanol and 0.3% FA, gradient, 8 min	72.8–97	10 µg/kg for honey, 11 µg/kg for milk and pork	34 µg/kg for honey, 36 µg/kg for milk and pork	[27]
HILIC-MS/MS	human serum	kanamycin	**SPE**, Oasis MCX cartridge (30 mg)	ZIC-HILIC	HILIC column, SeQuant ZIC-HILIC (100 × 2.1 mm)	A (5/95/0.2, *v/v/v*) and B (95/5/0.2, *v/v/v*) each being a mixture of ACN: 2 mM ammonium acetate: FA, gradient	-	-	100 ng/mL	[37]
LC-MS/MS	veal muscle	kanamycin disulfate salt	**LLE**, defatting using hexane	ZIC-HILIC	HILIC column, ZIC-HILIC (50 × 2.1 mm, 5 µm)	0.4% formic acid inwater/ACN, gradient, 15 min	-	6 ng/g	-	[22]

Table 4. Reaction schemes of kanamycin with different derivatization regents.

Reaction Scheme	Refs
NO_2, F_3C, Cl, NO_2 **CNBF** + NH_2, HO, HO, OH, H_2N, O, NH_2, HO, O, OH, OH, HO, NH_2 $\xrightarrow{OH^-}$ O_2N, CF_3, NH, NO_2, HO, HO, OH, H_2N, O, NH_2, HO, O, OH, OH, HO, NH_2	[31]
Ph–N=C=O + RNH_2 → Ph–N(H)–C(=O)–NH–R **PIC** **N-aryl-N'-phenyl urea**	[48]
N=C=S (naphthyl) + RNH_2 → NH–C(=S)–NHR (naphthyl) **NITC**	[13]
Cl–C(=O)–O– (fluorenylmethyl) + HN(R_1)(R_2) → (fluorenylmethyl)–O–C(=O)–N(R_1)(R_2) + HCl **FMOC-Cl**	[12]

Table 4. *Cont.*

Reaction Scheme	Refs
OPA + RNH_2 + $HOCH_2CH_2SH$ → (N-R, SCH_2CH_2OH) + H_2O	[36]
$B_4O_7^{2-}$ + 9 H_2O ⇌ $4B(OH)_4^-$ + $2H^+$ HO, OH, B, HO, OH + HO, HO, NH_2, O, OH ⇌ HO, O, B, HO, O, NH_2, O, OH	[9]

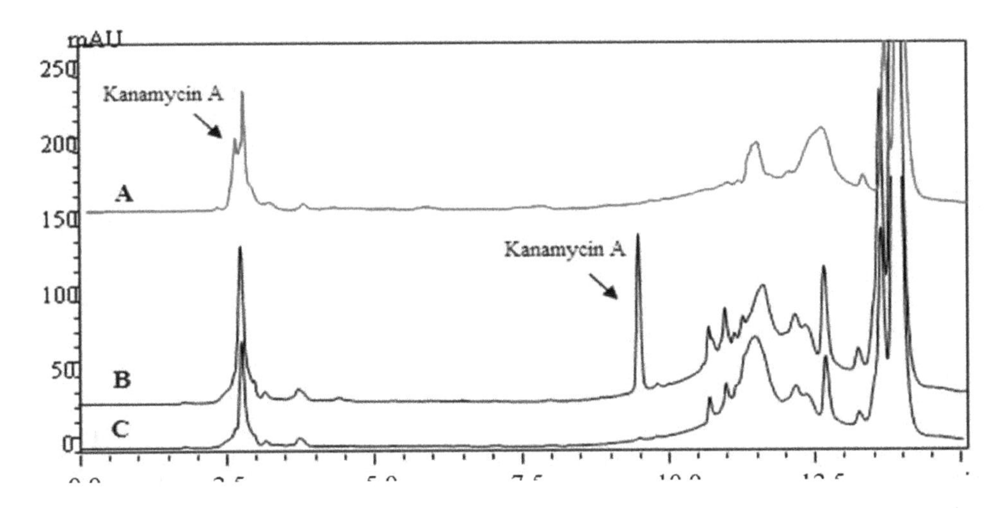

Figure 3. The HPLC-UV Chromatogram of CNBF-kanamycin A derivative. (A) The retention time of CNBF-kanamycin A derivative was 2.71 min without TFA in the mobile phase. The derivative could not be separated completely with interference. (B) The 0.1% TFA could improve separation efficiency. A perfect separation of CNBF-kanamycin A derivative was obtained with retention time of 9.58 min. (C) Blank soil sample.

PIC could react easily with primary or secondary amines, forming the stable *N*-aryl-*N′*-phenyl urea derivative, which was detected by UV at 242 nm. In Patel's study, a corresponding derivative through reaction of KANA with PIC (5 mg/mL in ACN) was formed in the presence of TEA for 10 min, followed by the RPLC method. The derivatives were separated on a Phenomenex C18 column (250 × 4.6 mm, 5 μm). Linearity range was 5–15 μg/mL. LOD was 0.597 μg/mL. The reaction scheme of PIC with kanamycin is presented in Table 4. The HPLC-UV Chromatogram of the kanamycin-PIC derivative is shown in Figure 4 [48].

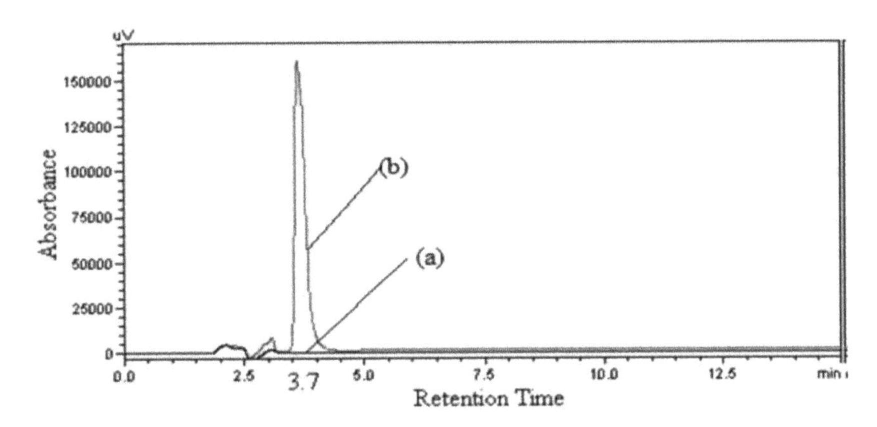

Figure 4. The HPLC-UV Chromatogram of the kanamycin-PIC derivative. (a) Blank; (b) Kanamycin-PIC derivative, 10 mg/mL showing retention time at 8.5 min.

NITC was used as a pre-column derivatization reagent to detect kanamycin A in human plasma by UV at 230 nm. The mixture containing kanamycin A was reacted in pyridine for 1 h. Methylamine was added to eliminate the remnant NITC after derivatization. The stationary phase was a Purospher STAR RP-18 column (55 × 4 mm, 3 μm). Linearity range was 1.2–40 μg/mL, and LOD was 0.3 μg/mL. The reaction scheme of NITC with kanamycin is presented in Table 4. The HPLC-UV Chromatogram of the kanamycin-NITC derivative is shown in Figure 5 [13].

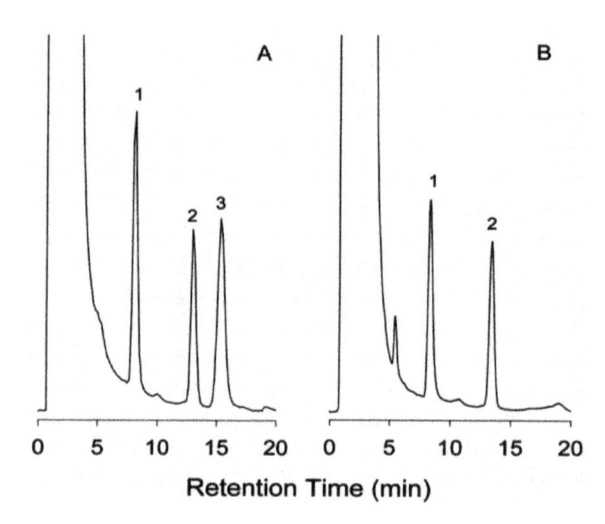

Figure 5. HPLC-UV chromatogram of the kanamycin-NITC derivative. (A) Separation of kanamycin A from kanamycin B, each at 40 μg/mL; (B) Determination of kanamycin A in commercial capsule sample. Peaks: 1, kanamycin A-NITC derivative; 2, acenaphthene (IS), 3, kanamycin B-NITC derivative.

FMOC-Cl was commonly used as a pre-column derivatization reagent of kanamycin, and the following detection was conducted by fluorescence. Kanamycin in human plasma reacted with FMOC-Cl in borate buffer solution (pH 8.5) for 30 min at room temperature, then separated by an Eclipse XDB C8 column (150 × 4.6 mm, 5 μm). LOD was 0.01 μg/mL, fluorescence wavelength was set at excitation of 268 nm and emission 318 nm. The reaction mechanism is shown in Table 4. The HPLC-FL Chromatogram of the kanamycin-FMOC derivative is shown in Figure 6 [12]. Similarly, pre-column FMOC-Cl derivatization of kanamycin was performed in swine tissue. The sample tissue was purified with the MCX SPE column. The derivatives were separated on a Waters symmetry C18 column (150 × 3.9 mm, 5 μm). LOD was 0.03 mg/kg for muscle, 0.06 mg/kg for liver and 0.18 mg/kg for kidney. The fluorescence measurements were set as excitation wavelength at 260 nm and emission wavelength at 315 nm. LOQ was 0.025 μg/mL, which was far lower than that reported by other researchers [36]. Another FMOC-Cl derivatization was prepared in wastewater and soil using a Diamonsil C18 column (250 × 4.6 mm, 5 μm). This is the first reported analysis that reduced the kanamycin derivative with the DLLME-SFO procedure. The fluorescence was measured at excitation wavelength 265 nm and emission wavelength 315 nm [23].

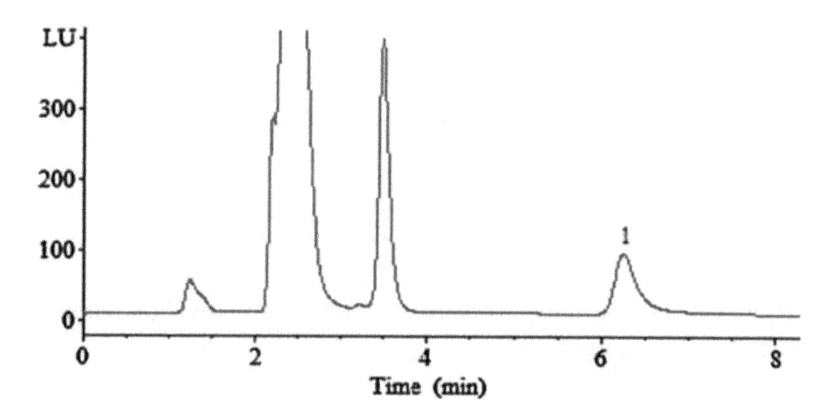

Figure 6. The HPLC-FL Chromatogram of the kanamycin-FMOC derivative. Kanamycin extracted from plasma from the same person 1.5 h after oral administration of 0.75 g of the drug. Peak 1, kanamycin-FMOC derivative.

OPA is a widely used derivatization reagent that introduces chromophores in HPLC methods using UV or fluorescence detection. A typical example is a pre-column derivatization of kanamycin with OPA in animal feeds; the reaction scheme is presented in Table 4 [35]. Oasis MCX SPE was used for cleanup. Chromatographic separation was implemented on a XTerra C18 column (250 × 4.6 mm, 5 μm). LOD was 5 g/ton in animal feeds with fluorescence measurement at excitation wavelength of 230 nm and emission wavelength of 389 nm. The HPLC-FL Chromatogram of the kanamycin-OPA pre-column derivative is shown in Figure 7 [35].

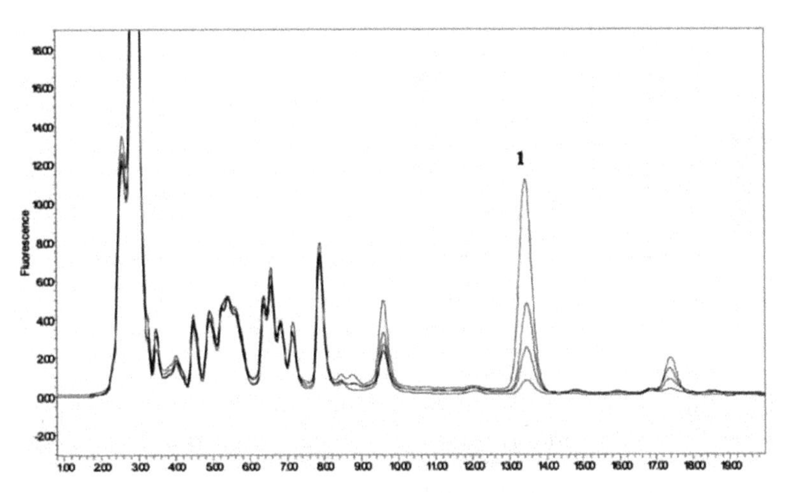

Figure 7. The HPLC-FL Chromatogram of kanamycin-OPA pre-column derivative. Peak 1: kanamycin-OPA derivative, with kanamycin in poultry feeds at levels of 10 mg/g, 40 mg/g, 80 mg/g, and 200 mg/g.

Although the pre-column derivatization methods can avoid using ion pair reagent (IPR), IPR is still needed under certain conditions. The derivatization of kanamycin using borate complexation is an example of this [9]; with reaction scheme is shown in Table 4. The HPLC-UV chromatogram of the kanamycin A-borate derivative is shown in Figure 8 [9]. Borate ion was obtained by dissolving borax in water. After borate complexation formation, the derivatives were analyzed with a XBridge C18 column (250 × 4.6 mm, 5 μm), using sodium octanesulphonate as IPR, and with UV detection at 205 nm. Baseline separation from kanamycins B, C, and D were achieved.

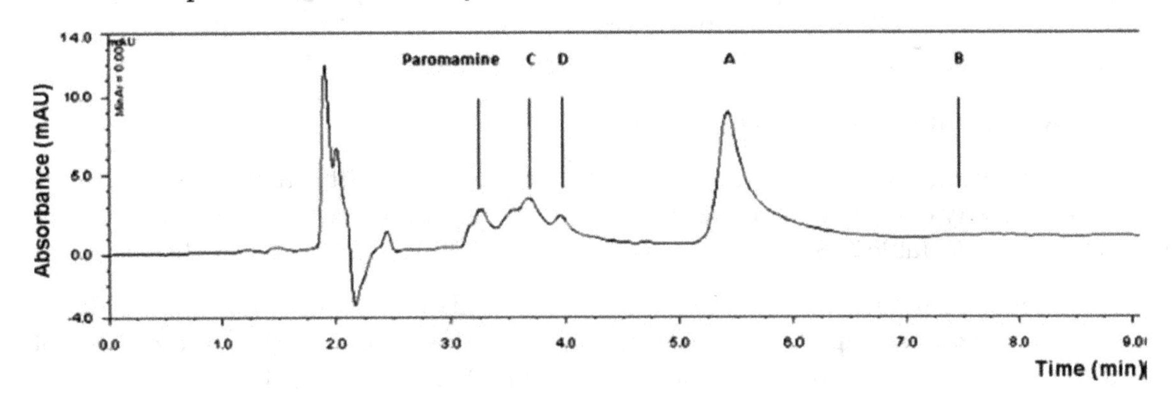

Figure 8. The HPLC-UV chromatogram of kanamycin A borate complexation. Chromatogram obtained after injection of kanamycin A solution (1 g/L) spiked with kanamycins B, C, and D, and paromamine (0.1 g/L each).

3.1.2. Post-Column Derivatization

Post-column derivatization requires more complicated instruments [47] and is confined by reaction time and the solvent system. However, the chemical reaction does not need to be complete since it is repeatable, and long-term stability of the derivative is not a concern [47].

OPA could be used as both pre-column [35] and post-column [19] derivatization agent. Post-column derivatization of kanamycin using OPA was achieved after RPLC with a C8 TSK ODS 120T (150 × 4.6 mm, 5 μm) or Hypersil ODS column (150 × 3.2 mm, 5 μm). Both columns led to good results. The HPLC-FL chromatogram of the kanamycin-OPA post-column derivative is shown in Figure 9 [19]. LOD was 0.2 mg/L in pig feeds, detected with fluorescence measurement at excitation wavelength of 355 nm and emission wavelength of 415 nm.

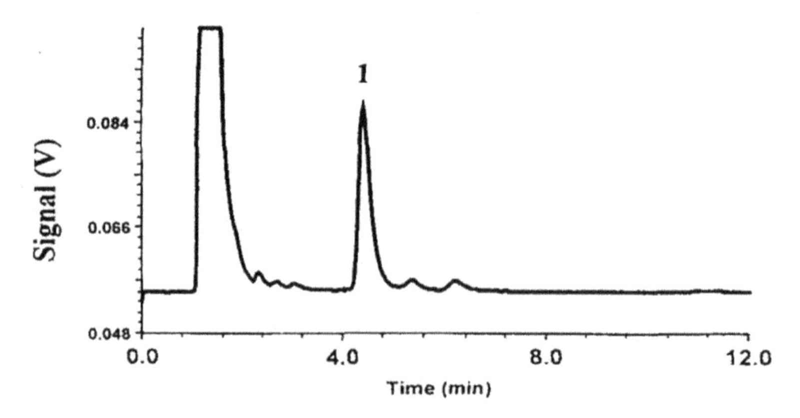

Figure 9. The HPLC-FL chromatogram of the kanamycin-OPA post-column derivative. Peak 1: kanamycin-OPA derivative, with kanamycin in swine feed at a level of 120 mg/kg.

3.2. ELSD and PED-Ion Pair Liquid Chromatography

In ion-pair liquid chromatography (IPLC) methods, the ion pairing reagent (IPR) is used as a mobile phase additive, which interacts with the RPLC stationary phase [47] and allows separating of the ionic and highly polar compounds on RP-HPLC columns. Alkyl sulfonates compounds like octanesulfonate could be used as IPR [10]. Meanwhile, volatile TFA and heptafluorobutyric acid (HFBA) [50,52] could also be used as IPR when coupled with MS detection. Since the high potency of IPR (>20 mM) is harmful to the column packing material, it is ideal to minimize the potency so as to achieve appropriate retention and peak shape [47].

In the IPLC method, an extra buffer system is required to maintain a stable pH of the mobile phase [47]. Ammonium acetate and phosphate are the most frequently used buffer solutions. Phosphate buffer is compatible with UV but not with an MS or ELSD detector. Meanwhile, ammonium acetate buffer is incompatible with UV but compatible with an MS detector [47].

3.2.1. Evaporative Light Scattering Detection (ELSD)

ELSD is increasingly being applied in IPLC for compounds without chromophores, because it eliminates the necessity of derivatization [50]. For HPLC applications in the analysis of kanamycin with ELSD detection, refer to Table 2. Some applications of the IPLC-ELSD methods are discussed hereinafter.

The separation of kanamycins A, B, and sulfate were validated through a novel IPLC-ELSD method without the derivatization step. Chromatographic separations were carried out with a Spherisorb ODS-2C18 column (250 × 4.6 mm, 5 μm) using 11.6 mM HFBA as IPR. The LODs were 0.20 μg/mL for kanamycin A, 1.4 μg/mL for kanamycin B and 2.3 μg/mL for kanamycin sulfates [50]. Another example of the IPLC-ELSD method was determination of kanamycin B and tobramycin impurities with HFBA as IPR. Kanamycin was separated on an Agilent SB-Aq C18 column (150 × 4.6 mm, 5 μm) after sample extraction on a weak acidic cation-exchange resin CD180 [52].

HILIC is a very important alternative approach for the separation of kanamycin. A new HILIC-coupled ELSD method was applied for kanamycin detection. In this research, a HILIC column Click TE-Cys (150 × 4.6 mm, 5 μm) was applied for selective separation of kanamycin. High buffer potency (≥50 mM) and low pH (2.7 or 3.0) are required for the mobile phase to improve peak shape and selectivity [51].

3.2.2. Pulsed Electrochemical Detection (PED)

HPLC together with pulsed electrochemical detector (PED) has been adopted in US Pharmacopoeia [50]. Analysis of kanamycin A and its related substances using IPLC coupled with PED has been reported [10,53]. For IPLC-ELSD applications in the analysis of kanamycin, refer to Table 2.

In Adams' work, octanesulfonate was selected as the IPR. To improve the sensitivity of PED detection, 0.5 M NaOH was added in the post-column effluent to adjust the pH to 13. The packing materials of column PLRP-S (250 × 4.6 mm, 8 μm) was poly (styrene-divinylbenzene). Eight components including kanamycin B and D were separated, and the method was applied to commercial samples [53].

Manyanga improved Adams' work [53] and applied the method to silica-based columns Platinum EPS (150 × 4.6 mm, 3 μm). The amount of salt in the mobile phase was reduced to improve stability, with the use of IPR of octanesulfonate remaining [10]. This method indicated better selectivity and sensitivity.

Nevertheless, the PED method has some disadvantages [54]. First, experience is important for repeatable quantitative results. Second, long equilibration time is required after washing of the electrodes of the electrochemical cell. Therefore, the PED method demands further improvement.

3.3. Liquid Chromatography-Mass Spectrometry

LC-MS/MS is a common analytical method in antibiotics residue analysis [33]. Applications of MS with RPLC, IPLC, HILIC or ZIC-HILIC in the analysis of kanamycin are discussed below; refer to Table 3. Mass spectral acquisition was performed in positive-ion mode by applying multiple reactions monitoring (MRM) using electrospray ionization (ESI) or atmospheric pressure chemical ionization (APCI) to detect kanamycin in this review. Kanamycin B produced $[M + H]^+$ ions at *m/z* 484, which is the precursor ion (Q1). The most abundant product ion (Q3) from the fragmentation was at *m/z* 324, and the relatively abundant product ions were *m/z* 205 and *m/z* 163. The three transition Q3 fragments of kanamycin were 163 for KANA1, and 324 or 205 for KANA2, respectively. The MS/MS spectra of kanamycin B was shown in Figure 10, and the fragmentation pathway of kanamycin B was shown in Figure 11 [55].

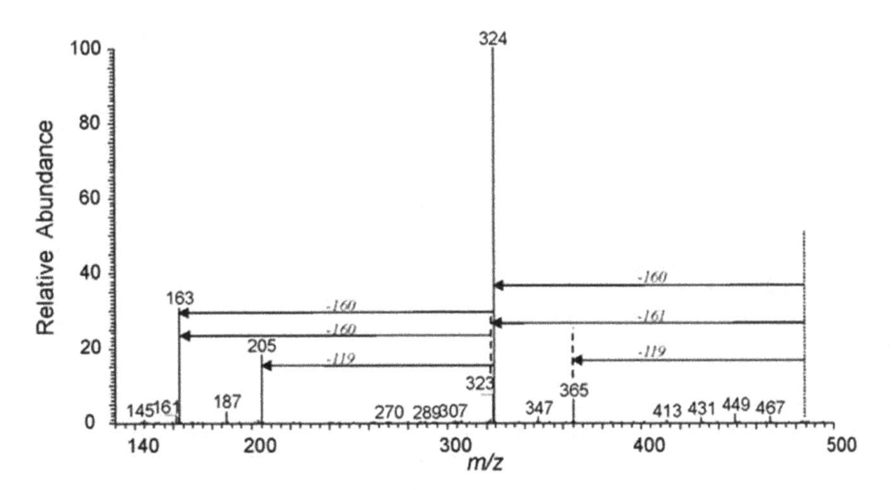

Figure 10. MS/MS spectra of $[M + H]^+$ ions of kanamycin B at *m/z* 484.

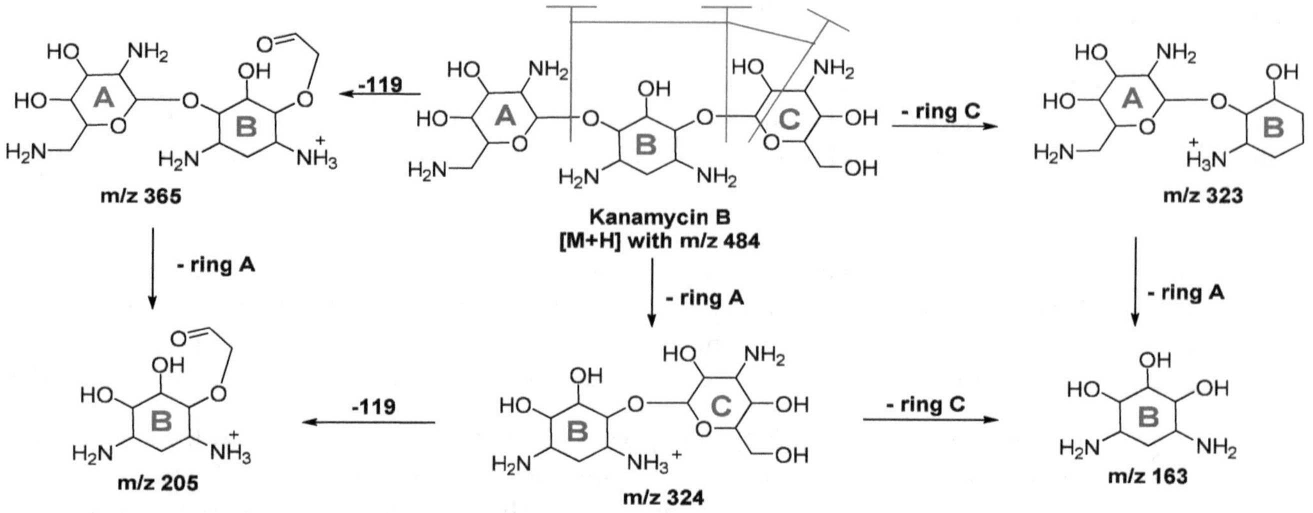

Figure 11. Summary of the fragmentation pathway of kanamycin B reference substances.

3.3.1. IPLC-MS/MS

The IPLC with MS/MS is a powerful tool commonly used in the separation of aminoglycosides [56,57]. The widely used IPRs in kanamycin IPLC-MS-MS analysis were HFBA [17,21,42], TCA [15] and Nonafluoropentanoic acid (NFPA) [20].

In a recent example, kanamycin along with other 12 aminoglycoside antibiotics (AGs) was determined in muscle, kidney, liver, honey, and milk [33]. Volatile HFBA was used as IPR, which was compatible with mass spectrometry and could cause strong retention on the reversed-phase column. Separation was performed using Capcell Pak C18 UG120 column (150 × 2.0 mm, 5 μm). Tobramycin was used as an internal standard (IS). Another rapid qualitative determination of 9 AGs including kanamycin in bovine matrix was realized by IPLC-MS/MS on a Waters BEH C18 (50 × 2.1 mm, 1.7 μm) column, using HFBA as IPR and tobramycin as IS. Since the column material particles were only 1.7 μm ID, the analysis time was shortened to 2.4 min [42]. In another multi-residue study, kanamycin together with 35 other antibiotics were detected in chicken meat on a Betasil phenyl hexyl column (50 × 2.1 mm, 3 μm) [21]. HFBA was chosen as an optimal IPR with minor or no ion suppression effect. For kanamycin detection, LOQ was 25 μg/kg, the decision limit CCα was 121 μg/kg, and detection capability CCβ was 143 μg/kg. Another example was the determination of kanamycin and amikacin in serum using IPLC-MS [17]. IPLC separation was achieved through a water-methanol gradient, containing 0.05% HFBA as IPR, on a Thermo ScientificTM HyPURITYTM C18 column (5.0 × 2.1 mm, 3 μm). Apramycin was used as IS solution [17].

Kanamycin, gentamicin and apramycin were quantified in rat plasma by Cheng et al. [15]. In this research, TCA acted as both a protein precipitator and an IPR, which only existed in the sample but not in the mobile phase; yet the system yielded better sensitivity. The absence of TCA in the mobile phase could reduce the contamination of ion source and result in good reproducibility [15]. The retention of AGs was improved on the Phenomenex Synergi C12 Max-RP column (50 × 2.0 mm, 4 μm), using tobramycin as the internal standard.

In a multi-residue analysis, kanamycin and nine other AGs were detected in bovine milk and bovine, swine and poultry muscle using a Waters X-Terra C18 column (100 × 2.1 mm, 3.5 μm) [20]. NFPA was used as IPR in the mobile phase, which improved kanamycin retention in the C18 column and improved its ionization, enhancing the MS/MS signal. Monitoring and screening was performed by LC-QTOF-MS and then confirmed by the LC-MS/MS method. LOQs for kanamycin were 37.5 ng/g for milk and 25 ng/g for muscle. The LODs for kanamycin was 15 ng/g in milk and muscle [20].

3.3.2. HILIC-MS/MS

HILIC shows a similar separation to normal phase liquid chromatography (NPLC), but it can also use water and volatile buffering solution as the mobile phases of RPLC, which are compatible with MS. Therefore, this technique can be applied to separate strong polar and hydrophilic chemical compounds [47].

Kanamycin is extremely hydrophilic because it has many amino and hydroxyl groups, so it has good solubility in the aqueous mobile phases of HILIC [58]. There is no need to use IPR in the mobile phase of HILIC, so it will cause less ion suppression and is fully compatible with MS systems. HILIC can provide higher sensitivity because the organic solvent-rich mobile phase is more volatile and can enhance desolvation and ionization efficiency of the ESI source [47].

3.3.3. ZIC-HILIC-MS/MS Method

In recent years, HILIC-coupled mass spectrometry has been successfully applied to the separation of AGs. The application of HILIC to quantify kanamycin and other 5 AGs in human serum was reported [37], with a zwitterionic ZIC-HILIC column (100 × 2.1 mm). LOQ of the method was 100 ng/mL for kanamycin [37].

Another application was reported in kidney and muscle tissues using a ZIC-HILIC column (100 × 2.1 mm, 5 μm) [44]. The LOQ of kanamycin was low—50 ng/g. It was observed that the high sorption affinity of kanamycin to polar surfaces required only polypropylene during sample preparation and storage, thus glass was avoided [44].

Kanamycin together with six other AGs was determined in veal muscle, and a ZIC-HILIC column (50 × 2.1 mm, 5 μm) was applied [22]. The ZIC-HILIC column (50 × 2.1 mm, 3.5 μm) was also used to determine kanamycin in honey, milk and pork samples [27].

Kumar et al. compared six kinds of HILIC stationary phases, including bare silica (anionic), amino phenol (cationic), amide (neutral), and zwitter ionic (ZIC) materials [39]. They concluded that the ZIC phase offered the best result, which might be attributed to the ZIC phase providing interaction with both the electropositive amino and the electronegative hydroxyl. The zwitterionic ZIC-HILIC column (150 × 2.1 mm, 3.5 μm) was used to determine Kanamycin A disulphate dihydrate in honey matrix. Amikacin was selected as the internal standard. The linearity range was 70–2000 μg/L. LOD and LOQ were 8 μg/L and 27 μg/L, respectively [39]. The year after that research, the above-mentioned method was improved and applied to the honey and kidney sample analysis for kanamycin, and validated according to Commission Decision 2002/657/EC. The CCα were 50 μg/kg for honey and 2733 μg/kg for kidney. LOQs were 41 μg/kg for honey and 85 μg/kg for kidney, respectively. The linearity was narrowed down to 70–495 μg/kg for honey and 200–4375 μg/kg for kidney [40].

In another similar study, kanamycin was detected in muscle, kidney (cattle and pig) and cow's milk using ZIC-HILIC column (100 × 2.1 mm, 5 μm) [29], and the internal standard tobramycin was used. The CCα ranges from 118 μg/kg to 2829 μg/kg, and the CCβ range from 153 μg/kg to 3401 μg/kg [29].

The usage of a new ZIC-HILIC column Obelisc R (100 × 2.1 mm, 5 μm) was also reported when detecting kanamycin in honey, milk and liver [41]. Obelisc R is a mixed-mode zwitterionic-type LiSC stationary phase, which has a similar structure to ZIC-HILIC column. However, Obelisc R is better than ZIC-HILIC because it has better sensitivity for AGs. The CCα ranges from 3 μg/kg to 793 μg/kg, and CCβ ranges from 5 μg/kg to 881 μg/kg [41].

3.3.4. Other HILIC-MS/MS Methods

The HILIC column CAPCELL PAK ST (150 × 2.0 mm, 4 μm) was applied in separation of 15 AGs residues including kanamycin in animal tissues, milk and eggs [43]. Measurement was carried out through a Thermo electron TSQ Quantum MS. The CCβ of kanamycin ranges from 17.4 μg/kg to 21.9 μg/kg, which was lower than the MRL defined by EU, USA and other countries [43].

In another analysis, kanamycin was separated through an Atlantis HILIC column (150 × 2.1 mm, 3 μm) [14], using apramycin as the internal standard. The calibration range was 100–2500 ng/mL for kanamycin in human plasma [14].

A new Click TE-Cys HILIC column (150 × 3 mm, 3 μm) was used to separate kanamycin in milk sample [30]. The LOD and LOQ were 6.1 μg/kg and 19.4 μg/kg, respectively, and the calibration range was 40 ng/mL to 4000 ng/mL [30].

The Phenomenex Kinetex HILIC column (100 × 2.1 mm, 1.7 μm) was applied to analyze kanamycin residues in different kinds of milk [28]. The LOD and LOQ were 13.6 μg/kg and 45.5 μg/kg, respectively, and the calibration range was 45.5 μg/k to 250 μg/kg kanamycin in milk [28].

Waters HSS T3 column (50 × 2.1 mm, 1.8 μm) was used to analyze kanamycin in serum, gentamicin as IS solution. The LOD and LOQ were 0.5 μg/mL and 2.5 μg/mL, respectively [16]. The LOD and LOQ were further expanded to 0.3 μg/mL and 5.0 μg/mL, respectively, and tested in dried blood spots (DBSs) samples in another study [18].

4. Conclusions

The extraction and clean-up methods play a very important role in the analysis of kanamycin. A series of information on methodologies for extraction and clean-up of kanamycin have been published. The extraction and clean-up methods for kanamycin have been applied to a variety of matrices, including animal feeds, liver and kidney tissues, and serum, among others. When the sample contains protein, as milk and serum, protein precipitation is an initial and key step. After protein precipitation, liquid-liquid extraction can be performed to remove fats by using hexane. SPE can be used to remove salts that might affect the ionization of the MS detector.

Much progress has been achieved in kanamycin detection. However, numerous problems still exist and need to be addressed. The UV and fluorescence derivatization methods are time consuming, and the reaction by-products often cause difficulties in quantitation. Therefore, simpler and direct detection methods are preferred, such as PED [10] and ELSD [50–52]. Nevertheless, ELSD is less sensitive than PED, needs to use volatile additives, and does not display a direct linear relation with the amount injected [10]. Some are semi-quantitative determination methods. LC-MS/MS methods can ensure good sensitivity and separation ability to detect kanamycin in animal-origin food [21,30]. However, the required instruments are not commonly available in many laboratories owing to their high cost. The IPLC is also suitable for MS-MS detector, while the IPR must be volatile and compatible with MS detector with low ionization suppression. Besides IPLC, HILIC is fully compatible with MS systems and free from IPR in the mobile phase. Meanwhile, the HILIC method can achieve lower detection limits [47]. Therefore, the HILIC-MS-MS offers further direction. Moreover, the MRLs of kanamycin residues defined by the EU Commission Decision is still not quite comprehensive, such as the absence of honey; thus, more sample materials needed to be included. We hope that this paper provides some help for kanamycin detection.

Author Contributions: X.Z. wrote the paper and organized data; J.W. revised the paper; Q.W. and L.L. collected relevant literature; Y.W. provided language modification; H.Y. provided the ideas about the paper and financial support.

Acknowledgments: The research was supported by the Science and Technology Research Project from Hubei Provincial Education Department (Q20181321), the Yangtze Fund for Youth Teams of Science and Technology Innovation (2016cqt02).

Abbreviations

ACN	Acetonitrile
AGs	Aminoglycoside antibiotics
anti-TB drug	anti-Tuberculosis drug
APCI	Atmospheric Pressure Chemical Ionization
CBA	Carboxylic Acid
CBX	Carboxypropyl
CCα	Decision Limit
CCβ	Detection Capability
CNBF	4-chloro-3,5-dinitrobenzotrifluoride
DBSs	Dried Blood Spots
DLLME-SFO	Dispersive Liquid-Liquid Microextraction Based on Solidification of Floating Organic Droplet
DPX	Disposable (or Dispersive) Pipet Extraction
ELSD	Evaporative light scattering detection
ESI	Electrospray ionization
FA	Fomic acid
FL	Fluorescence
FOMC-Cl	9-Fluorenylmethyl chloroformate
HCl	Hydrochloricacid
HFBA	Heptafluorobutyric acid
HILIC	Hydrophilic interaction chromatography
HLB	Hydrophilic-lipophilic balanced
HPLC	High Performance Liquid Chromatography
IPLC	Ion-pair liquid Chromatography
IPR	Ion-pairing agent
IS	Internal standard
KANA	Kanamycin
LC-MS	Liquid chromatography-mass spectrometry methods
LLE	Liquid-liquid extraction
LOD	Limit of detection
LOQ	Limit of quantification
ME	Mercaptoethanol
MIPs	Molecularly imprinted polymers
MPA	Mobile phase A
MPB	Mobile phase B
MRLs	Maximum residue limits
MRM	Multiple reactions monitoring
MS	Mass Spectrometry
NITC	1-Naphthyl isothiocyanate
NFPA	Nonafluoropentanoic acid
NPLC	Normal Phase liquid Chromatography
ODS	Octadecyl silane
OPA	*O*-phthaladehyde
PED	Pulsed Electrochemical Detection
PIC	Phenylisocyanate
Q1	Precursor ion
Q3	Product ion
RPLC	Reverse phase liquid chromatography
SPE	Solid-phase extraction
TCA	Trichloroacetic acid
TFA	Trifluoroacetic acid
UV	Ultraviolet
ZIC-HILIC	Zwitter ionic-hydrophilic interaction chromatography

References

1. Durante-Mangoni, E.; Grammatikos, A.; Utili, R.; Falagas, M.E. Do we still need the aminoglycosides? *Int. J. Antimicrob. Ag.* **2009**, *33*, 201–205. [CrossRef]
2. Salimizand, H.; Zomorodi, A.R.; Mansury, D.; Khakshoor, M.; Azizi, O.; Khodaparast, S.; Baseri, Z.; Karami, P.; Zamanlou, S.; Farsiani, H.; et al. Diversity of aminoglycoside modifying enzymes and 16S rRNA methylases in Acinetobacter baumannii and Acinetobacter nosocomialis species in Iran; wide distribution of aadA1 and armA. *Infect. Genet. Evol.* **2018**, *66*, 195–199. [CrossRef]
3. Lee, J.H.; Kim, H.J.; Suh, M.W.; Ahn, S.C. Sustained Fos expression is observed in the developing brainstem auditory circuits of kanamycin-treated rats. *Neurosci. Lett.* **2011**, *505*, 98–103. [CrossRef]
4. Jiang, M.; Karasawa, T.; Steyger, P.S. Aminoglycoside-Induced Cochleotoxicity: A Review. *Front. Cell Neurosci.* **2017**, *11*, 308. [CrossRef] [PubMed]
5. Shavit, M.; Pokrovskaya, V.; Belakhov, V.; Baasov, T. Covalently linked kanamycin-Ciprofloxacin hybrid antibiotics as a tool to fight bacterial resistance. *Bioorgan. Med. Chem.* **2017**, *25*, 2917–2925. [CrossRef] [PubMed]
6. Jiang, Y.F.; Sun, D.W.; Pu, H.B.; Wei, Q.Y. Ultrasensitive analysis of kanamycin residue in milk by SERS-based aptasensor. *Talanta* **2019**, *197*, 151–158. [CrossRef] [PubMed]
7. European Commission. European Commission Decision (2002/657/EC) of 12 August 2002 implementing council directive 96/23/EC concerning the performance of analytical methods and interpretation of results. *Off. J. Eur. Communities* **2002**, *221*, 8–36.
8. Isoherranen, N.; Soback, S. Chromatographic methods for analysis of aminoglycoside antibiotics. *J. AOAC Int.* **1999**, *82*, 1017–1045.
9. Blanchaert, B.; Poderós Jorge, E.; Jankovics, P.; Adams, E.; Van Schepdael, A. Assay of Kanamycin A by HPLC with Direct UV Detection. *Chromatographia* **2013**, *76*, 1505–1512. [CrossRef]
10. Manyanga, V.; Dhulipalla, R.L.; Hoogmartens, J.; Adams, E. Improved liquid chromatographic method with pulsed electrochemical detection for the analysis of kanamycin. *J. Chromatogr. A* **2010**, *1217*, 3748–3753. [CrossRef]
11. Khan, S.M.; Miguel, E.M.; de Souza, C.F.; Silva, A.R.; Aucelioa, R.Q. Thioglycolic acid-CdTe quantum dots sensing and molecularly imprinted polymer based solid phase extraction for the determination of kanamycin in milk, vaccine and stream water samples. *Sensor. Actuat. B-Chem.* **2017**, *246*, 444–454. [CrossRef]
12. Wang, L.; Peng, J. LC Analysis of Kanamycin in Human Plasma, by Fluorescence Detection of the 9-Fluorenylmethyl Chloroformate Derivative. *Chromatographia* **2008**, *69*, 519–522. [CrossRef]
13. Chen, S.H.; Liang, Y.C.; Chou, Y.W. Analysis of kanamycin A in human plasma and in oral dosage form by derivatization with 1-naphthyl isothiocyanate and high-performance liquid chromatography. *J. Sep. Sci.* **2006**, *29*, 607–612. [CrossRef]
14. Kim, H.J.; Seo, K.A.; Kim, H.M.; Jeong, E.S.; Ghim, J.L.; Lee, S.H.; Lee, Y.M.; Kim, D.H.; Shin, J.G. Simple and accurate quantitative analysis of 20 anti-tuberculosis drugs in human plasma using liquid chromatography-electrospray ionization-tandem mass spectrometry. *J. Pharmaceut. Biomed. Anal.* **2015**, *102*, 9–16. [CrossRef] [PubMed]
15. Cheng, C.; Liu, S.; Xiao, D.; Hansel, S. The Application of Trichloroacetic Acid as an Ion Pairing Reagent in LC–MS–MS Method Development for Highly Polar Aminoglycoside Compounds. *Chromatographia* **2010**, *72*, 133–139. [CrossRef]
16. Han, M.; Jun, S.H.; Lee, J.H.; Park, K.U.; Song, J.; Song, S.H. Method for simultaneous analysis of nine second-line anti-tuberculosis drugs using UPLC-MS/MS. *J. Antimicrob. Chemoth.* **2013**, *68*, 2066–2073. [CrossRef] [PubMed]
17. Dijkstra, J.A.; Sturkenboom, M.G.; Kv, H.; Koster, R.A.; Greijdanus, B.; Alffenaar, J.W. Quantification of amikacin and kanamycin in serum using a simple and validated LC-MS/MS method. *Bioanalysis* **2014**, *6*, 2125–2133. [CrossRef]
18. Lee, K.; Jun, S.H.; Han, M.; Song, S.H.; Park, J.S.; Lee, J.H.; Park, K.U.; Song, J. Multiplex Assay of Second-Line Anti-Tuberculosis Drugs in Dried Blood Spots Using Ultra-Performance Liquid Chromatography-Tandem Mass Spectrometry. *Ann. Lab. Med.* **2016**, *36*, 489–493. [CrossRef] [PubMed]

19. Morovjan, G.C.; Csokan, P.P.; Nemeth-Konda, L. HPLC Determination of Colistin and Aminoglycoside Antibiotics in Feeds by Post-Column Derivatization and Fluorescence Detection. *Chromatographia* **1998**, *48*, 32–36. [CrossRef]
20. Arsand, J.B.; Jank, L.; Martins, M.T.; Hoff, R.B.; Barreto, F.; Pizzolato, T.M.; Sirtori, C. Determination of aminoglycoside residues in milk and muscle based on a simple and fast extraction procedure followed by liquid chromatography coupled to tandem mass spectrometry and time of flight mass spectrometry. *Talanta* **2016**, *154*, 38–45. [CrossRef]
21. Bousova, K.; Senyuva, H.; Mittendorf, K. Quantitative multi-residue method for determination antibiotics in chicken meat using turbulent flow chromatography coupled to liquid chromatography-tandem mass spectrometry. *J. Chromatogr. A* **2013**, *1274*, 19–27. [CrossRef] [PubMed]
22. Martos, P.A.; Jayasundara, F.; Dolbeer, J.; Jin, W.; Spilsbury, L.; Mitchell, M.; Varilla, C.; Shurmer, B. Multiclass, multiresidue drug analysis, including aminoglycosides, in animal tissue using liquid chromatography coupled to tandem mass spectrometry. *J. Agr. Food Chem.* **2010**, *58*, 5932–5944. [CrossRef]
23. Hu, S.; Song, Y.h.; Bai, X.H.; Jiang, X.; Yang, X.L. Derivatization and Solidification of Floating Dispersive Liquid-phase Microextraction for the Analysis of Kanamycin in Wastewater and Soil by HPLC with Fluorescence Detection. *CLEAN-Soil Air Water* **2014**, *42*, 364–370. [CrossRef]
24. Jia, X.Y.; Gong, D.R.; Xu, B.; Chi, Q.Q.; Zhang, X. Development of a novel, fast, sensitive method for chromium speciation in wastewater based on an organic polymer as solid phase extraction material combined with HPLC-ICP-MS. *Talanta* **2016**, *147*, 155–161. [CrossRef] [PubMed]
25. Zhou, Q.X.; Lei, M.; Li, J.; Liu, Y.L.; Zhao, K.F.; Zhao, D.C. Magnetic solid phase extraction of N- and S-containing polycyclic aromatic hydrocarbons at ppb levels by using a zerovalent iron nanoscale material modified with a metal organic framework of type Fe@MOF-5, and their determination by HPLC. *Microchim. Acta* **2017**, *184*, 1029–1036. [CrossRef]
26. Wang, T.; Zhu, Y.; Ma, J.; Xuan, R.; Gao, H.; Liang, Z.; Zhang, L.; Zhang, Y. Hydrophilic solid-phase extraction of melamine with ampholine-modified hybrid organic-inorganic silica material. *J. Sep. Sci.* **2015**, *38*, 87–92. [CrossRef] [PubMed]
27. Yang, B.; Wang, L.; Luo, C.; Wang, X.; Sun, C. Simultaneous Determination of 11 Aminoglycoside Residues in Honey, Milk, and Pork by Liquid Chromatography with Tandem Mass Spectrometry and Molecularly Imprinted Polymer Solid Phase Extraction. *J. AOAC Int.* **2017**, *100*, 1869–1878. [CrossRef] [PubMed]
28. Moreno-Gonzalez, D.; Hamed, A.M.; Garcia-Campana, A.M.; Gamiz-Gracia, L. Evaluation of hydrophilic interaction liquid chromatography-tandem mass spectrometry and extraction with molecularly imprinted polymers for determination of aminoglycosides in milk and milk-based functional foods. *Talanta* **2017**, *171*, 74–80. [CrossRef]
29. Bohm, D.A.; Stachel, C.S.; Gowik, P. Validation of a method for the determination of aminoglycosides in different matrices and species based on an in-house concept. *Food Addit. Contam. A* **2013**, *30*, 1037–1043. [CrossRef]
30. Wang, Y.; Li, S.; Zhang, F.; Lu, Y.; Yang, B.; Zhang, F.; Liang, X. Study of matrix effects for liquid chromatography-electrospray ionization tandem mass spectrometric analysis of 4 aminoglycosides residues in milk. *J. Chromatogr. A* **2016**, *1437*, 8–14. [CrossRef]
31. Sun, Y.; Li, D.; He, S.; Liu, P.; Hu, Q.; Cao, Y. Determination and dynamics of kanamycin A residue in soil by HPLC with SPE and precolumn derivatization. *Int. J. Environ. Anal. Chem.* **2013**, *93*, 472–481. [CrossRef]
32. Mahrouse, M.A. Simultaneous ultraperformance liquid chromatography/tandem mass spectrometry determination of four antihypertensive drugs in human plasma using hydrophilic-lipophilic balanced reversed-phase sorbents sample preparation protocol. *Biomed. Chromatogr.* **2018**, *32*, e4362. [CrossRef]
33. Zhu, W.X.; Yang, J.Z.; Wei, W.; Liu, Y.F.; Zhang, S.S. Simultaneous determination of 13 aminoglycoside residues in foods of animal origin by liquid chromatography-electrospray ionization tandem mass spectrometry with two consecutive solid-phase extraction steps. *J. Chromatogr. A* **2008**, *1207*, 29–37. [CrossRef] [PubMed]
34. Casado, N.; Damian, P.Q.; Morante-Zarcero, S.; Sierra, I. Bi-functionalized mesostructured silicas as reversed-phase/strong anion-exchange sorbents. Application to extraction of polyphenols prior to their quantitation by UHPLC with ion-trap mass spectrometry detection. *Microchim. Acta* **2019**, *186*, 1–13. [CrossRef] [PubMed]

35. Zhou, Y.X.; Yang, W.J.; Zhang, L.Y.; Wang, Z.Y. Determination of Kanamycin A in Animal Feeds by Solid Phase Extraction and High Performance Liquid Chromatography with Pre-Column Derivatization and Fluorescence Detection. *J. Liq. Chromatogr. R. T.* **2007**, *30*, 1603–1615. [CrossRef]
36. Chen, Y.; Chen, Q.; Tang, S.; Xiao, X. LC method for the analysis of kanamycin residue in swine tissues using derivatization with 9-fluorenylmethyl chloroformate. *J. Sep. Sci.* **2009**, *32*, 3620–3626. [CrossRef]
37. Oertel, R.; Neumeister, V.; Kirch, W. Hydrophilic interaction chromatography combined with tandem-mass spectrometry to determine six aminoglycosides in serum. *J. Chromatogr. A* **2004**, *1058*, 197–201. [CrossRef]
38. Tran, N.H.; Hu, J.Y.; Ong, S.L. Simultaneous determination of PPCPs, EDCs, and artificial sweeteners in environmental water samples using a single-step SPE coupled With HPLC-MS/MS and isotope dilution. *Talanta* **2013**, *113*, 82–92. [CrossRef]
39. Kumar, P.; Rubies, A.; Companyo, R.; Centrich, F. Hydrophilic interaction chromatography for the analysis of aminoglycosides. *J. Sep. Sci.* **2011**, *35*, 498–504. [CrossRef] [PubMed]
40. Kumar, P.; Rubies, A.; Companyo, R.; Centrich, F. Determination of aminoglycoside residues in kidney and honey samples by hydrophilic interaction chromatography-tandem mass spectrometry. *J. Sep. Sci.* **2012**, *35*, 2710–2717. [CrossRef]
41. Diez, C.; Guillarme, D.; Staub Sporri, A.; Cognard, E.; Ortelli, D.; Edder, P.; Rudaz, S. Aminoglycoside analysis in food of animal origin with a zwitterionic stationary phase and liquid chromatography-tandem mass spectrometry. *Anal. Chim. Acta* **2015**, *882*, 127–139. [CrossRef] [PubMed]
42. Lehotay, S.J.; Mastovska, K.; Lightfield, A.R.; Nunez, A.; Dutko, T.; Ng, C.; Bluhm, L. Rapid analysis of aminoglycoside antibiotics in bovine tissues using disposable pipette extraction and ultrahigh performance liquid chromatography-tandem mass spectrometry. *J. Chromatog. A* **2013**, *1313*, 103–112. [CrossRef]
43. Tao, Y.; Chen, D.; Yu, H.; Huang, L.; Liu, Z.; Cao, X.; Yan, C.; Pan, Y.; Liu, Z.; Yuan, Z. Simultaneous determination of 15 aminoglycoside(s) residues in animal derived foods by automated solid-phase extraction and liquid chromatography-tandem mass spectrometry. *Food Chem.* **2012**, *135*, 676–683. [CrossRef] [PubMed]
44. Ishii, R.; Horie, M.; Chan, W.; MacNeil, J. Multi-residue quantitation of aminoglycoside antibiotics in kidney and meat by liquid chromatography with tandem mass spectrometry. *Food Addit. Contam. A* **2008**, *25*, 1509–1519. [CrossRef] [PubMed]
45. Figueiredo, L.; Erny, G.L.; Santos, L.; Alves, A. Applications of molecularly imprinted polymers to the analysis and removal of personal care products: A review. *Talanta* **2016**, *146*, 754–765. [CrossRef]
46. Tan, F.; Sun, D.; Gao, J.; Zhao, Q.; Wang, X.; Teng, F.; Quan, X.; Chen, J. Preparation of molecularly imprinted polymer nanoparticles for selective removal of fluoroquinolone antibiotics in aqueous solution. *J. Hazard. Mater.* **2013**, *244*, 750–757. [CrossRef]
47. Farouk, F.; Azzazy, H.M.; Niessen, W.M. Challenges in the determination of aminoglycoside antibiotics, a review. *Anal. Chim. Acta* **2015**, *890*, 21–43. [CrossRef]
48. Patel, K.N.; Limgavkar, R.S.; Raval, H.G.; Patel, K.G.; Gandhi, T.R. High-Performance Liquid Chromatographic Determination of Cefalexin Monohydrate and Kanamycin Monosulfate with Precolumn Derivatization. *J. Liq. Chromatogr. R. T.* **2015**, *38*, 716–721. [CrossRef]
49. Qian, K.; Tao, T.; Shi, T.; Fang, W.; Li, J.; Cao, Y. Residue determination of glyphosate in environmental water samples with high-performance liquid chromatography and UV detection after derivatization with 4-chloro-3,5-dinitrobenzotrifluoride. *Anal. Chim. Acta* **2009**, *635*, 222–226. [CrossRef]
50. Megoulas, N.C.; Koupparis, M.A. Direct determination of kanamycin in raw materials, veterinary formulation and culture media using a novel liquid chromatography–evaporative light scattering method. *Anal. Chim. Acta* **2005**, *547*, 64–72. [CrossRef]
51. Wei, J.; Shen, A.; Wan, H.; Yan, J.; Yang, B.; Guo, Z.; Zhang, F.; Liang, X. Highly selective separation of aminoglycoside antibiotics on a zwitterionic Click TE-Cys column. *J. Sep. Sci.* **2014**, *37*, 1781–1787. [CrossRef]
52. Zhang, Y.; He, H.M.; Zhang, J.; Liu, F.J.; Li, C.; Wang, B.W.; Qiao, R.Z. HPLC-ELSD determination of kanamycin B in the presence of kanamycin A in fermentation broth. *Biomed. Chromatogr.* **2015**, *29*, 396–401. [CrossRef]
53. Adams, E.; Dalle, J.; De, B.E.; De, S.I.; Roets, E.; Hoogmartens, J. Analysis of kanamycin sulfate by liquid chromatography with pulsed electrochemical detection. *J. Chromatogr. A* **1997**, *766*, 133–139. [CrossRef]
54. Chopra, S.; Vanderheyden, G.; Hoogmartens, J.; Van Schepdael, A.; Adams, E. Comparative study on the analytical performance of different detectors for the liquid chromatographic analysis of tobramycin. *J. Pharmaceut. Biomed. Anal.* **2010**, *53*, 151–157. [CrossRef]

55. Li, B.; Van Schepdael, A.; Hoogmartens, J.; Adams, E. Characterization of impurities in tobramycin by liquid chromatography-mass spectrometry. *J. Chromatogr. A* **2009**, *1216*, 3941–3945. [CrossRef]
56. Gremilogianni, A.M.; Megoulas, N.C.; Koupparis, M.A. Hydrophilic interaction vs ion pair liquid chromatography for the determination of streptomycin and dihydrostreptomycin residues in milk based on mass spectrometric detection. *J. Chromatogr. A* **2010**, *1217*, 6646–6651. [CrossRef]
57. Babin, Y.; Fortier, S. A high-throughput analytical method for determination of aminoglycosides in veal tissues by liquid chromatography/tandem mass spectrometry with automated cleanup. *J. AOAC Int.* **2007**, *90*, 1418–1426.
58. Buszewski, B.; Noga, S. Hydrophilic interaction liquid chromatography (HILIC)—A powerful separation technique. *Anal. Bioanal. Chem.* **2012**, *402*, 231–247. [CrossRef]

Quality Evaluation of *Gastrodia Elata* Tubers based on HPLC Fingerprint Analyses and Quantitative Analysis of Multi-Components by Single Marker

Yehong Li [2,†], Yiming Zhang [1,†], Zejun Zhang [1], Yupiao Hu [1], Xiuming Cui [1,3,4] and Yin Xiong [1,3,4,*]

1 Faculty of Life Science and Technology, Kunming University of Science and Technology, Kunming 650500, China; jr93586@163.com (Y.Z.); 18380802826@126.com (Z.Z.); hypflygo@163.com (Y.H.); cuisanqi37@163.com (X.C.)
2 School of Pharmacy, China Pharmaceutical University, Nanjing 210009, China; yhlcpu@163.com
3 Yunnan Key Laboratory of *Panax notoginseng*, Kunming University of Science and Technology, Kunming 650500, China
4 Institute of Biology Leiden, Leiden University, 2333BE Leiden, The Netherlands
* Correspondence: yhsiung@163.com
† These authors contributed equally to this work.

Academic Editors: Marcello Locatelli, Angela Tartaglia, Dora Melucci, Abuzar Kabir, Halil Ibrahim Ulusoy and Victoria Samanidou

Abstract: *Gastrodia elata* (*G. elata*) tuber is a valuable herbal medicine used to treat many diseases. The procedure of establishing a reasonable and feasible quality assessment method for *G. elata* tuber is important to ensure its clinical safety and efficacy. In this research, an effective and comprehensive evaluation method for assessing the quality of *G. elata* has been developed, based on the analysis of high performance liquid chromatography (HPLC) fingerprint, combined with the quantitative analysis of multi-components by single marker (QAMS) method. The contents of the seven components, including gastrodin, *p*-hydroxybenzyl alcohol, *p*-hydroxy benzaldehyde, parishin A, parishin B, parishin C, and parishin E were determined, simultaneously, using gastrodin as the reference standard. The results demonstrated that there was no significant difference between the QAMS method and the traditional external standard method (ESM) ($p > 0.05$, RSD $< 4.79\%$), suggesting that QAMS was a reliable and convenient method for the content determination of multiple components, especially when there is a shortage of reference substances. In conclusion, this strategy could be beneficial for simplifying the processes in the quality control of *G. elata* tuber and giving references to promote the quality standards of herbal medicines.

Keywords: *Gastrodia elata* tuber; quality evaluation; HPLC; QAMS

1. Introduction

Gastrodia elata (G. elata) Blume is a traditional medicinal herb that has been used in oriental countries, for centuries, to treat general paralysis, headaches, dizziness, rheumatism, convulsion, and epilepsy [1,2]. Modern pharmacological studies have demonstrated that the extracts of *G. elata* tuber and some compounds that originate from it, possesses wide-reaching biological activities, including anti-tumor, anti-virus, memory-improving, anti-oxidation, and anti-aging actions [3–5]. Nowadays, it is also widely used as a sub-material in food and Chinese Patent Medicines (CPM) [6], and this herbal medicine is also listed as one of the functional foods approved by the Ministry of Health in China [7,8]. As the wild *G. elata* is not sufficient enough for commercial large-scale exploitation, its artificial cultivation in medicine has become essential, to meet the increasing requirement of markers [6].

Due to their high medicinal value, *G. elata* tubers have been cultivated and produced in many areas of Asia, like China and Korea, which could lead to great differences in quality and, possibly, could lead to differences in the following clinical efficacies. Many studies have indicated that the efficacy and quality of herbal medicines are somewhat different depending on the cultivation soil and climate, based on the geographic origin, even when coming from the same species [9,10]. Therefore, a reasonable and effective method for the quality evaluation of *G. elata* tuber, plays an important role in its medication safety.

Gastrodin and its aglycone (*p*-hydroxybenzyl alcohol) are major components of the *G. elata* tuber, which are also markers for the quality control of this herbal medicine [11]. However, over 81 compounds from *G. elata* tuber have been currently isolated and identified. Along with the above two marker components, others like *p*-hydroxy benzaldehyde, parishin A, parishin B, parishin C, parishin E, and so on have also been reported to be correlated with the bioeffects of the *G. elata* tuber [12,13]. Accordingly, a qualitative analysis and quantification of one or two compounds, could be insufficient for a complete profile of the chemical characterization of the *G. elata* tuber, due to its complex compositions. In recent years, the chromatographic fingerprint analysis has been accepted as a strategy for the quality assessment of herbal medicines and preparations by the US Food and Drug Administration [14], State Food and Drug Administration of China [15], and the European Medicines Agency [16]. Since the fingerprint is characterized by more chemical information, the method is often used for the origin identification, species authentication, and quality control for herbal medicines, by observing the presence or absence of a limited number of peaks in the chromatographic fingerprints [17,18]. Therefore, the fingerprint analysis of high performance liquid chromatography (HPLC) was developed for the qualitative analysis of *G. elata* tuber.

A single standard to determine multiple components, also known as the quantitative analysis of multi-components by single marker (QAMS) [19], is a novel method designed for the quality evaluation of herbal medicines and related products [20]. Researchers have used QAMS to determine three components in Fructus Evodiae, simultaneously, by using rutaecarpine as the internal reference compound to calculate the relative correction factor of evodin and evodiamine [21]. To make up for the limitations of the fingerprint which cannot be quantified accurately, a QAMS method using berberine as the standard, was developed and validated for a simultaneous quantitative analysis of fourteen components [22]. This strategy could not only reduce the cost of the experiment and time of detection but could also be independent of the availability of all target ingredients [19]. Thus, the QAMS method was applied for a quantitative analysis of *G. elata* tuber.

This study aimed to establish a reliable and practical method, realizing both qualitative and quantitative analyses for *G. elata* tuber, via HPLC fingerprinting, combined with QAMS. The differences and similarities of the HPLC fingerprints were visually compared, using a hierarchical cluster analysis (HCA) and similarity analysis. The contents of seven major active constituents were accurately determined by both the QAMS method and external standard method (ESM), through which we hoped to offer a suitable and efficient approach for assessing the quality of *G. elata* tuber.

2. Results and Discussion

2.1. Optimization of the Chromatographic Conditions

As the components of *G. elata* tuber are very intricate, it is critical to optimize the chromatographic conditions, including favorable mobile phase systems, gradient elution systems, and the detection wavelength, to obtain an efficient separation of the target components. Lei [23] indicated that the HPLC fingerprints of *G. elata* tubers were the most informative, while the UV wavelength was 220 nm from HPLC-DAD-3D spectrum of *G. elata* tuber. So in this case, we chose the UV wavelength of 220 nm, to determinate the selected components. We chose acetonitrile-water containing 0.1% phosphoric acid system. The samples were dissolved in 60% methanol and ultrasound, for 60 min. We optimized the gradient elution system as Section 3.5, and 35 °C was selected as the proper temperature for analysis, while the flow rate was set at 1.0 mL/min. The S1 sample of *G. elata* tuber and the mixed standards containing seven reference substances were analyzed to obtain the HPLC fingerprints (Figure 1) under the conditions of Section 3.5, producing sharp and symmetrical chromatographic peak shapes, good separation, and preventing the peak tailing.

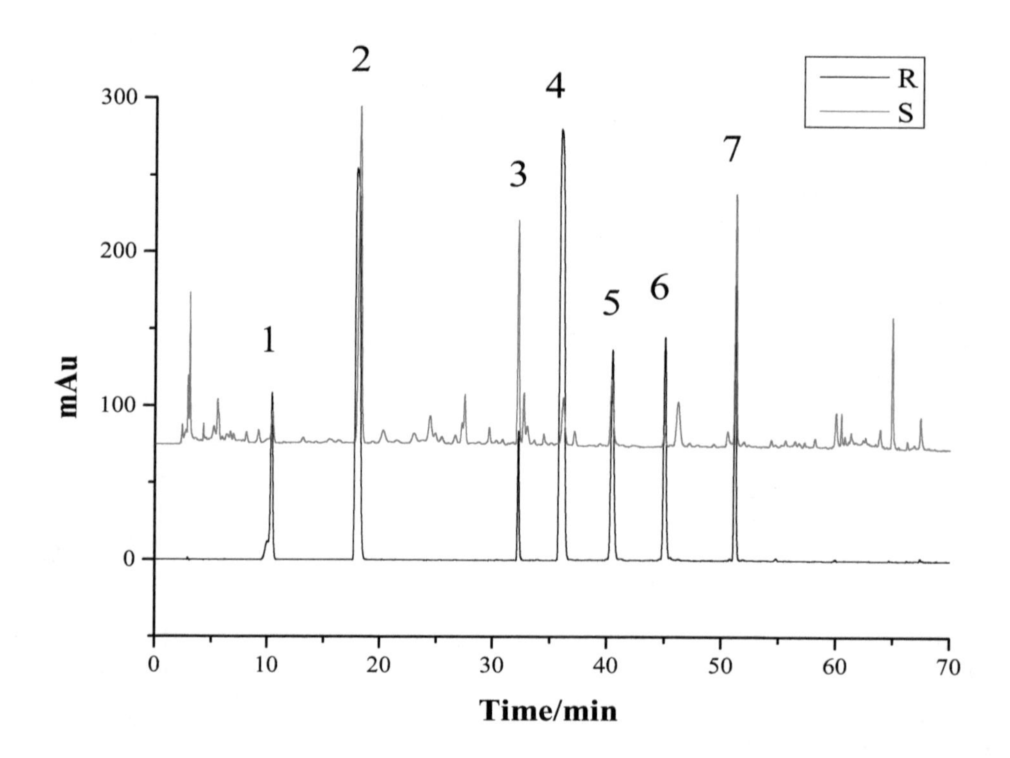

Figure 1. The HPLC fingerprints of the *Gastrodia elata* tuber sample and the mixed standards. R: The mixed standards; S: The *G. elata* tuber sample. 1—Gastrodin; 2—*p*-Hydroxy benzyl alcohol; 3—Parishin E; 4—*p*-Hydroxy benzaldehyde; 5—Parishin B; 6—Parishin C; 7—Parishin A.

According to the retention time of each peak in the chromatogram [24], the peaks of 1, 2, 3, 4, 5, 6, and 7 were identified to be gastrodin, *p*-hydroxybenzyl alcohol, parishin E, *p*-hydroxy benzaldehyde, parishin B, parishin C, and parishin A. The separation degree of each peak was greater than 1.5, in the present HPLC system, indicating the peaks were well-separated, under the chromatographic conditions.

2.2. Method Validation

2.2.1. Linearity

The mixed reference solution containing all the reference substances was diluted in series, with 60% methanol, to obtain six different concentrations for the seven reference curves. The linearity of each analyte was assessed by plotting its calibration curve with different concentrations and the corresponding peak areas. The results were shown in Table 1. The high correlation coefficient values indicated that there was a good correlation between the concentration and peak area of the seven compounds, at a relatively wide range of concentrations. The correlation coefficient of more than 0.9990, indicated a satisfactory linearity. The calibration curve could be utilized for the quantitative analysis in the given concentration range. The standard solution of the individual analyte was diluted gradually, to determine its Limit of Detection (LOD) and Limit of Quantity (LOQ) with signal-to-noise ratio of 3:1 and 10:1, respectively. LOD and LOQ values for the analytes are also listed in Table 1.

Table 1. The regression equations, Limit of Detection (LODs) and Limit of Quantity (LOQs) of seven components.

Analytes	Regression Equations	Linear Ranges (mg/mL)	R^2	LOD (mg/mL)	LOQ (mg/mL)
Gastrodin	$Y = 18634X - 264.07$	1.906~6.483	0.9997	0.042	0.139
p-Hydroxybenzyl alcohol	$Y = 39300X + 42.955$	0.075~1.773	0.9995	0.001	0.003
Parishin E	$Y = 14141X + 142.93$	2.273~7.052	0.9997	0.037	0.122
p-Hydroxy benzaldehyde	$Y = 52536X + 7.9174$	0.079~2.588	1.0000	0.001	0.005
Parishin B	$Y = 20791X + 6.7746$	1.450~5.190	1.0000	0.004	0.015
Parishin C	$Y = 31240X - 335.24$	0.286~0.356	0.9997	0.005	0.015
Parishin A	$Y = 11769X - 100.83$	0.181~19.301	0.9995	0.020	0.070

2.2.2. Precision, Stability, Repeatability, and Accuracy

The precision was evaluated according to the assay of S1, in which the solution was analyzed for six times in a day, to evaluate the intra-day precision, and was analyzed on three consecutive days, to evaluate the inter-day precision. Calculating the RSDs of each chromatographic peak, the results showed that the RSDs of gastrodin, *p*-hydroxybenzyl alcohol, parishin E, *p*-hydroxy benzaldehyde, parishin B, parishin C, and parishin A were 1.93%, 1.10%, 1.29%, 2.30%, 2.03%, 2.63%, and 0.89% (n = 6), respectively, indicating that the precision of the method was good.

The stability was tested with the S1 solution that was stored at room temperature (25 ± 5 °C) and analyzed at 0, 2, 4, 6, 8, 12, and 24 h, to calculate the RSDs. The results showed that the RSDs of gastrodin, *p*-hydroxybenzyl alcohol, parishin E, *p*-hydroxy benzaldehyde, parishin B, parishin C, and parishin A were 1.15%, 2.04%, 1.51%, 2.37%, 2.10%, 1.12%, and 2.25%, respectively, suggesting that the method was stable within 24 h.

In the repeatability test, six duplicates of S1 were extracted and analyzed, according to the sample preparation procedure, and the HPLC method. The RSDs of the peak areas were calculated. The results showed that the RSDs of gastrodin, *p*-hydroxybenzyl alcohol, parishin E, *p*-hydroxy benzaldehyde, parishin B, parishin C, and parishin A were 1.25%, 2.15%, 1.60%, 1.81%, 1.72%, 1.84%, and 1.60% (n = 6), respectively, indicating that the repeatability of the method was good.

In the accuracy test, certain amounts of the seven analytes' standards were added to the *G. elata* tuber samples (S1), with the six replicates. Then, these seven mixed samples were treated, as in the method described above. Recovery rate was used as the evaluation index and calculated as Recovery rate (%) = (Found amount − Known amount) × 100%/Added amount. The RSD of the accuracy values of the seven components are shown in Table 2, respectively.

Table 2. RSD of precision, stability, repeatability and accuracy for determination of seven components.

Analyte RSD (%)	Precision RSD (%)	Stability RSD (%)	Repeatability RSD (%)	Accuracy	
				Mean (%)	RSD (%)
Gastrodin	1.93	1.15	1.25	92.05%	2.02%
p-Hydroxybenzyl alcohol	1.10	2.04	2.15	95.78%	1.09%
Parishin E	1.29	1.51	1.60	98.05%	2.90%
p-Hydroxy benzaldehyde	2.30	2.37	1.81	92.44%	0.25%
Parishin B	2.03	2.10	1.72	93.33%	1.32%
Parishin C	2.63	1.12	1.84	92.91%	2.10%
Parishin A	0.89	2.25	1.60	91.80%	1.36%

The HPLC method was validated in terms of precision, repeatability, stability, and accuracy, as shown in Table 2. The RSD of the precision values of the seven components were less than 2.63%. RSD values for the stability and the repeatability were less than 2.37% and 2.15%, respectively. The recovery rates of the analytes ranged from 91.80% to 98.05%, with the RSD values being lower than 2.90%. All results indicated that the developed method was stable, accurate, and repeatable. This established HPLC method could be applied for a simultaneous determination of gastrodin, *p*-hydroxybenzyl alcohol, parishin E, *p*-hydroxy benzaldehyde, parishin B, parishin C, and parishin A, in the *G. elata* tuber samples.

2.3. HPLC Fingerprints Analysis

The 21 batches of *G. elata* tuber samples from the different producing areas were prepared according to Section 3.3, and 10 μL of S1 sample solution was injected into the HPLC system according to the chromatographic conditions in Section 3.5, to obtain the fingerprints. The retention time was the horizontal axis and the peak area was the vertical axis; the 3D fingerprints of the 21 batches of *G. elata* tuber samples were established by the software Origin 9.0, as shown in Figure 2.

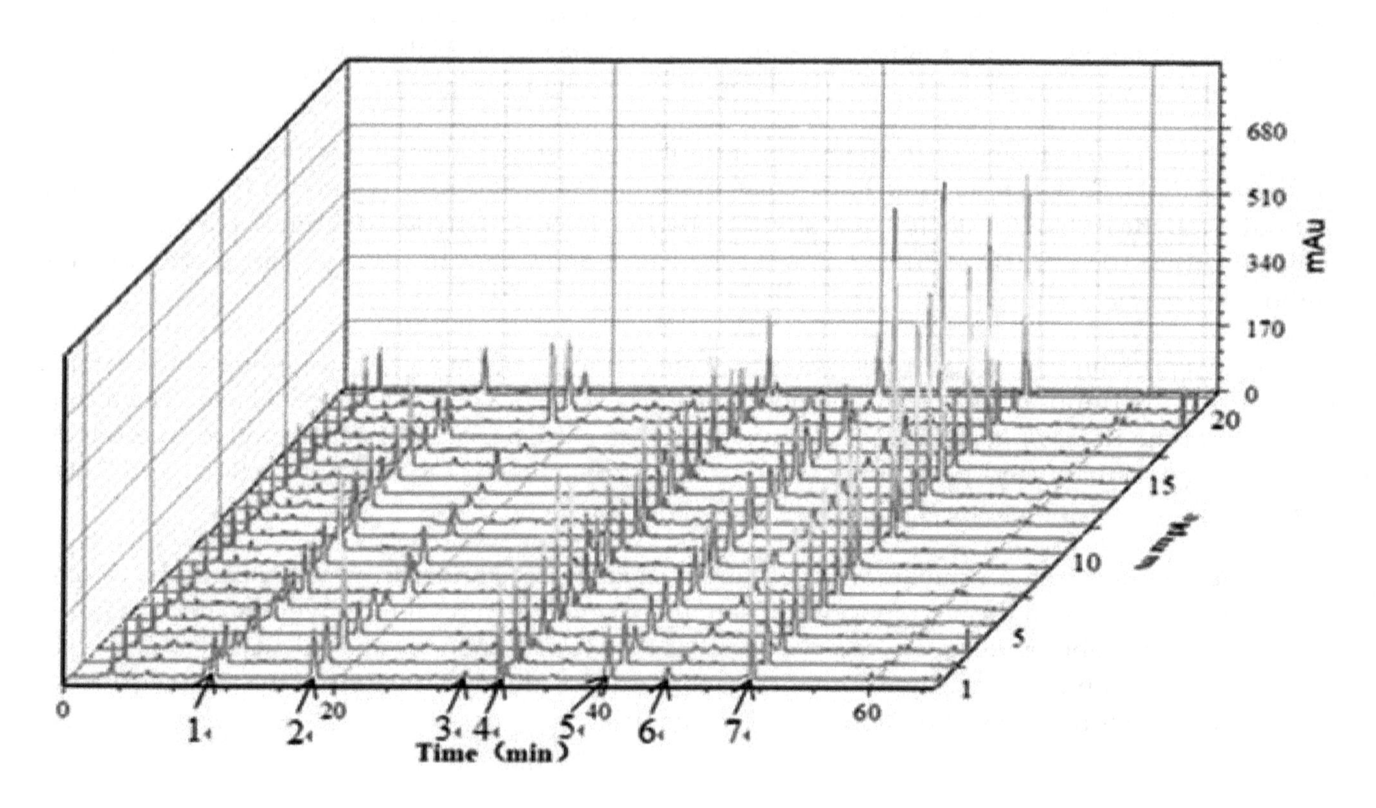

Figure 2. HPLC fingerprints of the 21 batches of *G. elata* tuber samples. 1—Gastrodin; 2—*p*-Hydroxy benzyl alcohol; 3—Parishin E; 4—*p*-Hydroxy benzaldehyde; 5—Parishin B; 6—Parishin C; 7—Parishin A.

According to Figure 2, the seven peaks with stable and better shape were determined to be the major ones for the HPLC fingerprints of *G. elata* tubers. The peak areas of the seven peaks are shown in Table 3. The variance coefficients of the peak area were greater than 32.2 percent, indicating that the content of each marker component varied greatly from place to place.

Table 3. The information and peak areas of the seven characteristic peaks in HPLC fingerprints of *G. elata* tubers.

No.	Peak Area of Seven Characteristic Peaks						
	Gastrodin	*p*-Hydroxy Benzyl Alcohol	Parishin E	*p*-Hydroxy Benzaldehyde	Parishin B	Parishin C	Parishin A
S1	1797.1	2249.5	2337.8	217.4	2263.6	420.7	4340.3
S2	1470.2	2144.3	2523.4	227.6	2526.6	462.2	4561.9
S3	623.8	4536.4	1528.4	301.7	1017	280.3	1487.8
S4	1325.1	1516.1	1412.1	116.7	1906.9	402.8	3150
S5	1659.3	2123.3	1991.3	111.9	2141.7	383.4	4006.4
S6	1161	1463.7	3734.3	108.1	1867.1	390.2	3167.6
S7	1492.8	663.8	2991.6	85.8	1818.7	392.5	3473.5
S8	1470.9	823.8	1573.2	127.1	2231.8	546.3	5104.3
S9	1898	1876.6	2572.7	82	2629.1	595.9	4430.8
S10	3816.6	136.2	1316.2	110.1	2663.3	441.9	3383.6
S11	2353.9	970.8	1563.7	131.9	3073.4	789.5	9224.6
S12	1794	830	2577.2	45.7	2141.8	101.1	3845.8
S13	2344.5	572.4	2363.1	57.6	2039.1	499.2	5019.1
S14	1369.4	427.8	1961.6	41.9	2408.9	622.1	5512.4
S15	2177.5	1270.2	2076.6	56.6	3133.4	791.2	8184.9
S16	3322.1	108.1	1240.9	73.1	1935.1	357.8	2127.8
S17	1081.8	322.8	2365	104.7	2363.6	500.7	5062.6
S18	1893.7	270.9	1719.4	78.3	2475.6	823	6072.7
S19	380.4	4012.7	1414.3	617.3	781.5	136.4	1789.1
S20	300.9	3287.7	878.9	564	479.5	102.3	687.1
S21	2175.1	1076.8	2057.2	143.7	2826.4	94.7	6278.5
C.V. (%) [1]	49.7	85.2	33.3	96.5	32.2	50.1	47.9

[1] C.V. (%) = $\delta/\mu \times 100$, δ—The standard deviation of peak area and μ—The average value of each peak area.

2.4. Similarity Analysis

According to the data of HPLC fingerprints in Figure 2, the similarity of HPLC fingerprints from the different producing regions were evaluated using the Similarity Evaluation System for chromatographic fingerprint of traditional Chinese medicines (TCM) (Version 2012), with correlation coefficient (median) on behalf of the similarity of HPLC fingerprints. We utilized the average correlation coefficient method of 21 batches of the samples for the multipoint correction, and the time window width was set to 0.5 [25], while the establishment of a common model was to generate a control fingerprints of the *G. elata* tuber. Compared with the reference fingerprint chromatogram (R), the similarities of the 21 batches of samples were higher than 0.96, indicating that the batch-to-batch consistency was good. The results suggested that those samples of *G. elata* tuber had a similar chemical composition, and the samples were collected from the same genus, even though they were from different producing countries or were produced under different processing conditions (Table 4). Therefore, the developed fingerprint by HPLC could be used as a practical tool for the qualitative identification of the *G. elata* tuber.

Table 4. Similarity of the *G. elata* tuber samples.

No.	Similarity	No.	Similarity	No.	Similarity	No.	Similarity
S1	0.983	S7	0.970	S13	0.988	S19	0.990
S2	0.987	S8	0.988	S14	0.982	S20	0.988
S3	0.983	S9	0.989	S15	0.979	S21	0.988
S4	0.975	S10	0.982	S16	0.989	*R*	1.000
S5	0.983	S11	0.987	S17	0.980		
S6	0.975	S12	0.990	S18	0.964		

2.5. Hierarchical Cluster Analysis (HCA)

Using the peak areas of the seven compounds from the 21 *G. elata* tuber samples as the clustering variable, the HCA of the standardized data was performed with the heat map software of Heml 1.0. The graph in Figure 3 illustrated that the samples could be categorized into three groups. Group 1 contained S1 and S2 from Zhaotong, Yunnan in China; Group 2 contained S19 and S20 tubers from South Korea; and Group 3 contained the rest of samples. From the result, the samples from the same producing area were not always classified into the same group. For example, Zhaotong has been considered as the Daodi production area (area which produces authentic and superior medicinal materials) of the *G. elata* tuber in China. However, samples 1 to 6 from Zhaotong, showed different levels and ratios of chemical components, which could be due to the variations in harvesting time, planting patterns, dying methods, and other factors. Additionally, the preliminary processing method also contributes to the differences in the chemical composition. For instance, *G. elata* tubers and slices from South Korea were classified into different categories. Therefore, it is insufficient to determine the quality of the *G. elata* tubers by only their producing areas or any other single factor. Although the HCA could be used to classify the *G. elata* tubers on the basis of the peak areas of the seven components, it was hard to tell which group had a better quality. Therefore, other methods for the quantitative analysis of *G. elata* tubers should be developed, to reflect the quality difference.

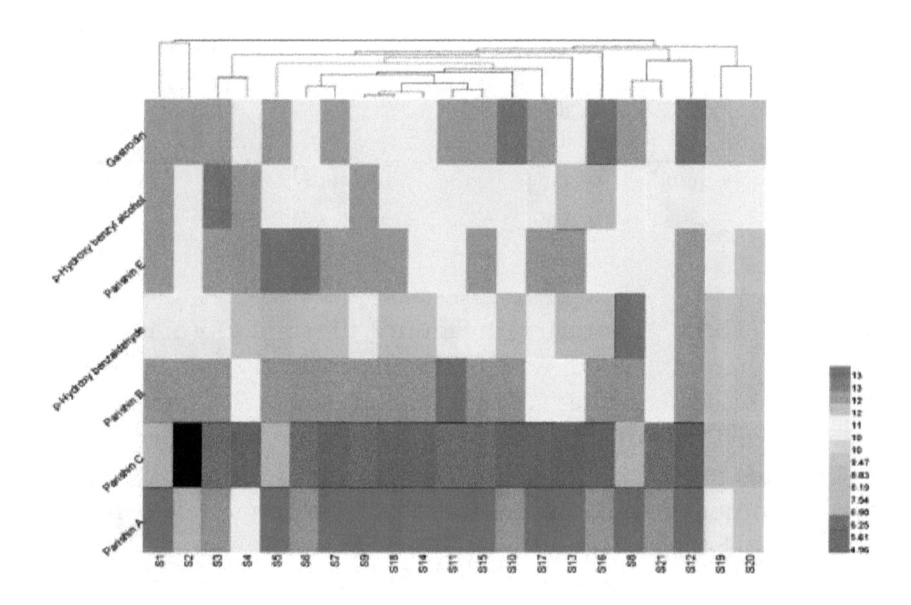

Figure 3. Clustering analysis graph of the 21 *G. elata* tuber samples.

2.6. Quantitative Analysis of Multiple Components by Single Marker

Theoretically, the quantity (mass or concentration) of an analyte is in direct proportion of the detector response. Then, in multi-component quantitation, a typical botanical compound (readily available) might be selected as an internal standard and the relative correction factor (RCF) of this marker, and the other components can be calculated.

2.6.1. Calculation of RCFs

It is of vital importance to select a proper internal referring standard for the accurate assay of multiple components in TCM. The component chosen as the internal referring substance should be stable, easily obtainable, and have relatively clear pharmacologic effects related to the clinical efficacy of the herbal medicine [26]. In this work, the gastrodin was used as an internal referring substance for its easy availability, lower cost, moderate retention value, and good stability.

In order to simultaneously determine the contents of the seven components in the *G. elata* tuber, by using the QAMS method, the relative correction factors (RCFs, f_x) were first determined, according to the ratio of the peak areas and the ratio of the concentration between the gastrodin and other compounds, as described in Section 3.6. We calculated the RCFs of six components (shown in Table 5).

Table 5. Relative correction factor (RCF) values of six components of the *G. elata* tuber.

Instrument	Chromatogram Column	RCF Values	
Agilent 1260	YMC-Tyiart C18 (250 × 4.6 mm, 5 μm)	$f_{\text{P-hydroxy benzyl alcohol/gastrodin}}$	2.1090
		$f_{\text{parishin E/gastrodin}}$	0.7589
		$f_{\text{P-hydroxy benzaldehyde/gastrodin}}$	2.8194
		$f_{\text{parishin B/gastrodin}}$	1.1156
		$f_{\text{parishin C/gastrodin}}$	1.6771
		$f_{\text{parishin A/gastrodin}}$	0.6316

2.6.2. Results from the QAMS Method

After preparing the sample solutions of *G. elata* tubers, they were injected into the HPLC system to obtain the peak areas. The contents of seven compounds were calculated, according to the calibration curves. Those scattered in the vicinity of the lowest concentration point on the standard curve were determined with a one point ESM. Meanwhile, the contents of the seven components of the *G. elata* tuber calculated according to QAMS method, are shown in Table 6.

The validated traditional ESM and QAMS method were employed to test the 21 batches of *G. elata* tuber samples from the different producing areas, which were based on the principle of the linear relationship between a detector response and the levels of components within certain concentration ranges. The validation of the QAMS method might be implemented, based on *t*-test, correlation coefficient [27], RSD [28], and relative error [29], through a comparison with an external standard. Correlation coefficient, as a statistical parameter, ranging from 0 (no correlation) to 1 (complete correlation), reflecting the closeness of two variables, is often used in similarity assessments of traditional Chinese medicine fingerprints [30]. As shown in Table 7, Correlation coefficients of the assay results obtained from the two methods were calculated here; all coefficients were found to be >0.998.

The data showed that the results of the two methods were highly correlated. Then, a *t*-test was performed for the calculated results, by the QAMS method, and the on detected results, by an external standard method. *p*-values of gastrodin, *p*-hydroxy benzyl alcohol, parishin E, *p*-hydroxy benzaldehyde, parishin B, parishin C and parishin A, were all >0.05. The relative error and RSD values were all lower than 5%. Above all, the results indicated that there was no significant difference between the data from the QAMS and the ESM method, indicating that the present QAMS method was reliable for the simultaneous quantification of the seven components of the *G. elata* tuber.

Table 6. Contents of the seven components in *G. elata* tubes determined by the external standard method (ESM) and the quantitative analysis of multi-components by single marker (QAMS) methods ($mg \cdot g^{-1}$) [1].

No.	Gastrodin	*p*-Hydroxy Benzyl Alcohol		Parishin E		*p*-Hydroxy Benzaldehyde		Parishin B		Parishin C		Parishin A		Total
		ESM	QAMS	ESM	QAMS	ESM	QAMS	ESM	QAMS	ESM	QAMS	ESM	QAMS	
S1	5.23 ± 0.16	1.77 ± 0.05	1.82 ± 0.05	5.14 ± 0.01	5.35 ± 0.03	0.24 ± 0.01	0.25 ± 0.06	3.54 ± 0.14	3.60 ± 0.05	0.18 ± 0.00	0.19 ± 0.01	11.58 ± 0.45	11.71 ± 0.49	27.68
S2	4.51 ± 0.38	1.61 ± 0.08	1.64 ± 0.06	2.27 ± 0.14	2.38 ± 0.08	0.24 ± 0.01	0.25 ± 0.01	3.43 ± 0.21	3.47 ± 0.01	0.13 ± 0.00	0.14 ± 0.00	0.18 ± 0.00	0.00 ± 0.00	12.38
S3	2.44 ± 0.28	1.07 ± 0.08	1.09 ± 0.10	4.82 ± 0.25	5.00 ± 0.04	0.26 ± 0.01	0.26 ± 0.02	2.84 ± 0.10	2.88 ± 0.04	0.16 ± 0.02	0.16 ± 0.02	8.33 ± 0.31	8.32 ± 0.11	19.91
S4	1.35 ± 0.02	3.36 ± 0.12	3.39 ± 0.08	2.62 ± 0.06	2.62 ± 0.23	0.23 ± 0.01	0.24 ± 0.04	1.45 ± 0.04	1.46 ± 0.09	0.16 ± 0.00	0.16 ± 0.10	4.23 ± 0.12	4.10 ± 0.31	13.41
S5	3.41 ± 0.38	1.55 ± 0.10	1.58 ± 0.07	2.31 ± 0.22	2.36 ± 0.10	0.12 ± 0.02	0.13 ± 0.01	2.49 ± 0.12	2.51 ± 0.11	0.20 ± 0.00	0.20 ± 0.01	8.94 ± 0.63	8.92 ± 0.29	19.03
S6	1.91 ± 0.12	1.03 ± 0.07	1.06 ± 0.10	7.05 ± 0.13	7.26 ± 0.23	0.23 ± 0.02	0.24 ± 0.05	2.69 ± 0.03	2.73 ± 0.17	0.17 ± 0.00	0.17 ± 0.02	7.64 ± 0.31	7.63 ± 0.61	20.72
S7	3.12 ± 0.01	0.51 ± 0.00	0.531 ± 0.00	6.02 ± 0.14	6.20 ± 0.09	0.21 ± 0.01	0.21 ± 0.00	2.71 ± 0.02	2.73 ± 0.01	0.15 ± 0.00	0.16 ± 0.00	8.87 ± 0.02	8.86 ± 0.08	21.58
S8	3.06 ± 0.10	0.62 ± 0.01	0.64 ± 0.01	3.00 ± 0.18	3.13 ± 0.17	2.59 ± 0.02	2.60 ± 0.05	3.25 ± 0.15	3.27 ± 0.01	0.15 ± 0.01	0.14 ± 0.04	12.77 ± 0.58	12.75 ± 0.54	25.44
S9	2.85 ± 0.37	1.22 ± 0.18	1.24 ± 0.20	4.54 ± 0.03	4.69 ± 0.06	0.28 ± 0.01	0.28 ± 0.01	3.61 ± 0.06	3.63 ± 0.15	0.17 ± 0.01	0.16 ± 0.05	10.78 ± 0.15	10.77 ± 0.08	23.44
S10	5.89 ± 0.22	0.10 ± 0.01	0.10 ± 0.01	3.37 ± 0.24	3.52 ± 0.23	0.11 ± 0.01	0.11 ± 0.06	3.84 ± 0.13	3.91 ± 0.05	0.15 ± 0.02	0.16 ± 0.00	7.90 ± 0.67	7.91 ± 0.63	21.36
S11	4.74 ± 0.37	0.69 ± 0.08	0.71 ± 0.08	3.40 ± 0.22	3.55 ± 0.22	0.26 ± 0.02	0.26 ± 0.04	5.191 ± 0.09	5.23 ± 0.02	0.18 ± 0.01	0.19 ± 0.00	26.70 ± 0.46	26.93 ± 0.54	41.15
S12	7.10 ± 0.27	0.65 ± 0.04	0.66 ± 0.01	4.88 ± 0.23	5.04 ± 0.11	0.08 ± 0.12	0.08 ± 0.02	3.03 ± 0.16	3.18 ± 0.01	0.15 ± 0.02	0.15 ± 0.00	9.43 ± 0.54	9.80 ± 0.10	25.32
S13	4.03 ± 0.03	0.37 ± 0.01	0.39 ± 0.01	3.69 ± 0.11	3.86 ± 0.11	0.16 ± 0.01	0.17 ± 0.02	2.37 ± 0.06	2.41 ± 0.06	0.15 ± 0.00	0.15 ± 0.00	10.27 ± 0.24	10.33 ± 0.24	21.04
S14	2.59 ± 0.03	0.19 ± 0.16	0.20 ± 0.00	2.97 ± 0.08	3.10 ± 0.07	0.20 ± 0.00	0.21 ± 0.02	2.88 ± 0.06	2.90 ± 0.05	0.15 ± 0.01	0.15 ± 0.00	11.61 ± 0.37	11.58 ± 0.33	20.59
S15	4.29 ± 0.15	0.92 ± 0.04	0.94 ± 0.04	3.76 ± 0.16	3.90 ± 0.16	0.31 ± 0.02	0.31 ± 0.07	4.27 ± 0.17	4.30 ± 0.10	0.13 ± 0.00	0.14 ± 0.00	19.30 ± 0.83	19.42 ± 0.84	32.98
S16	6.48 ± 0.21	0.08 ± 0.00	0.08 ± 0.00	2.35 ± 0.11	2.41 ± 0.13	0.11 ± 0.00	0.11 ± 0.05	2.75 ± 0.11	2.78 ± 0.04	0.15 ± 0.01	0.15 ± 0.01	5.24 ± 0.15	5.16 ± 0.20	17.15
S17	5.09 ± 0.39	0.13 ± 0.22	0.13 ± 0.26	4.20 ± 0.05	4.39 ± 0.05	0.24 ± 0.00	0.25 ± 0.07	1.54 ± 0.17	1.56 ± 0.08	0.15 ± 0.01	0.14 ± 0.01	11.65 ± 0.22	11.77 ± 0.23	23.01
S18	3.71 ± 0.05	0.20 ± 0.01	0.20 ± 0.01	4.22 ± 0.09	4.41 ± 0.13	0.20 ± 0.00	0.20 ± 0.04	3.47 ± 0.08	3.52 ± 0.03	0.15 ± 0.00	0.16 ± 0.01	14.60 ± 0.21	14.76 ± 0.24	26.54
S19	0.42 ± 0.01	1.23 ± 0.02	1.26 ± 0.02	1.10 ± 0.01	1.14 ± 0.01	0.18 ± 0.01	0.18 ± 0.00	0.46 ± 0.01	0.47 ± 0.01	0.38 ± 0.01	0.39 ± 0.04	1.99 ± 0.06	1.98 ± 0.05	5.76
S20	0.36 ± 0.02	1.01 ± 0.01	1.04 ± 0.01	0.64 ± 0.00	0.65 ± 0.00	0.12 ± 0.00	0.11 ± 0.00	0.28 ± 0.00	0.29 ± 0.00	0.30 ± 0.01	0.30 ± 0.03	0.83 ± 0.09	0.82 ± 0.00	3.52
S21	1.50 ± 0.14	0.33 ± 0.00	0.34 ± 0.00	1.70 ± 0.40	1.71 ± 0.04	0.68 ± 0.02	0.69 ± 0.01	1.75 ± 0.02	1.78 ± 0.01	0.17 ± 0.00	0.16 ± 0.01	6.62 ± 0.03	6.61 ± 0.03	12.75
Mean	3.53		0.91		3.65		0.34		2.79		0.18		9.53	20.70

[1] ESM—external standard method, and its content was determined by the calibration equation method; QAMS—quantitative analysis multi-components by single marker, and its content was determined by RCFs; RSD—relative standard deviation; Total—the sum of the six alkaloid contents in each batch.

Table 7. The relative error, RSD, correlation coefficient, and p values of the contents from the ESM and the QAMS [1].

No.	*p*-Hydroxy Benzyl Alcohol		Parishin E		*p*-Hydroxy Benzaldehyde		Parishin B		Parishin C		Parishin A	
	Relative Error	RSD	Relative Error	RSD	Relative Error	RSD	Relative Error	RSD	Relative Error	RSD	Relative Error	RSD
S1	2.38%	1.70%	4.04%	2.92%	2.39%	1.71%	1.76%	1.25%	1.47%	1.05%	1.15%	0.82%
S2	1.78%	1.27%	4.56%	3.30%	1.72%	1.23%	1.10%	0.78%	1.55%	1.11%	0.00%	0.00%
S3	2.38%	1.70%	3.72%	2.68%	1.94%	1.39%	1.37%	0.97%	1.56%	1.11%	0.02%	0.10%
S4	0.72%	0.51%	0.10%	0.07%	1.11%	0.79%	0.60%	0.43%	0.94%	0.67%	0.18%	2.13%
S5	1.52%	1.08%	2.13%	1.52%	2.14%	1.53%	0.90%	0.64%	1.11%	0.79%	0.03%	0.18%
S6	2.50%	1.79%	2.95%	2.12%	2.01%	1.44%	1.38%	0.98%	0.66%	0.47%	0.01%	0.07%
S7	3.39%	2.43%	2.84%	2.03%	1.70%	1.21%	0.97%	0.69%	3.09%	2.22%	0.02%	0.12%
S8	2.38%	1.70%	4.27%	3.08%	0.32%	0.23%	0.37%	0.26%	3.81%	2.74%	0.02%	0.09%
S9	1.65%	1.17%	3.18%	2.29%	1.17%	0.83%	0.60%	0.43%	0.74%	0.52%	0.02%	0.10%
S10	1.29%	0.92%	4.24%	3.06%	2.81%	2.01%	1.59%	1.13%	3.05%	2.19%	0.21%	0.15%
S11	2.57%	1.84%	4.20%	3.03%	1.38%	0.98%	0.74%	0.53%	4.84%	3.51%	0.88%	0.62%
S12	0.77%	0.54%	3.22%	2.32%	3.59%	2.59%	4.72%	3.42%	0.31%	0.22%	3.81%	2.75%
S13	4.77%	3.45%	4.54%	3.29%	2.41%	1.73%	1.48%	1.05%	0.90%	0.64%	0.61%	0.43%
S14	4.79%	3.47%	4.31%	3.12%	1.15%	0.82%	0.39%	0.27%	0.66%	0.47%	0.27%	0.19%
S15	1.95%	1.39%	3.72%	2.68%	1.10%	0.78%	0.59%	0.42%	4.61%	3.34%	0.61%	0.44%
S16	4.71%	3.41%	2.10%	1.50%	2.69%	1.93%	1.15%	0.82%	0.39%	0.28%	1.44%	1.02%
S17	0.76%	0.54%	4.31%	3.12%	2.23%	1.60%	1.61%	1.15%	3.77%	2.72%	0.99%	0.70%
S18	0.35%	0.25%	4.31%	3.11%	2.23%	1.59%	1.43%	1.02%	4.95%	3.59%	1.14%	0.81%
S19	2.54%	1.82%	3.76%	2.71%	2.50%	1.79%	2.33%	1.67%	0.94%	0.67%	0.38%	0.27%
S20	3.25%	2.33%	1.43%	1.02%	3.59%	2.59%	3.34%	2.40%	0.07%	0.05%	1.13%	0.80%
S21	2.53%	1.81%	0.70%	0.50%	1.75%	1.24%	1.70%	1.21%	2.10%	1.50%	0.20%	0.14%
Correlation coefficient	0.999 **		0.999 **		0.999 **		1.000 **		0.998 **		0.999 **	
***p* values**	0.940		0.802		0.978		0.923		0.960		0.986	

[1] RSD—relative standard deviation; p values—the paired t-test results; ESM—external standard method, and its content was determined by the calibration equation method; QAMS—quantitative analysis multi-components by single marker, and its content was determined by RCFs; ** $p < 0.01$.

The results from the QAMS determination of the 21 batches of *G. elata* tuber samples showed the mean contents of 3.5275 mg·g^{-1}, 0.9060 mg·g^{-1}, and 0.3398 mg·g^{-1} for gastrodin, *p*-hydroxy benzyl alcohol, and *p*-hydroxy benzaldehyde; and 3.6511 mg·g^{-1}, 9.5303 mg·g^{-1}, 2.7901 mg·g^{-1}, and 0.1766 mg·g^{-1} for the parishin E, parishin A, parishin B, and parishin C, respectively (Table 4). It was obvious that parishin A is one of the most abundant components in *G. elata* tuber, thus, is well-deserved as a reference substance and index for quality assessment and control of the *G. elata* tuber. Obvious inter-batch content variations could be found for all these components with the mean ranging from 0.1766 mg·g^{-1} to 9.5303 mg·g^{-1}; these seven components in total averaged 20.7031 mg·g^{-1} in the *G. elata* tuber, for the 21 batches of samples. The data in Table 4 shows differences among various samples. To show the clear classification of the *G. elata* tuber samples, the QAMS method with chemometrics analysis was performed in the subsequent analyses.

Meanwhile, the results (Table 6) illustrated that there were remarkable differences in the contents of the seven components, in *G. elata* tubers from different regions, which could be attributed to the variations of genetics, plant origins, environmental factors, drying process, storage conditions, and so on. It was obvious that gastrodin is one of the most abundant components in *G. elata* tuber. Combined with its activities related to the efficacies of *G. elata* tuber [31], gastrodin is well-deserved as a reference substance and index for quality assessment and control of *G. elata* tuber.

In the Chinese Pharmacopoeia of 2015 edition, gastrodin and *p*-hydroxy benzyl alcohol are determined as the marker components for the quality control and evaluation of *G. elata* tuber. Despite their close correlation with the efficacies of *G. elata* tuber, gastrodin can transform to *p*-hydroxybenzyl alcohol, which is the aglycone and metabolite of gastrodin [32]. Fresh *G. elata* tubers have to be processed before being traded as materia medica in the market. During the steaming process, the change trend of the gastrodin content was often contrary to the one of *p*-hydroxybenzyl alcohol. When the content of gastrodin was increased, the content of *p*-hydroxybenzyl alcohol was generally decreased, and vice versa. Additionally, different processing methods will result in different variation of the contents of the two components. Choi et al. [33] applied drying methods of freeze drying, hot air, infrared ray, and steaming, to process *G. elata* tuber. The results showed that after steaming, the content of gastrodin in *G. elata* tuber processed by freeze drying was decreased, whereas, the content of *p*-hydroxybenzyl alcohol was increased. However, tubers processed by hot-air and infrared ray drying showed the opposite results. Such transformations between gastrodin and *p*-hydroxybenzyl alcohol might be due to the deglycosylation or glycosylation, during the processing. Since the herbal medicine in the global market is often processed or dried by different methods, which results in the fluctuation in the content of single component, it is relatively stable and more comprehensive to reflect on the quality of *G. elata* tuber by monitoring multiple components, instead of a single one.

3. Materials and Methods

3.1. Plant Material

Samples of *G. elata* tuber from different producing areas were collected, as shown in Table 8.

Table 8. The information of *G. elata* tubers from different producing areas.

No.	Sample	Producing Areas	No.	Sample	Producing Areas
S1	*G. elata* tubers	Zhaotong, Yunnan, China	**S12**	*G. elata* tubers	Enshi, Hubei, China
S2	*G. elata* tubers	Zhaotong, Yunnan, China	**S13**	*G. elata* tubers	Yichang, Hubei, China
S3	*G. elata* tubers	Zhaotong, Yunnan, China	**S14**	*G. elata* tubers	Hanzhong, Shanxi, China
S4	*G. elata* tubers	Zhaotong, Yunnan, China	**S15**	*G. elata* tubers	Qinling, Shanxi, China
S5	*G. elata* tubers	Zhaotong, Yunnan, China	**S16**	*G. elata* tubers	Qinchuan, Sichuang, China

S6	*G. elata* tubers	Zhaotong, Yunnan, China
S7	*G. elata* tubers	Lijiang, Yunnan, China
S8	*G. elata* tubers	Bijie, Guizhou, China
S9	*G. elata* tubers	Zhengyuan, Guizhou, China
S10	*G. elata* tubers	Qiandongnan, Guizhou, China
S11	*G. elata* tubers	Bijie, Guizhou, China
S17	*G. elata* tubers	Longnan, Gansu, China
S18	*G. elata* tubers	Anhui, China
S19	*G. elata* tubers	Moju, South Korea
S20	*G. elata* tubers	Chun chuan, South Korea
S21	*G. elata* tuber slices	Yingyang, South Korea

3.2. Chemicals

The reference standards of gastrodin (no. B21243, purity HPLC ≥ 98%), *p*-hydroxybenzyl alcohol (no. B20326, purity HPLC ≥ 98%), *p*-hydroxy benzaldehyde (no. B20327, purity HPLC ≥ 99%), parishin A (no. BP1063, purity HPLC ≥ 98%), parishin B (no. BP1064, purity HPLC ≥ 98%), parishin C (no. B20913, purity HPLC ≥ 98%), parishin E (no. BP1648, purity HPLC ≥ 98%) were purchased from Sichuan Victory Biological Technology Co., Ltd. (Sichuan, China), and their structures are shown in Figures 4 and 5. Methyl alcohol was purchased from the Tianjin Fengchuan Chemical Reagent Technology Co. Ltd. Acetonitrile (HPLC grade) was purchased from Sigma-Aldrich, Inc. (St. Louis, MO, USA). Phosphoric acid was purchased from the Tianjin JinDongTianZheng Precision Chemical Reagent Factory. Ultrapure water was generated with an UPT-I-20T ultrapure water system (Yunnan Ultrapure Technology, Inc., Yunnan, China). All other chemicals used were of analytical grade.

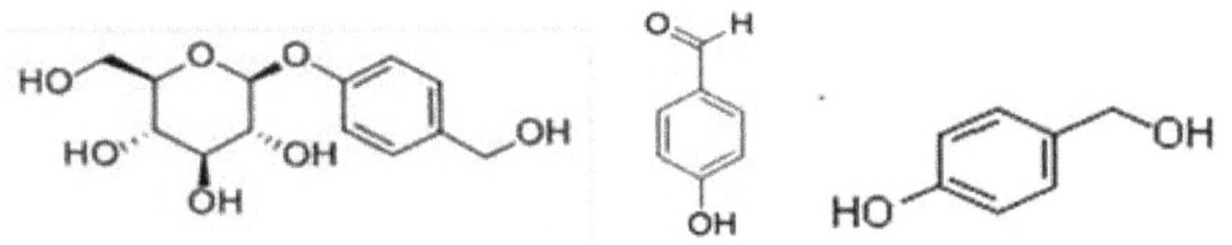

Figure 4. The structures of some compounds in the *G. elata* tuber. (**a**) Gastrodin [34], (**b**) *p*-hydroxy benzaldehyde [35], and (**c**) *p*-hydroxybenzyl alcohol [36].

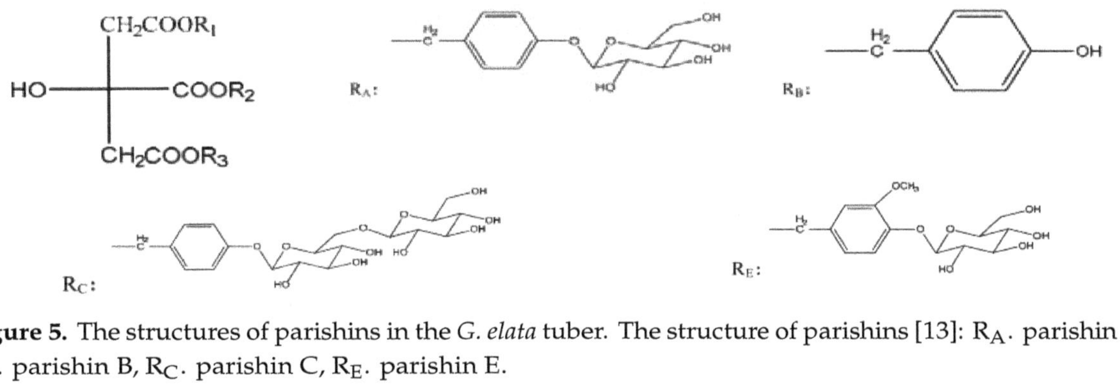

Figure 5. The structures of parishins in the *G. elata* tuber. The structure of parishins [13]: R_A. parishin A, R_B. parishin B, R_C. parishin C, R_E. parishin E.

3.3. Preparation of the Sample Solution

The 21 batches of dried *G. elata* tubers from different producing areas were crushed by a Wiggling high-speed Chinese medicine shredder, then powdered and sieved through a 40-mesh sieve. The sample solution of *G. elata* tuber was precisely absorbed (2.0 mg) and immersed in 25 mL volumetric flask, with 60% methanol. Additional 60% methanol was added to compensate for the weight loss after ultrasonic extraction for 60 min, and shaking it well. All solutions were filtered through 0.22 μm filter membranes, before being precisely injected into the HPLC system.

3.4. Reference Solution Preparation

The reference solution of *G. elata* tuber was prepared by accurately dissolving weighed samples of each compound in 60% methanol, making a mixture of 0.8 mg/mL of parishin A, 0.9 mg/mL of parishin B, 0.5 mg/mL of parishin E, 1.5 mg/mL of *p*-hydroxy benzaldehyde, 3.4 mg/mL of *p*-hydroxybenzyl alcohol, 0.9 mg/mL of gastrodin, 1.3 mg/mL of parishin C, mixed evenly. All the standard solutions were stored in a refrigerator at 4 °C, before use.

3.5. Chromatographic Procedures

The HPLC analysis of the *G. elata* tuber were done on an Agilent 1260 series system (Agilent Technologies, Santa Clara, CA, USA) consisting of a G1311B pump, a G4212B DAD detector, and a G1329B auto-sampler. The YMC-Tyiart C18 column (250 × 4.6 mm, 5 μm) was adopted for the analysis. The mobile phase consisted of A (0.1% phosphate solution) and B (acetonitrile). The gradient mode was as follows: 3–5% B for 0–11 min; 5% B for 11–18 min; 5–14% B for 18–31 min; 14% B for 31–38 min; 14–20% B for 38–48min; 20–24% B for 48–55 min; 24–80% B for 55–75 min; 80–100% B for 75–80 min; 100% B for 80–95 min; 100–70% B for 95–100 min; 70–50% B for 100–105 min; 50–30% B for 105–110 min; 30–3% B for 110–115 min; 3% B for 115–130 min. The flow rate was set at 1.0 mL/min. The detection wavelength was 220 nm. The column temperature was set at 35 °C and sample volume was 10 μL.

3.6. Theory of the QAMS Method

Methods for calculating the RCFs have been previously reported [24,37]. First, gastrodin was selected as the internal standard, and a multipoint method (Equation (1)) was used to calculate the relative correction factors (RCF) for *p*-hydroxy benzaldehyde, *p*-hydroxybenzyl alcohol, parishin A, parishin B, parishin E, and parishin C. Then the content of the measured component was calculated according to Equation (2) [38].

The RCFs were calculated using the calibration curves as follows:

$$f_{k/s} = \frac{a_k}{a_s} \quad (1)$$

The content of the measured component was calculated as follows:

$$C_k = \frac{A_k}{\left(A_s \times f_{k/s}\right)} \quad (2)$$

where, a_s is the ratio of the slope of internal standard reference calibration equations; a_k is the ratio of the slope of measured component calibration equations; A_k is the peak area of the measured component; and A_s is the peak area of the internal standard reference [37].

The content of the multi-marker components measured by QAMS was compared with results from ESM, to validate the methods of QAMS.

3.7. Data Analysis

We used the ESM and QAMS to calculate the seven components in 21 batches of *G. elata* tuber, to verify the feasibility of QAMS. At the same time, HCA was performed using the heat map software of Heml 1.0, to further investigate the difference among the *G. elata* tuber samples. The data were analyzed and evaluated by the Similarity Evaluation System for the chromatographic fingerprint of TCM (Version 2012), to evaluate similarities of the chromatographic profiles of the *G. elata* tuber.

4. Conclusions

In this study, the quality assessment method of *G. elata* tubers were established using QAMS methods, in combination with HPLC fingerprints analyses. The *G. elata* tubers from different areas were analyzed by HPLC fingerprints and the contents of the seven components in *G. elata* tuber samples was determined by the QAMS method. On the basis of these results, the quality of *G. elata* tubers could be quantified and better identified comprehensively by HCA of synthesis and similarity analysis. HPLC fingerprint analyses, combined with the QAMS methods, could be a powerful and reliable way to provide both qualitative insight and quantitative data for comprehensive quality assessment of the complex multi-component systems. QAMS combined with the HPLC fingerprint might offer a holistic phytochemical profile of botanicals, along with similarity analysis and HCA of synthesis, and the quality of *G. elata* tubers would be evaluated and better and more comprehensively identified. Moreover, in subsequent analyses, it is also necessary to combine the chemical analysis, biological evaluation, pharmacological activity, and other methods to evaluate the quality of *G. elata* tubers for better studying the clinical effect.

Author Contributions: Y.X. supervised the project and designed the experimental works; Y.L. performed the chemical analyses and wrote the paper; Y.Z., Z.Z., Y.H., and X.C. contributed to sample process and data analyses; Y.X. revised the paper. All authors read and approved the final manuscript.

Abbreviations

The following abbreviations have been used in this manuscript.

HPLC	High performance liquid chromatography
QAMS	Qualitative and quantitative analysis of multi-component by single marker
ESM	External standard method
RSD	Relative standard deviation
HCA	Hierarchical cluster analysis
TCM	Traditional Chinese medicine
RCF	Relative correction factor

References

1. Zhan, H.D.; Zhou, H.Y.; Sui, Y.P.; Du, X.L.; Wang, W.H.; Dai, L.; Sui, F.; Huo, H.R.; Jiang, T.L. The rhizome of *Gastrodia elata* Blume–An ethnopharmacological review. *J. Ethnopharmacol.* **2016**, *189*, 361–385. [CrossRef] [PubMed]
2. Li, Z.F.; Wang, Y.W.; Ouyang, H.; Lu, Y.; Qiu, Y.; Feng, Y.L.; Jiang, H.L.; Zhou, X.; Yang, S.L. A novel dereplication strategy for the identification of two new trace compounds in the extract of *Gastrodia elata* using UHPLC/Q-TOF-MS/MS. *J. Chromatogr. B* **2015**, *988*, 45–52. [CrossRef] [PubMed]
3. Hu, M.; Yan, H.; Fu, Y.; Jiang, Y.; Yao, W.; Yu, S.; Zhang, L.; Wu, Q.; Ding, A.; Shan, M. Optimal extraction study of gastrodin-type components from *Gastrodia Elata* tubers by response surface design with integrated phytochemical and bioactivity evaluation. *Molecules* **2019**, *24*, 547. [CrossRef] [PubMed]
4. Heo, J.C.; Woo, S.U.; Son, M.; Park, J.Y.; Choi, W.S.; Chang, K.T.; Kim, S.U.; Yoon, E.K.; Shin, S.H.; Lee, S.H. Anti-tumor activity of Gastrodia elata Blume is closely associated with a GTP-Ras-dependent pathway. *Oncol. Rep.* **2007**, *18*, 849–853. [CrossRef]
5. Hu, Y.H.; Li, C.Y.; Shen, W. Gastrodin alleviates memory deficits and reduces neuropathology in a mouse model of Alzheimer's disease. *Neuropathology* **2014**, *34*, 370–377. [CrossRef] [PubMed]
6. Zuo, Y.; Deng, X.; Wu, Q. Discrimination of gastrodia elata from different geographical origin for quality evaluation using newly-build near infrared spectrum coupled with multivariate analysis. *Molecules* **2018**, *23*, 1088. [CrossRef]
7. Zhao, Y.; Kang, Z.J.; Zhou, X.; Yang, S.L. An edible medicinal plant-Gastrodia elata Bl. J. *Guizhou Norm. Univ.* **2013**, *31*, 9–12.

8. Kang, C.; Lai, C.J.; Zhao, D.; Zhou, T.; Liu, D.H.; Lv, C.; Wang, S.; Kang, L.; Yang, J.; Zhan, Z.L.; et al. A practical protocol for comprehensive evaluation of sulfur-fumigation of Gastrodia Rhizoma using metabolome and health risk assessment analysis. *J. Hazard. Mater.* **2017**, *340*, 221–230. [CrossRef] [PubMed]
9. Zervakis, G.I.; Koutrotsios, G.; Katsaris, P. Composted versus raw olive mill waste as substrates for the production of medicinal mushrooms: An assessment of selected cultivation and quality parameters. *Biomed. Res. Int.* **2013**, *2013*, 546830. [CrossRef]
10. Chen, W.C.; Lai, Y.S.; Lu, K.H.; Lin, S.H.; Liao, L.Y.; Ho, C.T.; Sheen, L.Y. Method development and validation for the high-performance liquid chromatography assay of gastrodin in water extracts from different sources of *Gastrodia elata* Blume. *J. Food Drug Anal.* **2015**, *23*, 803–810. [CrossRef] [PubMed]
11. Chinese Pharmacopoeia Commission. *Pharmacopoeia of the People's Republic of China. Part I*; Chinese Medical Science and Technology Press: Beijing, China, 2015.
12. Ojemann, L.M.; Nelson, W.L.; Shin, D.S.; Rowe, A.O.; Buchanan, R. Tian ma, an ancient Chinese herb, offers new options for the treatment of epilepsy and other conditions. *Epilepsy Behav.* **2006**, *8*, 376–383. [CrossRef] [PubMed]
13. Xie, M.; Sha, M.; Zhai, Q.C.; Yu, H.; Guo, L.; Wei, Y.Q.; Li, Y. Research Progress on Parishins from *Gastrodia Elata*. *Guangdong Chem. Ind.* **2016**, *22*, 93–95.
14. FDA. *Guidance for Industry-Botanical Drug Products*; U. S. Food and Drug Administration: Silver Spring, MD, USA, 2004.
15. SFDA. *Technical Requirements for Studying Fingerprint of Traditional Chinese Medicine Injections (Draft)*; Drug Administration Bureau of China: Beijing, China, 2000.
16. EMEA. *Guidance on Quality of Herbal Medicinal Products/Traditional Herbal Medicinal Products*; European Medicines Agency: London, UK, 2006.
17. Cui, L.L.; Zhang, Y.; Shao, W.; Gao, D. Analysis of the HPLC fingerprint and QAMS from *Pyrrosia* species. *Ind. Crop. Prod.* **2016**, *85*, 29–37. [CrossRef]
18. Schaneberg, B.T.; Crockett, S.; Khan, I.A. The role of chemical fingerprinting: Application to Ephedra. *Phytochemistry* **2003**, *62*, 911–918. [CrossRef]
19. Bauer, R. Quality criteria and phytopharmaceuticals: Can acceptable drug standards be achieved? *Drug Inf. J.* **1998**, *31*, 101–110. [CrossRef]
20. Wang, X.; Tan, Y.; Wang, D.J.; Qing, D.S.; Yao, L.C.; Li, L.Y.; Luo, W.Z. Application and advance of QAMS in quality control of traditional Chinese medicines. *Chin. Tradit. Pat. Med.* **2016**, *38*, 395–402.
21. Li, D.W.; Zhu, M.; Shao, Y.D.; Shen, Z.; Weng, C.C.; Yan, W.D. Determination and quality evaluation of green tea extracts through qualitative and quantitative analysis of multi-components by single marker (QAMS). *Food Chem.* **2016**, *197*, 1112–1120. [CrossRef]
22. Song, Y.; Wang, Z.; Zhu, J.; Yan, L.; Zhang, Q.; Gong, M.; Wang, W. Assay of evodin, evodiamine and rutaecarpine in Fructus Evodiae by QAMS. *Zhongguo Zhong Yao Za Zhi* **2009**, *34*, 2781–2785. [PubMed]
23. Lei, Y.C. *Authenticity Identification and Quality Assessment of Gastrodia tuber (Tianma) Based on Chemical Characteristics*; Chengdu University of TCM: Chengdu, China, 2015; p. 5.
24. Gao, X.Y.; Jiang, Y.; Lu, J.Q.; Tu, P.F. One single standard substance for the determination of multiple anthraquinone derivatives in rhubarb using high-performance liquid chromatography-diode array detection. *J. Chromatogr. A* **2009**, *1216*, 2118–2123. [CrossRef]
25. Peng, Y.; Dong, M.H.; Zou, J.; Liu, Z.H. Analysis of the HPLC Fingerprint and QAMS for Sanhuang Gypsum Sou. *J. Anal. Methods Chem.* **2018**. [CrossRef]
26. Wang, N.; Li, Z.Y.; Zheng, X.L.; Li, Q.; Yang, X.; Xu, H. Quality Assessment of Kumu Injection, a Traditional Chinese Medicine Preparation, Using HPLC Combined with Chemometric Methods and Qualitative and Quantitative Analysis of Multiple Alkaloids by Single Marker. *Molecules* **2018**, *23*, 856. [CrossRef] [PubMed]
27. He, B.; Liu, Y.; Tian, J.; Li, C.H.; Yang, S.Y. Study on quality control of Houttuynia Cordata, a tradtional Chinese medicine by fingerprint combined with quantitative analysis of multi-components by single marker. *China J. Chin. Mater. Med.* **2013**, *38*, 2682–2689.
28. Ding, L.Y.; Zhou, L.; Wang, L.N.; Guo, Y.H.; Wang, H. Multi-components quantitation by one marker for simultaneous content determination of four components in Psoralea corylifolia. *Chin. J. Exp. Tradit. Med Formulae* **2013**, *19*, 152–154.
29. Dou, Z.H.; Qiao, J.; Bian, L.; Hou, J.Y.; Mao, C.F.; Chen, Z.X.; Shi, Z. Combinational quality control method of Rhei Radix et Rhizoma based on fingerprint and QAMS. *J. Chin. Pharm. Sci.* **2015**, *50*, 442–448.

30. Yan, D.M.; Chang, Y.X.; Kang, L.Y.; Gao, X.M. Quality evaluation and regional analysis of Psoraleae Fructus by HPLC-DAD-MS/MS plus chemometrics. *Chin. Herb. Med.* **2010**, *2*, 216–223.
31. Chen, W.H.; Luo, D. Research Development on Pharmacological Action in *Gastrodia elata blume* and *Gastrodia Elata Polysaccharide*. *China J. Drug Eval.* **2013**, *30*, 132–141.
32. Lu, G.W.; Zou, Y.J.; Mo, Q.Z. Kinetic aspects of absorption, distribution, metabolism and excretion of ~3H-gastrodin in rats. *Yao Xue Xue Bao* **1985**, *20*, 167–172. [PubMed]
33. Choi, S.R.; Jang, I.; Kim, C.S.; You, D.H.; Kim, J.Y.; Kim, Y.G.; Ahn, Y.S.; Kim, J.M.; Kim, Y.S.; Seo, K.W. Changes of components and quality in gastrodiae rhizoma by different dry methods. *Korean J. Med. Crop Sci.* **2011**, *19*, 354–361. [CrossRef]
34. KPC Pharmaceuticals, Inc. The Invention Relates to Gastrodin Compound, Preparation Method, Preparation and Application. CN107056853A, 2017.
35. Wang, W.Y. *Study on Solubility of p-Hydroxybenzaldehyde, m-Hydroxybenzaldehyde and Their Mixture in Supercritical Carbon Dioxide*; Beijing University of Chemical Technology: Beijing, China, 2014.
36. Wang, W.J. *Study on the Aerobic Oxidation of o/p-Cresol to o/p-Hydroxybenzaldehyde Catalyzed by Metallopoprhyrinns*; Beijing University of Chemical Technology: Beijing, China, 2013.
37. Hou, J.J.; Wu, W.Y.; Da, J.; Yao, S.; Long, H.L.; Yang, Z.; Cai, L.Y.; Yang, M.; Liu, X.; Jiang, B.H.; et al. Ruggedness and robustness of conversion factors in method of simultaneous determination of multi-components with single reference standard. *J. Chromatogr. A* **2011**, *1218*, 5618–5627. [CrossRef]
38. Huang, J.; Yin, L.; Dong, L.; Quan, H.F.; Chen, R.; Hua, S.Y.; Ma, J.H.; Guo, D.Y.; Fu, X.Y. Quality evaluation for Radix Astragali based on fingerprint, indicative components selection and QAMS. *Biomed. Chromatogr.* **2018**, *32*, e4343. [CrossRef]

12

Liquid Phase and Microwave-Assisted Extractions for Multicomponent Phenolic Pattern Determination of Five Romanian *Galium* Species Coupled with Bioassays

Andrei Mocan [1], Alina Diuzheva [2], Sabin Bădărău [3], Cadmiel Moldovan [1], Vasil Andruch [2], Simone Carradori [4], Cristina Campestre [4], Angela Tartaglia [4], Marta De Simone [4], Dan Vodnar [5], Matteo Tiecco [6], Raimondo Germani [6], Gianina Crişan [1] and Marcello Locatelli [4,*]

[1] Department of Pharmaceutical Botany, "Iuliu Haţieganu" University of Medicine and Pharmacy, 400012 Cluj-Napoca, Romania; mocan.andrei@umfcluj.ro (A.M.); moldovan.cadmiel@yahoo.com (C.M.); gcrisan@umfcluj.ro (G.C.)
[2] Department of Analytical Chemistry, Pavol Jozef Šafárik University, SK-04180 Košice, Slovakia; adyuzheva@gmail.com (A.D.); vasil.andruch@gmail.com (V.A.)
[3] Department of Environmental Sciences, Babeş-Bolyai University, 400084 Cluj-Napoca, Romania; alexandru@transsilvanica.net
[4] Department of Pharmacy, University "G. D'Annunzio" of Chieti-Pescara, 66100 Chieti, Italy; simone.carradori@unich.it (S.C.); cristina.campestre@unich.it (C.C.); angela.tartaglia@unich.it (A.T.); marta.desimone@studenti.unich.it (M.D.S.)
[5] Department of Food Science, University of Agricultural Sciences and Veterinary Medicine, 400372 Cluj-Napoca, Romania; dan.vodnar@usamvcluj.ro
[6] Department of Chemistry, Biology and Biotechnology, University of Perugia, 06132 Perugia, Italy; matteotiecco@gmail.com (M.T.); raimondo.germani@unipg.it (R.G.)
* Correspondence: m.locatelli@unich.it

Abstract: Background: Galium is a plant rich in iridoid glycosides, flavonoids, anthraquinones, and small amounts of essential oils and vitamin C. Recent works showed the antibacterial, antifungal, antiparasitic, and antioxidant activity of this plant genus. **Methods**: For the determination of the multicomponent phenolic pattern, liquid phase microextraction procedures were applied, combined with HPLC-PDA instrument configuration in five Galium species aerial parts (G. verum, G. album, G. rivale, G. pseudoaristatum, and G. purpureum). Dispersive Liquid–Liquid MicroExtraction (DLLME) with NaCl and NAtural Deep Eutectic Solvent (NADES) medium and Ultrasound-Assisted (UA)-DLLME with β-cyclodextrin medium were optimized. **Results**: The optimal DLLME conditions were found to be: 10 mg of the sample, 10% NaCl, 15% NADES or 1% β-cyclodextrin as extraction solvent—400 μL of ethyl acetate as dispersive solvent—300 μL of ethanol, vortex time—30 s, extraction time—1 min, centrifugation at 12000× g for 5 min. **Conclusions**: These results were compared with microwave-assisted extraction procedures. *G. purpureum* and *G. verum* extracts showed the highest total phenolic and flavonoid content, respectively. The most potent extract in terms of antioxidant capacity was obtained from *G. purpureum*, whereas the extract obtained from *G. album* exhibited the strongest inhibitory effect against tyrosinase.

Keywords: dispersive liquid-liquid microextraction; microwave-assisted extraction; natural deep eutectic solvent; β-cyclodextrin; *Galium* species; tyrosinase inhibition

1. Introduction

The use of plants for the treatment of human diseases is a centuries-old tradition, based on phytotherapy research as well as on ethnopharmacological knowledge. Recently, the use of herbal medicines applied for the prevention and/or preservation of health covers a central role in modern medicine related to the fact that these plant-derived materials avoid the classical side effects of synthetic drugs. Additionally, there are benefits of their long-term historic use—safety, accessibility, and efficacy with a wide range of therapeutic actions [1]. *Galium* is a well-known genus with many medicinal representatives that are rich sources of iridoid glycosides [2–4], flavonoids [5], anthraquinones [6], and small amounts of essential oils and vitamin C [7]. Recent studies showed the antibacterial, antifungal, antiparasitic, and antioxidant activities of representatives of this plant genus [7,8].

G. verum L., also known as Lady's Bedstraw, is an herbaceous perennial plant, native to Europe and Asia, and used commonly in many countries' folk medicine for a large variety of treatments. The dried plants' aerial parts were used to stuff mattresses, and the flowers were also used to coagulate milk for cheese production [9]. The cut and dried aerial parts of the plant, *'Herba gallii verii'*, are used for homeopathic purposes. These are still used for exogenous treatment of psoriasis or as a tea with diuretic effect for the cure of pyelitis or cystitis [10]. Moreover, *G. verum* L. has been used as a diuretic for bladder and kidney irritation, externally for poorly healing wounds, as well as for epilepsy and hysteria in Montenegro's traditional medicine [11]. Regarding Turkish folk medicine, it has been used for its diuretic, choleretic, antidiarrheal, and sedative effects [4]. In Romania, the plant is used in traditional medicine mainly for its diuretic, depurative, laxative, sedative, and antirheumatic effects. Additionally, in the Romanian traditional medicine, several *Galium* species are used as components of different cosmetic formulations [12]. *G. album* Mill., the "white bedstraw" or "hedge bedstraw", is an herbaceous annual plant, cited in traditional Albanian pharmacopoeias and folk medicine for healing wounds and gingival inflammations [13]. *G. rivale*, *G. pseudoaristatum*, and *G. purpureum* (syn. *Asperula purpurea*) are less-known species, and to the extent of our knowledge, they have not been investigated yet in terms of chemical composition and antioxidant capacity, nor in terms of enzyme inhibitory potential.

Generally, the use of different extraction procedures on plant-derived material yields different biological activities. In this field, the availability of an efficient, fast, exhaustive, and reproducible extraction procedure allows obtaining a standardized starting material for food additives, nutraceuticals, and phytoformulations. For the extraction of bioactive compounds from *Galium* maceration in methanol [7] or ethanol [14], percolation in methanol [8], and ultrasound-assisted extraction [12] were applied, wherein the extraction time was varied from 30 min to one week. In order to reduce the extraction time and retain or increase the extraction efficiency, new extraction methods are required.

Liquid phase microextraction techniques are positioned as 'green' chemistry methodologies, which require small amounts of organic solvents. In order to make the procedures more environmentally-friendly, ionic liquids (ILs) or natural deep eutectic solvents (NADESs) are frequently used. Comparing ILs with NADESs, more advantages are on the side of NADESs due to their natural original (the main components can be sugars and organic acids), which may vary depending on analysis purpose, making them nontoxic, biodegradable, and incombustible. In comparison with NADESs, most ILs are toxic, have low biodegradability, and have high cost. Either IL or NADES can have high viscosity, so their extracts are limited for direct analysis using HPLC or GC systems [15–18].

Regarding biological activities, in the current work, a key enzyme was considered in order to further evaluate the extracts. Particularly, pigmentation is one of the most obvious phenotypical characteristics in the natural world. Between the pigments, melanin is one of the most widely distributed and is found in bacteria, fungi, plants, and animals. Melanins are heterogeneous polyphenol-like biopolymers with complex structure and color varying from yellow to black. The synthesis of melanin plays an important role in skin color and pigmentation. Tyrosinase, a copper-containing mono-oxygenase, is a key enzyme in melanin biosynthesis [19]. Skin disorders, such as melasma (facial pigmentation), scarce, and freckles, are related to excessive melanin biosynthesis. Thus, tyrosinase inhibitors are used to control or treat pigmentation disorders and are widely used in the cosmetic industry. In fact, some tyrosinase inhibitors, such as kojic acid and hydroquinones, are nowadays commercially produced, but they can present severe side effects, such as skin inflammation [20]. Hence, in recent years, more attention has been paid to the use of natural plant extracts as a safe and alternative source of tyrosinase inhibitors for cosmetic purposes.

In the present study, following our research on innovative microextraction procedures [21–27], different microextraction procedures were examined for the analysis of phenolic compounds in *G. verum* aerial parts, and then applied for the determination of the phenolic pattern of four other *Galium* species (*G. album*, *G. rivale*, *G. pseudoaristatum*, and *G. purpureum*). As an alternative, the microwave-assisted extraction (MAE) technique was used as a reference method [27–29]. To the best of our knowledge, it is the first time that microextraction techniques have been applied for the recovery and the establishment of phenolic compounds in *Galium* species.

2. Results and Discussion

2.1. Preliminary Examinations

Several liquid phase extraction techniques, such as DLLME, UA-DLLME in water, 10% NaCl, NADES, and 1% β-CD media, SA-LLE, and SULLE, were performed in order to select the procedure providing the better quali-quantitative multicomponent profile of phenolic compounds. The extractions were carried out as described in the experimental section. Figure 1 shows that the best results were achieved in the case of DLLME in 10% NaCl and 10% NADES media and UA-DLLME in 1% β-CD. In UA-DLLME, the phase separation was observed only with β-CD, whereas no phase separation was observed using the other additives. The notable increasing of extraction recovery using UA-DLLME with β-CD could be explained because β-CD was able to better dissolve the metabolites in the extraction solvent, contributing to an increased inclusion in its cavity of a higher amount of phenolic compounds. Therefore, DLLME in NaCl and NADES, UA-DLLME in β-CD media were selected for optimization.

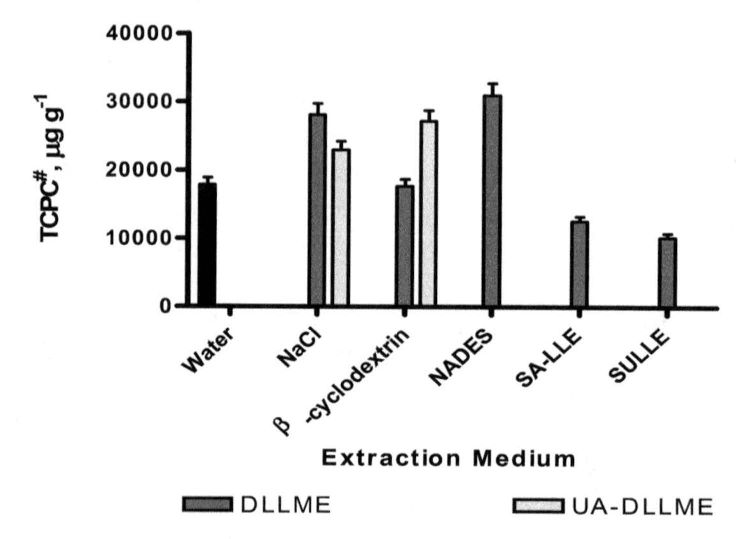

Figure 1. Selection of microextraction procedure. # TCPC—Total concentration of phenolic compounds. Values expressed are means ± S.D. of three measurements. All the values were statistically significant ($p < 0.001$). Raw data regarding the statistical analyses were reported in *Supplementary Materials* section S1.

2.2. Optimization of DLLME and UA-DLLME

Several parameters that could influence the extraction efficiency, such as solid:liquid ratio, extraction and dispersive solvent types, and volume, were selected for optimization. For UA-DLLME, ultrasonication time was also optimized.

2.2.1. Optimization of Extraction Medium Concentration

The extraction medium can significantly affect the extraction yields; therefore, a series of experiments were carried out by adding 5–15% NaCl or NADES solution, or 0.5–1.5% β-CD solution into the vessel containing 10 mg of the dry herbal material. For β-CD, the concentration was lower due to their low water solubility. With 10% NaCl, 15% NADES, and 1% β-CD, the best extraction recoveries were achieved (Figure 2a). Thus, these conditions were applied in further experiments.

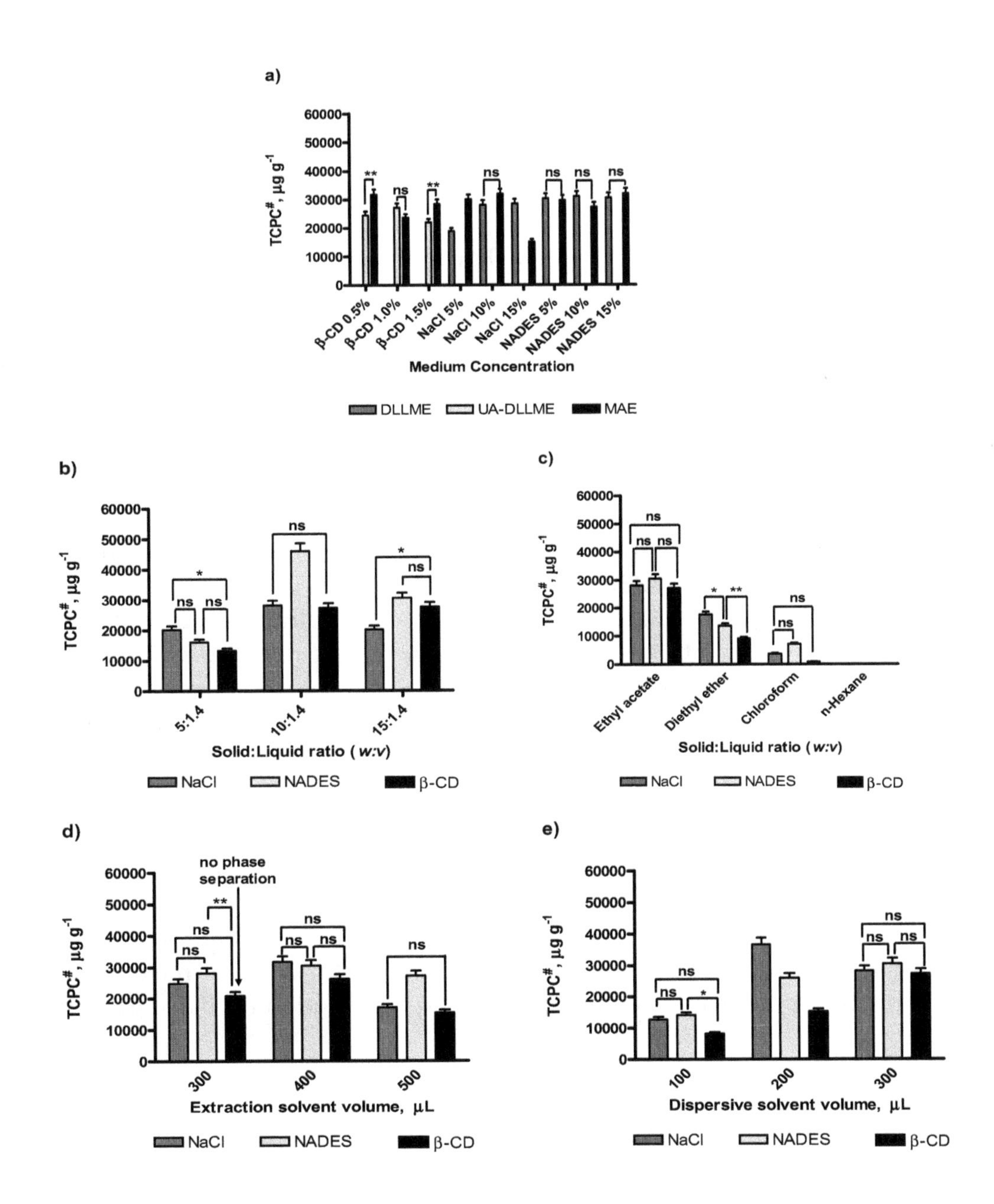

Figure 2. Optimization of DLLME, UA-DLLME, and MAE. (**a**) Effect of medium concentration; (**b**) Effect of solid:liquid ratio; (**c**) Selection of extraction solvent; (**d**) Effect of extraction solvent volume; (**e**) Effect of dispersive solvent volume. # TCPC—Total concentration of phenolic compounds. Values expressed are means ± S.D. of three measurements. All the values were statistically significant ($p < 0.001$), unless otherwise indicated as n.s. (not statistically significant), ** (statistically significant at $p < 0.01$), or * (statistically significant at $p < 0.05$). Raw data regarding the statistical analyses were reported in *Supplementary Materials* section S1.

2.2.2. Optimization of Solid:Liquid Ratio

Three solid:liquid ratios, expressed as mg/mL (5:1.4, 10:1.4, 15:1.4 *w:v*), were examined for their impact on the extraction efficiency. The experimental results showed that the tendency for NaCl and NADES was similar, and the maximum of the extraction recovery was reached with the ratio 10:1.4. For β-CD, with the ratios 10:1.4 and 15:1.4 (*w:v*), no significant differences were observed. Therefore, the optimal solid:liquid ratio was established as 10:1.4 (*w:v*) (Figure 2b). In fact, in the analytical chemistry workflow, if two different systems show similar data, the lower ratio is generally used because it can get the same analytical performances using a lower amount of solvents, raw material, chemicals, etc.

2.2.3. Selection of Extraction Solvent Type and Volume

n-Hexane, ethyl acetate, chloroform, and diethyl ether were tested as potential extractants. The experiments revealed that a higher amount of phenolic compounds was extracted using ethyl acetate (Figure 2c). This could be explained by the different polarities of the extraction solvents and by the interaction with polar phenolic compounds. For instance, with *n*-hexane, a nonpolar solvent, the phenolic compounds were poorly extracted. Diethyl ether and chloroform showed similar extraction efficiency with NaCl and NADES additives, whereas the addition of β-CD did not provide an exhaustive extraction. Taking into account the high volatility of diethyl ether, it was easier to work with ethyl acetate. Therefore, ethyl acetate was selected as appropriate solvent for all samples.

To determine the optimal volume of the extraction solvent, 200, 300, 400, and 500 μL were examined. When the volume is less than 300 μL, the phase separation was not achieved in DLLME and UA-DLLME, while phase separation was reached with 300 μL or more of NaCl and NADES. In order to apply this volume amount to other solvent media, the extraction procedure was modified as follows: the extraction solvent was added in two steps, firstly 200 μL were added in order to achieve an emulsion, then an additional 100 μL of ethyl acetate were rapidly injected. The phase separation was achieved after 5 min in the rest. Applying this procedure, no phase separation in UA-DLLME in β-CD media was observed; therefore, the UA-DLLME was carried out with 400 and 500 μL of ethyl acetate. It was found that with the increase of volume of ethyl acetate, the extraction of total content of phenolics decreased. Therefore, 400 μL was selected for further study on solid samples (Figure 2d).

2.2.4. Selection of Dispersive Solvent Volume

Commonly, ethanol, methanol, and acetonitrile are reported as dispersive solvents in DLLME. In this study, ethanol was selected as dispersive solvent because some food supplements, not considered in this study, of *Galium* are in ethanolic solution. Therefore, the effect of its volume (100–500 μL) on the extraction yields was tested. The results showed that the extraction efficiency was enhanced with the increase of the ethanol volume in the solution until 300 μL, while for higher volumes, phase separation was not achieved (Figure 2e).

2.2.5. Optimization of Ultrasonication Time in UA-DLLME

The cyclodextrins (α, β, γ) show amphiphilic characteristics related to a hydrophilic shell and a hydrophobic cavity and could be usefully used as emulsifiers in order to enhance the extraction recovery for the target analytes. Their capacity to improve the extraction efficiency is related to their ability to reduce the interfacial tension between the two phases by an organic solvent/cyclodextrin complex located in the liquid–liquid interface. In this way, an increased contact area between the two phases was observed [30–33]. The aid of ultrasonication was generally required in order to enhance the solubility, as discussed by Saokham et al. [34] in a recent review paper. Different times have been investigated in the range of 2 to 10 min. Since 5 and 10 min showed similar responses, 5 min was selected as optimal in order to reduce the time of analysis.

2.3. *Reference Method: Microwave-Assisted Extraction*

In order to evaluate the performances of the proposed procedure, as in the comparison method, MAE was selected and carried out in the same media as LPME procedures at different concentration levels of the solvents (5–15% solution of NaCl and NADES, and 0.5–1.5% solution of β-CD). Figure 2a shows that the extraction efficiencies obtained in 10% NaCl and 15% NADES were comparable to DLLME and UA-DLLME. Therefore, the recovery of total phenolics, using LPME and MAE, was also comparable.

Following our experimental data, the optimized DLLME conditions found were: 10 mg of the sample, 10% NaCl, 15% NADES or 1% β-cyclodextrin, extraction solvent—400 μL of ethyl acetate, dispersive solvent—300 μL of ethanol, vortex time—30 s, extraction time—1 min, centrifugation at 12000× g for 5 min. In the case of UA-DLLME, 5 min of ultrasonication was required.

2.4. *Total Phenolic and Flavonoid Content, Antioxidant Capacity, and Tyrosinase Inhibitory Activity*

2.4.1. Total Phenolic Content (TPC) by Spectrophotometric Assay

The Folin–Ciocâlteau assay was employed to determine the TPC of *Galium* extracts. The maximum TPC was registered in the ethanolic extract of *G. purpureum* (10.3 ± 0.8 mg GAE/g extract), whereas the lowest concentration was present in the ethanolic extract of *G. rivale* (1.3 ± 0.2 mg GAE/g extract). A recent study by Lakić et al. showed similar results regarding the low phenolic content of *G. verum* (2.4–5.2 mg GAE/g extract), using different extraction solvents [7].

2.4.2. Total Flavonoid Content (TFC) by Spectrophotometric Assay

Results of the total flavonoid content (TFC) of the different plant materials are presented in Table 1. The highest amount for the TFC was obtained for *G. verum* extract, with a value of 8.60 ± 0.07 mg QE/g d.w., comparable with the value obtained for *G. purpureum* extract, containing 8.50 ± 0.04 mg QE/g (d.w.). According to the results of the present study, Vlase et al. reported a TPC of 5.2 ± 0.2 g/100 g for a *G. verum* extract [12] and, additionally, Lakić et al. reported values of 6.4–17.9 mg QE/g (d.w.), for *G.·verum* extracts, using different solvents and extraction times [7] confirming the results herein presented.

2.4.3. Antioxidant Potential Assays

The ferric reducing antioxidant power (FRAP), scavenging of DPPH, and ABTS free radical assays were used to evaluate the antioxidant capacity of *Galium* species (Table 1). These methods are simple and widely used for the evaluation of antioxidant capacity of herbal extracts/pure compounds. Moreover, the values regarding the total phenolic content (TPC) and total flavonoid content (TFC) are in accordance with antioxidant capacity values of the extracts. In the DPPH assay, *G. purpureum* (6.3 ± 0.7 mg TE/g extract) exhibited a higher DPPH scavenging capacity than any other considered species (0.4–1.9 mg TE/g extract). The ABTS value for *G. purpureum* (16.7 ± 0.8 mg TE/g extract) was higher in comparison with the values obtained for the other considered species, which ranged from 4.5 to 7.6 mg TE/g extract.

2.4.4. Tyrosinase Inhibitory Activity

Galium extracts had good tyrosinase inhibitory activities (4.66–70.98% at 8 mg/mL), as reported in Table 1. The extract of *G. album* presented the highest tyrosinase inhibition, with a value of 70.98%. Despite the highest concentration of rutin and chlorogenic acid, the ethanolic extract of *G. rivale* showed no inhibitory effect against tyrosinase. This shows that the synergic effect of the compounds from the tested *Galium* samples have no or low inhibitory effects in some cases, although it was demonstrated by many studies that phenolic and flavonoid compounds are, in general, good inhibitors of tyrosinase [19]. The modest tyrosinase inhibitory activity for *Galium* species is confirmed by other studies as well.

For example, Chiocchio et al. reported no tyrosinase inhibitory activity for *G. album* [35]. The low inhibitory activity of these extracts can be explained by the presence of other nondetected compounds, which might block or interfere with the enzyme.

2.5. Quantitative Analysis of Galium Species

The dry extracts were analyzed to establish the fingerprint of phenolic compounds in five *Galium* species. Table 2 summarizes the results obtained by means of a validated HPLC-PDA method for phenolics determination. All measurements were performed in triplicate in order to obtain standard deviation. It can be observed that the amount of the phenolic compounds for all *Galium* species is in the range from 2526.2–11345.1 $\mu g\ g^{-1}$. The major biologically active compounds are chlorogenic acid and rutin. The highest number of the detected phenolic compounds was found in *G. rivale* (11345.1 $\mu g\ g^{-1}$), where the main compound was chlorogenic acid (10192 $\pm$ 34 $\mu g\ g^{-1}$), but the fingerprint was poorer in comparison with other species. The richest multicomponent pattern was observed in *G. pseudoaristatum*, but the quantity of phenolic compounds was the lowest (2526.2 $\mu g\ g^{-1} \pm 46.21$). Chromatograms for each *Galium* sp. were reported in Supplementary Materials section S2.

p-OH benzoic acid, vanillic acid, epicatechin, syringic acid, 3-OH-4-MeO benzaldehyde, *p*-coumaric acid, *t*-ferulic acid, naringin, 2.3-diMeO benzoic acid, benzoic acid, *o*-coumaric acid, harpagoside, *t*-cinnamic acid, and naringenin were not reported into the table because they were not detected by the HPLC-PDA method.

Table 1. Total phenolic and flavonoid content, antioxidant capacity, and enzyme inhibitory effects of the extracts of *Galium* spp. (values expressed are means ± S.D. of three measurements).

	TPC (mg GAE/g Extract)	TFC (mg QE/g Extract)	DPPH Scavenging (mg TE/g Extract)	ABTS Scavenging (mg TE/g Extract)	FRAP (mg TE/g Extract)	Tyrosinase Inhibit. (mg KAE/g Extract)	Tyrosinase Inhibit. (% Inhibition)
G. verum	3.1 ± 0.3 ^{a,e}	8.60 ± 0.07 ^{a,e}	1.9 ± 0.5 ^{a,e}	6.15 ± 0.02 ^{a,e}	21.9 ± 0.9 ^{a,c,e}	7.3 ± 0.2 ^{a,e}	36.96
G. album	2.7 ± 0.1 ^{a,e}	4.88 ± 0.03 ^{a,e}	1.1 ± 0.2 ^{a,e}	6.1 ± 0.1 ^{a,e}	17.19 ± 0.04 ^{b,e}	13.8 ± 3.4 ^{e}	70.98
G. rivale	1.3 ± 0.2 ^{a,d,e}	4.01 ± 0.08 ^{a,c,e}	0.4 ± 0.25 ^{a,e}	4.5 ± 0.5 ^{a,d,e}	12.6 ± 0.2 ^{a,e}	na	na
G. pseudoaristatum	4.5 ± 0.6 ^{a,d,e}	6.7 ± 0.3 ^{a,c}	1.9 ± 0.1 ^{a,e}	7.6 ± 0.2 ^{a,d,e}	19.4 ± 0.9 ^{a,b,c,e}	2.5 ± 1.6 ^{e}	4.66
G. purpureum	10.3 ± 0.8 ^{e}	8.50 ± 0.04 ^{a,e}	6.3 ± 0.7 ^{e}	16.7 ± 0.8 ^{e}	45.2 ± 1.1 ^{e}	6.3 ± 1.4 ^{a,e}	29.71
KA (0.1 mg/mL)							62.52

Na—not active; for tyrosinase inhibition (in percentages), the extracts concentration was 8 mg/mL. TFC = Total Flavonoid Content by spectrophotometric assay; TPC = Total Phenolic Content by spectrophotometric assay; ABTS = 2,2′-azino-bis(3-ethylbenzothiazoline-6-sulphonic acid); DPPH = 2,2-diphenyl-1-picryl-hydrazyl-hydrate; FRAP = Ferric Reducing Antioxidant Power; GAE = Gallic Acid Equivalents; QE = Quercetin Equivalents; TE = Trolox Equivalents; KAE = Kojic Acid Equivalents; KA = Kojic Acid. Raw data regarding the statistical analyses were reported in *Supplementary Materials* section S1. Data marked with different letters indicates significant difference ($p < 0.05$).

Table 2. Phenolic compounds quantified in Galium spp. using DLLME in 15% NADES (values expressed are means ± S.D. of three measurements).

Metabolite	***Galium* Species**				
	G. verum	***G. album***	***G. rivale***	***G. pseudoaristatum***	***G. purpureum***
	Conc., µg/g				
	Mean (±S.D.)	**Mean (±S.D.)**	**Mean (±S.D.)**	**Mean (±S.D.)**	**Mean (±S.D.)**
Gallic acid				109 (±7) ^{a,b,e}	23.2 (±1.5) ^{a,b,e}
Catechin		380 (±96) ^{a,e}	63.5 (±2.5) ^{a,e}	203 (±30) ^{a,e}	
Chlorogenic acid	2986 (±75) ^{e}	8310 (±231) ^{e}	10192 (±34) ^{e}	1640 (±30) ^{e}	5572 (±205) ^{e}
3-OH benzoic acid	853 (±184) ^{e}			87.4 (±12.6) ^{a,c}	374 (±16) ^{b,c,d,e}
Rutin	3624 (±97) ^{e}	275 (±38) ^{a,c,e}	987 (±24) ^{e}	283 (±5) ^{a,b,c,d,e}	137 (±13) ^{a,d,e}
Sinapinic acid				55.7 (±0.9) ^{a,d,e}	203 (±81) ^{a,b}
Quercetin	89.6 (±15.5) ^{a,e}	84.1 (±8.3) ^{a,c,e}		67.9 (±11.6) ^{a,c,e}	
Carvacrol	101 (±2) ^{a,e}	84.1 (±0.9) ^{a,c,e}	102.6 (±0.9) ^{a,e}	81.8 (±2.1) ^{a,c,e}	162 (±26) ^{a,c,e}
Total µg/g	**7654 (±222)**	**9133 (±253)**	**11345 (±42)**	**2526 (±47)**	**6471 (±223)**

Data marked with different letters indicate significant difference ($p < 0.05$).

3. Materials and Methods

3.1. Materials

Chemical standards of phenolic compounds (benzoic acid, carvacrol, catechin, chlorogenic acid, *t*-cinnamic acid, 8-cinnamoyl harpagide (harpagoside), *o*-coumaric acid, *p*-coumaric acid, 2,3-dimethoxybenzoic acid, epicatechin, *t*-ferulic acid, gallic acid, 3-hydroxybenzoic acid, 4-hydroxybenzoic acid, 3-hydroxy-4-methoxybenzaldehyde, naringin, naringenin, quercetin, rutin, sinapinic acid, syringic acid, vanillic acid (all purity > 98%)), β-cyclodextrin ($\geq$97%), *n*-hexane (HPLC-grade), diethyl ether ($\geq$99%), and chloroform (HPLC-grade) were purchased from Sigma-Aldrich (Milan, Italy).

Ethyl acetate ($\geq$99%), acetonitrile (HPLC-grade), methanol (HPLC-grade), ethanol (HPLC-grade), acetic acid ($\geq$99%) as well as D-(+)-glucose were obtained from Carlo Erba Reagents (Milan, Italy). Sodium chloride ($\geq$99%) was obtained from Honeywell (Seelze, Germany). NADES (glycolic acid/betaine mixture) was newly synthesized and supplied by University of Perugia. It was chosen between differently structured novel DES and NADES mixtures for its suitable properties (low freezing point and low viscosity, absence of aromatic compounds in its composition, low cost and natural source of the molecules forming it). Ultra-pure water was obtained using a Millipore Milli-Q Plus water treatment system (18 MΩ cm at 23 °C, Millipore Bedford Corp., Bedford, MA, USA).

3.2. Sampling and Sample Preparation

Samples of *Galium* species were collected from different locations from Romania, as follows: *G. verum* L. from Apuseni mountains region, Sartăș, Alba County, Transylvania, Romania in June 2017, *G. album* Mill. from Podeni, Cluj Coutry, Romania and from Rimetea, Alba Coutry, Romania, *G. purpureum* L. and *G. pseudoaristatum* Schur from Băile Herculane, Caraş-Severin Coutry, in August 2014. All species were authenticated by Dr. Sabin Bădărău and Dr. Andrei Mocan, and voucher specimens were deposited at the herbarium of the Department of Pharmaceutical Botany, "Iuliu Hațieganu" University of Medicine and Pharmacy. Fresh herbal material was dried at room temperature until reaching a constant mass. Afterwards, the plant material was ground into a fine powder using a laboratory mill, mixed to obtain homogenous sample, and kept at 4 °C, for further analyses. All assays were carried out three times (three separate samplings) and in triplicate, and the values reported are represented by average and the standard deviation (S.D.).

3.3. Apparatus

3.3.1. HPLC Analysis

The quantitative analysis of phenolic compounds was performed according to the reported method [36]. The chromatographic system consisted of HPLC Waters liquid chromatograph instrument (model 600 solvent pump, 2996 PDA). Mobile phase was directly on-line degassed by using a Biotech 4CH DEGASI Compact (Onsala, Sweden). For separation of phenolic compounds, C18 reversed-phase column (Prodigy ODS(3), 4.6 × 150 mm, 5 μm; Phenomenex, Torrance, CA), thermostated at 30 °C (±1 °C) was used. The collection and analysis of the data were performed by Empower v.2 software (Waters Corporation, Milford, MA, USA). The mobile phase was a mixture of solution A (3% solution of acetic acid in water) and solution B (3% solution of acetic acid in acetonitrile) in a ratio 93:7 and the gradient mode was applied. The total separation was completed in 1 h (the chemical standards chromatograms, retention times and maximum wavelengths are shown in *Supplementary Materials* section S3).

3.3.2. Auxiliary Equipment

As auxiliary equipment for the extraction procedures, centrifuge model NF048 (Nuve, Ankara, Turkey), vortex (VELP Scientifica Srl, Usmate, Italy), and ultrasonic bath (Falc Instruments, Treviglio,

Italy) were used. MAE was performed using an automatic Biotage Initiator™ 2.0 (Uppsala, Sweden) characterized by 2.45 GHz high-frequency microwaves and power range 0–300 W. An IR sensor probe controlled the internal vial temperature.

3.4. Extraction Procedures

Extraction optimization was carried out using *G. verum* and after, under optimized conditions, the microextraction procedure was applied for the other four *Galium* species. The following microextraction procedure were investigated: DLLME, ultrasound-assisted dispersive liquid-liquid microextraction (UA-DLLME), Salting-out liquid-liquid extraction (SA-LLE), and Sugaring-out liquid-liquid extraction (SULLE). The general procedure for the extractions reported in the following paragraphs was described in Figure 3.

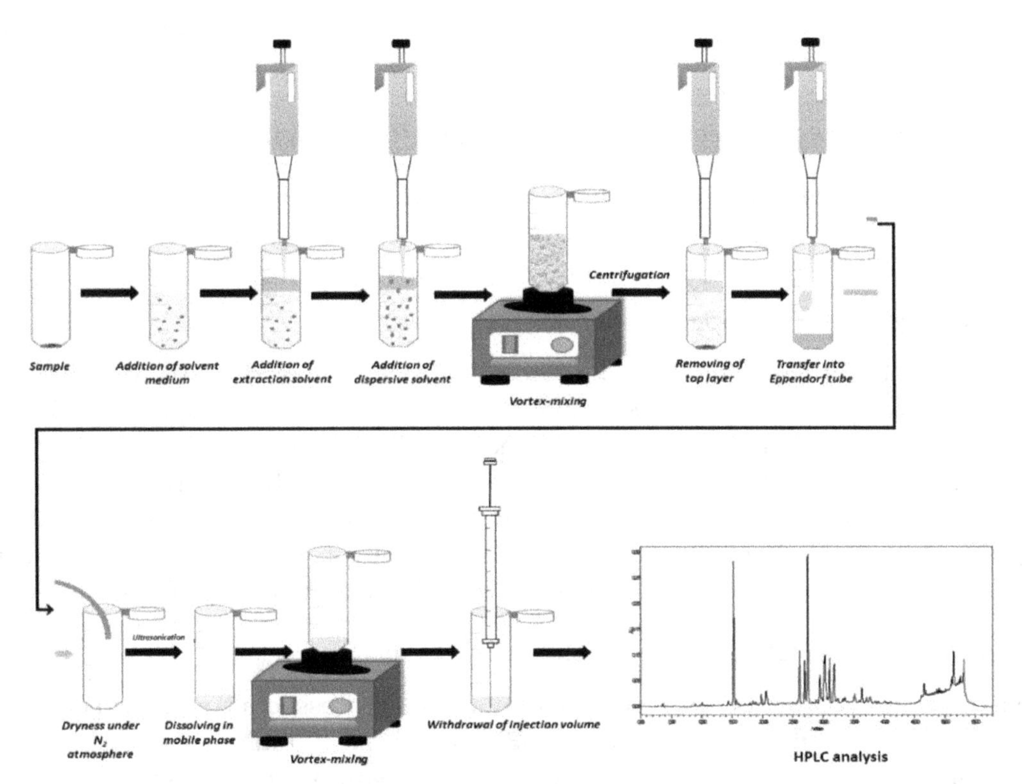

Figure 3. General extraction procedure.

3.4.1. DLLME and UA-DLLME

10 mg of the dry plant material of *G. verum* were accurately weighted and placed into a 2 mL Eppendorf tube. Subsequently, 700 μL of solvent medium (water, 10% NaCl, NADES, IL or 1% β-cyclodextrin (β-CD)), 400 μL of ethyl acetate, and 300 μL of ethanol were added to the Eppendorf tube by automatic pipette. The solution was vortexed during 30 s until a cloudy solution was formed. In the case of UA-DLLME, after those steps, the test tube was placed into the ultrasound bath for 5 min. Then, the solution was kept at rest for 1 min, for the analytes to distribute into the extraction solvent. For the phase separation, the solution was centrifuged at 12000× *g* for 5 min. The extraction solvent was found on the top of the Eppendorf tube, and its whole volume was collected using a microsyringe and transferred to the new Eppendorf tube, and then dried under a gentle stream of nitrogen. The dried residue was redissolved in 50 μL of mobile phase under ultrasonication for 5 min and 20 μL of the obtained solution were injected into the HPLC system.

3.4.2. Salting-Out-LLE

For salting-out-DLLME (SA-LLE), 10 mg of the dry herbal material of *G. verum* were placed into a 2 mL Eppendorf tube. Then, 200 μL of water and 400 μL of acetonitrile were added. To obtain the phase separation, 200 μL of 300 g L^{-1} solution of NaCl were rapidly injected into the Eppendorf tube. The mixture was vortexed for 1 min and a cloudy solution was formed. The next procedures were the same as in the Section 3.4.1.

3.4.3. Sugaring-Out-LLE

For the sugaring-out-LLE (SULLE), the procedure was similar to the SA-LLE. Instead of aqueous NaCl, 200 μL of glucose solution (600 g L^{-1}) was utilized for phase separation.

3.4.4. MAE

10 mg of the dry plant material were placed into a 2 mL sealed vessel suitable for an automatic single-mode microwave reactor and 1 mL of appropriate solvent medium (see Section 3.3.) was added, forming a yellow-green emulsion. MAE was carried out heating by microwave irradiation for 13 min 8 s at 80 °C (which correspond approximatively to 24 h of maceration at 25 °C), and then cooling to room temperature by pressurized air. Then, the homogenate was centrifuged at 12000× *g* for 5 min and 20 μL of the solution were directly injected into the HPLC system.

3.5. Total Phenolic and Flavonoid Content, Antioxidant Capacity, and Tyrosinase Inhibitory Activity

3.5.1. Antioxidant Assays

Total Phenolic Content (TPC) by Spectrophotometric Assay

The TPC was determined using the Folin–Ciocâlteu method described by Mocan et al. [37]. For a high throughput of samples, a SPECTROstar Nano Multi—Detection Microplate Reader with 96-well plates (BMG Labtech, Ortenberg, Germany) was used. Briefly, a mixture solution consisting of 20 μL of extract, 100 μL of Folin-Ciocâlteu reagent, and 80 μL of sodium carbonate (Na_2CO_3, 7.5% *w*/*v*) was homogenized and incubated at room temperature in the dark for 30 min. Afterwards, the absorbance of the samples was measured at 760 nm. Gallic acid was used as a reference standard, and the TPC was expressed as gallic acid equivalents (GAE) in mg/g dry weight (d.w.) of plant material.

Total Flavonoid Content (TFC) by Spectrophotometric Assay

The total flavonoid content (TFC) was calculated and expressed as quercetin equivalents using a method previously described by Mocan et al. [38]. Briefly, a 100 μL aliquot of 2% $AlCl_3$ aqueous solution was mixed with 100 μL of sample. After an incubation time of 15 min, the absorbance of the sample was measured at 420 nm. Quercetin was used as a reference standard, and the TFC was expressed as quercetin equivalents (QE) in mg/g dry weight (d.w.) of plant material.

DPPH Radical Scavenging Assay

The capacity to scavenge the "stable" free radical DPPH, monitored according to the method described by Martins et al. [39], with some modifications, was performed by using a SPECTROstar Nano microplate reader (BMG Labtech, Offenburg, Germany). The reaction mixture in each of the 96-wells consisted of 30 μL of sample solution (in an appropriated dilution) and a 0.004% methanolic solution of DPPH. The mixture was further incubated for 30 min in the dark, and the reduction of the DPPH radical was determined at 515 nm. Trolox was used as a standard reference and the results were expressed as Trolox equivalents per g of dry weight herbal extract (mg TE/g d.w. of herbal extract).

Trolox Equivalent Antioxidant Capacity (TEAC) Assay

In the Trolox equivalent antioxidant capacity (TEAC) assay, the antioxidant capacity is reflected in the ability of the *Galium* extracts to decrease the color, reacting directly with the ABTS radical. The latter was obtained by oxidation of ABTS (2,2′-azinobis(3-ethylbenzothiazoline-6-sulfonic acid)) with potassium peroxydisulfate ($K_2S_2O_8$). The amount of ABTS radical consumed by the tested compound was measured at 760 nm after 6 min of reaction time. The evaluation of the antioxidant capacity was obtained using the total change in absorbance at this wavelength. The percentage of ABTS consumption was transformed in Trolox equivalents (TE) using a calibration curve.

FRAP Assay

In FRAP assays, the reduction of Fe^{3+}-TPTZ to blue-colored Fe^{2+}-TPTZ complex was monitored by the method described by Damiano et al., (2017) with slight modifications [40]. The FRAP reagent was prepared by mixing ten volumes of acetate buffer (300 mM, pH 3.6), one volume of TPTZ solution (10 mM TPTZ in 40 mM HCl) and one volume of $FeCl_3$ solution (20 mM $FeCl_3 \cdot 6H_2O$ in 40 mM HCl). Reaction mixture (25 μL sample and 175 μL FRAP reagent) was incubated in the dark for 30 min at room temperature and the absorbance of each solution was measured at 593 nm using a SPECTROstar Nano Multi-Detection Microplate Reader with 96-well plates (BMG Labtech, Ortenberg, Germany). A Trolox^{TM} calibration curve (0.01–0.10 mg/mL) was plotted as a function of blue-colored Fe^{2+}-TPTZ complex formation, and the results were expressed as milligrams of trolox equivalents (TE) per milligram of extract (mg TE/mg extract).

3.5.2. Tyrosinase Inhibitory Activity

Tyrosinase inhibitory activity of each sample was determined by the method previously described by Likhitwitayawuid and Sritularak, (2001) and Masuda *et al.*, (2005) [41,42] using a SPECTROstar Nano Multi-Detection Microplate Reader with 96-well plates (BMG Labtech, Ortenberg, Germany). Samples were dissolved in water containing 5% DMSO; for each sample four wells were designated as A, B, C, D; each one contained the reaction mixture (200 μL) as follows: (A) 120 μL of 0.66 M phosphate buffer solution (pH = 6.8) (PBS) and 40 μL of mushroom tyrosinase in PBS (46 U/mL) (Tyr), (B) 160 μL of PBS, (C) 80 μL of PBS, 40 μL of Tyr, and 40 μL of sample, and (D) 120 μL of PBS and 40 μL of sample. The plate was then incubated at room temperature for 10 min; after incubation, 40 μL of 2.5 mM L-DOPA in PBS solution were added in each well and the mixtures were incubated again at room temperature for 20 min. The absorbance of each well was measured at 475 nm, and the inhibition percentage of the tyrosinase activity was calculated by the following equation, using as positive control a kojic acid solution (0.10 mg/mL):

$$\%I = \frac{(A - B) - (C - D)}{(A - B)} \times 100 \quad (1)$$

The results were also expressed as mg kojic acid equivalents per gram of dry weight extract (mg KAE/g extract) using a calibration curve between 0.01–0.10 mg kojic acid per milliliter of solution.

3.6. Statistical Analysis

All experiments were performed in triplicate and the results were expressed as the mean value ± standard deviation (S.D.). All comparisons were determined by using two-way ANOVA followed by Bonferroni post-test and GraphPad Prism v.4 for data elaboration. Raw data regarding the statistical analyses were reported in *Supplementary Materials* section S1.

4. Conclusions

In this work, a microextraction procedure was developed and applied for the establishment of the multicomponent phenolic pattern of aerial parts from *G. verum*, *G. album*, *G. rivale*, *G. pseudoaristatum*, and *G. purpureum*. The DLLME procedure in NADES solvent medium could provide high extraction efficiency within a short extraction time and with good correspondence with the MAE procedure. The biological results showed that *G. purpureum* and *G. verum* extracts contained the highest total phenolic and flavonoid contents, respectively. *G. purpureum* extract was the most active extract in terms of antioxidant capacity, whereas the *G. album* extract exhibited the strongest inhibitory effect against tyrosinase, an enzyme involved in several skin disorders. The results indicate that *Galium* extracts have the potential to be used as an alternative source of multifunctional agents and are a promising starting point for development of new bioactive formulations. Further studies are essential for the isolation of pure bioactive compounds and investigation of their molecular mechanisms of action.

Author Contributions: Data curation, A.T., A.D., and M.D.S.; Investigation, A.T., A.D., and M.D.S.; Methodology, M.L., A.M., S.B., C.M., V.A., S.C., C.C., M.T., R.G., D.V. and G.C.; Project administration, M.L., A.M., S.B., C.M., V.A., S.C., D.V. and G.C.; Supervision, M.L., A.M., S.B., C.M., V.A., S.C., D.V. and G.C.

Acknowledgments: A.D. gratefully thanks the University "G. d'Annunzio" of Chieti—Pescara, Dept. of Pharmacy and Erasmus+ program for giving her the opportunity of the research project implementation.

Abbreviations

ABTS	2,2'-azino-bis(3-ethylbenzothiazoline-6-sulphonic acid)
β-CD	β-cyclodextrin
DLLME	Dispersive Liquid–Liquid MicroExtraction
DMSO	Dimethyl sulfoxide
DPPH	2,2-diphenyl-1-picryl-hydrazyl-hydrate
d.w.	dry weight
FRAP	Ferric Reducing Antioxidant Power
GAE	Gallic Acid Equivalents
GC	Gas Chromatography
HPLC	High Performance Liquid Chromatography
HPLC-PDA	High Performance Liquid Chromatography-Photodiode array detector
ILs	ionic liquids
KA	Kojic Acid
KAE	Kojic Acid Equivalents
LPME	Liquid Phase MicroExtraction
MAE	Microwave-Assisted Extraction
NADES	NAtural Deep Eutectic Solvent
QE	Quercetin Equivalents
SA-LLE	Salting-out Liquid-liquid extraction
SULLE	Sugaring-out Liquid-liquid extraction
TCPC	Total Concentration of Phenolic Compounds
TE	Trolox Equivalents
TEAC	Trolox Equivalent Antioxidant Capacity
TFC	Total Flavonoid Content by spectrophotometric assay
TPC	Total Phenolic Content by spectrophotometric assay
UA-DLLME	Ultrasound-Assisted Dispersive Liquid–Liquid MicroExtraction

References

1. Efferth, T. Perspectives for globalized natural medicines. *Chin. J. Nat. Med.* **2011**, *9*, 1–6. [CrossRef]
2. Böjthe-Horváth, K.; Kocsis, A.; Párkány, L.; Simon, K. A new iridoid glycoside from *Galium verum* L. First X-ray analysis of a tricyclic iridoid glycoside. *Tetrahedron Lett.* **1982**, *23*, 965–966. [CrossRef]
3. Mitova, M.I.; Anchev, M.E.; Handjieva, N.V.; Popov, S.S. Iridoid patterns in *Galium* L. and some phylogenetic considerations, Zeitschrift Fur Naturforsch. *J. Biosci.* **2002**, *57*, 226–234. [CrossRef]
4. Demirezer, L.O.; Gürbüz, F.; Güvenalp, Z.; Ströch, K.; Zeeck, A. Iridoids, flavonoids and monoterpene glycosides from *Galium verum* subsp. *Verum*. *Turk J. Chem.* **2006**, *30*, 525–534.
5. Zhao, C.C.; Shao, J.H.; Li, X.; Kang, X.D.; Zhang, Y.W.; Meng, D.L.; Li, N. Flavonoids from *Galium verum* L. *J. Asian Nat. Prod. Res.* **2008**, *10*, 611–615. [CrossRef]
6. Zhao, C.C.; Shao, J.H.; Li, X.; Xu, J.; Wang, J.H. A new anthraquinone from *Galium verum* L. *Nat. Prod. Res.* **2006**, *20*, 981–984. [CrossRef]

7. Lakić, N.S.; Mimica-Dukić, N.M.; Isak, J.M.; Božin, B.N. Antioxidant properties of *Galium verum* L. (Rubiaceae) extracts. *Cent. Eur. J. Biol.* **2010**, *5*, 331–337. [CrossRef]
8. Khalili, M.; Ebrahimzadeh, M.A.; Safdari, Y. Antihaemolytic activity of thirty herbal extracts in mouse red blood cells. *Arh. Hig. Rada Toksikol.* **2014**, *65*, 399–406. [CrossRef]
9. Il'ina, T.V.; Kovaleva, A.M.; Goryachaya, O.V.; Aleksandrov, A.N. Essential oil from *Galium verum* flowers. *Chem. Nat. Compd.* **2009**, *45*, 587–588. [CrossRef]
10. Schmidt, M.; Scholz, C.; Gavril, G.; Otto, C.; Polednik, C.; Roller, J.; Hagen, R. Effect of *Galium verum* aqueous extract on growth, motility and gene expression in drug-sensitive and -resistant laryngeal carcinoma cell lines. *Int. J. Oncol.* **2014**, *44*, 745–760. [CrossRef] [PubMed]
11. Menković, N.; Šavikin, K.; Tasić, S.; Zdunić, G.; Stešević, D.; Milosavljević, S.; Vincek, D. Ethnobotanical study on traditional uses of wild medicinal plants in Prokletije Mountains (Montenegro). *J. Ethnopharmacol.* **2011**, *133*, 97–107. [CrossRef]
12. Vlase, L.; Mocan, A.; Hanganu, D.; Gheldiu, A.; Crişan, G. Comparative study of polyphenolic content, antioxidant and antimicrobial activity of four *Galium* species (Rubiaceae). *Dig. J. Nanomater. Biostruct.* **2014**, *9*, 1085–1094.
13. Pieroni, A.; Quave, C.L. Traditional pharmacopoeias and medicines among Albanians and Italians in southern Italy: A comparison. *J. Ethnopharmacol.* **2005**, *101*, 258–270. [CrossRef]
14. Schmidt, M.; Polednik, C.; Roller, J.; Hagen, R. *Galium verum* aqueous extract strongly inhibits the motility of head and neck cancer cell lines and protects mucosal keratinocytes against toxic DNA damage. *Oncol. Rep.* **2014**, *32*, 1296–1302. [CrossRef] [PubMed]
15. Płotka-Wasylka, J.; Rutkowska, M.; Owczarek, K.; Tobiszewski, M.; Namieśnik, J. Extraction with environmentally friendly solvents. *TrAC.* **2017**, *91*, 12–25. [CrossRef]
16. Craveiro, R.; Aroso, I.; Flammia, V.; Carvalho, T.; Viciosa, M.T.; Dionísio, M.; Barreiros, S.; Reis, R.L.; Duarte, A.R.C.; Paiva, A. Properties and thermal behavior of natural deep eutectic solvents. *J. Mol. Liq.* **2016**, *215*, 534–540. [CrossRef]
17. Khezeli, T.; Daneshfar, A.; Sahraei, R. A green ultrasonic-assisted liquid-liquid microextraction based on deep eutectic solvent for the HPLC-UV determination of ferulic, caffeic and cinnamic acid from olive, almond, sesame and cinnamon oil. *Talanta* **2016**, *150*, 577–585. [CrossRef]
18. Shishov, A.; Bulatov, A.; Locatelli, M.; Carradori, S.; Andruch, V. Application of deep eutectic solvents in analytical chemistry. A review. *Microchem. J.* **2017**, *135*, 33–38. [CrossRef]
19. Kim, Y.J.; Uyama, H. Tyrosinase inhibitors from natural and synthetic sources: Structure, inhibition mechanism and perspective for the future. *Cell. Mol. Life Sci.* **2005**, *62*, 1707–1723. [CrossRef] [PubMed]
20. Loizzo, M.R.; Tundis, R.; Menichini, F. Natural and synthetic tyrosinase inhibitors as antibrowning agents: An update. *Compr. Rev. Food Sci. Food Saf.* **2012**, *11*, 378–398. [CrossRef]
21. Zengin, G.; Menghini, L.; Di Sotto, A.; Mancinelli, R.; Sisto, F.; Carradori, S.; Cesa, S.; Fraschetti, C.; Filippi, A.; Angiolella, L.; et al. Chromatographic analyses, in vitro biological activities and cytotoxicity of *Cannabis sativa* L. essential oil: A multidiscipl.inary study. *Molecules* **2018**, *23*, 3266. [CrossRef] [PubMed]
22. Di Sotto, A.; Di Giacomo, S.; Amatore, D.; Locatelli, M.; Vitalone, A.; Toniolo, C.; Rotino, G.L.; Lo Scalzo, R.; Palamara, A.T.; Marcocci, M.E.; et al. A Polyphenol Rich Extract from *Solanum melongena* L. DR2 Peel Exhibits Antioxidant Properties and Anti-Herpes Simplex Virus Type 1 Activity In Vitro. *Molecules* **2018**, *23*, 2066. [CrossRef]
23. Boutaoui, N.; Zaiter, L.; Benayache, F.; Benayache, S.; Cacciagrano, F.; Cesa, S.; Secci, D.; Carradori, S.; Giusti, A.M.; Campestre, C.; et al. *Atriplex mollis* Desf. Aerial Parts: Extraction Procedures, Secondary Metabolites and Color Analysis. *Molecules* **2018**, *23*, 1962. [CrossRef]
24. Locatelli, M.; Macchione, N.; Ferrante, C.; Chiavaroli, A.; Recinella, L.; Carradori, S.; Zengin, G.; Cesa, S.; Leporini, L.; Leone, S.; et al. Graminex pollen: Phenolic pattern, colorimetric analysis and protective effects in immortalized prostate cells (PC3) and rat prostate challenged with LPS. *Molecules* **2018**, *23*, 1145. [CrossRef] [PubMed]
25. Boutaoui, N.; Zaiter, L.; Benayache, F.; Benayache, S.; Carradori, S.; Cesa, S.; Giusti, A.M.; Campestre, C.; Menghini, L.; Innosa, D.; et al. Qualitative and Quantitative Phytochemical Analysis of Different Extracts from *Thymus algeriensis* Aerial Parts. *Molecules* **2018**, *23*, 463. [CrossRef]

26. Melucci, D.; Locatelli, M.; Locatelli, C.; Zappi, A.; De Laurentiis, F.; Carradori, S.; Campestre, C.; Leporini, L.; Zengin, G.; Picot, C.M.N.; et al. A Comparative Assessment of Biological Effects and Chemical Profile of Italian *Asphodeline lutea* Extracts. *Molecules* **2018**, *23*, 461. [CrossRef] [PubMed]
27. Tartaglia, A.; Locatelli, M.; Kabir, A.; Furton, K.G.; Macerola, D.; Sperandio, E.; Piccolantonio, S.; Ulusoy, H.I.; Maroni, F.; Bruni, P.; et al. Comparison between Exhaustive and Equilibrium Extraction Using Different SPE Sorbents and Sol-Gel Carbowax 20M Coated FPSE Media. *Molecules* **2019**, *24*, 382. [CrossRef] [PubMed]
28. Hemwimon, S.; Pavasant, P.; Shotipruk, A. Microwave-assisted extraction of antioxidative anthraquinones from roots of *Morinda citrifolia*. *Sep. Purif. Technol.* **2007**, *54*, 44–50. [CrossRef]
29. Mollica, A.; Locatelli, M.; Macedonio, G.; Carradori, S.; Sobolev, A.P.; De Salvador, R.F.; Monti, S.M.; Buonanno, M.; Zengin, G.; Angeli, A.; et al. Microwave-assisted extraction, HPLC analysis, and inhibitory effects on carbonic anhydrase I, II, VA, and VII isoforms of 14 blueberry Italian cultivars. *J. Enzyme Inhib. Med. Chem.* **2016**, *6366*, 1–6. [CrossRef]
30. Hashizaki, K.; Kageyama, T.; Inoue, M.; Taguchi, H.; Ueda, H.; Saito, Y. Study on preparation and formation mechanism of *n*-alkanol/water emulsion using α-cyclodextrin. *Chem. Pharm. Bull.* **2007**, *55*, 1620–1625. [CrossRef]
31. Inoue, M.; Hashizaki, K.; Taguchi, H.; Saito, Y. Preparation and characterization of n-alkane/water emulsion stabilized by cyclodextrin. *J. Oleo. Sci.* **2009**, *58*, 85–90. [CrossRef]
32. Duchêne, D.; Bochot, A.; Yu, S.C.; Pépin, C.; Seiller, M. Cyclodextrins and emulsions. *Int. J. Pharm.* **2003**, *266*, 85–90. [CrossRef]
33. Diuzheva, A.; Carradori, S.; Andruch, V.; Locatelli, M.; De Luca, E.; Tiecco, M.; Germani, R.; Menghini, L.; Nocentini, A.; Gratteri, P.; et al. Use of Innovative (Micro)Extraction Techniques to Characterise *Harpagophytum procumbens* Root and its Commercial Food Supplements. *Phytochem. Anal.* **2018**, *3*, 233–241. [CrossRef] [PubMed]
34. Saokham, P.; Muankaew, C.; Jansook, P.; Loftsson, T. Solubility of Cyclodextrins and Drug/Cyclodextrin Complexes. *Molecules* **2018**, *23*, 1161. [CrossRef]
35. Chiocchio, I.; Mandrone, M.; Sanna, C.; Maxia, A.; Tacchini, M.; Poli, F. Industrial Crops & Products screening of a hundred plant extracts as tyrosinase and elastase inhibitors, two enzymatic targets of cosmetic interest. *Ind. Crop. Prod.* **2018**, *122*, 498–505. [CrossRef]
36. Zengin, G.; Menghini, L.; Malatesta, L.; De Luca, E.; Bellagamba, G.; Uysal, S.; Aktumsek, A.; Locatelli, M. Comparative study of biological activities and multicomponent pattern of two wild Turkish species: *Asphodeline anatolica* and *Potentilla speciosa*. *J. Enzyme Inhib. Med. Chem.* **2016**, *31*, 203–208. [CrossRef]
37. Mocan, A.; Schafberg, M.; Crisan, G.; Rohn, S. Determination of lignans and phenolic components of *Schisandra chinensis* (Turcz.) Baill. using HPLC-ESI-ToF-MS and HPLC-online TEAC: Contribution of individual components to overall antioxidant activity and comparison with traditional antioxidant assays. *J. Funct. Food.* **2016**, *24*, 579–594. [CrossRef]
38. Mocan, A.; Crişan, G.; Vlase, L.; Crişan, O.; Vodnar, D.C.; Raita, O.; Gheldiu, A.M.; Toiu, A.; Oprean, R.; Tilea, I. Comparative studies on polyphenolic composition, antioxidant and antimicrobial activities of *Schisandra chinensis* leaves and fruits. *Molecules* **2014**, *19*, 15162–15179. [CrossRef] [PubMed]
39. Martins, N.; Barros, L.; Dueñas, M.; Santos-Buelga, C.; Ferreira, I.C.F.R. Characterization of phenolic compounds and antioxidant properties of *Glycyrrhiza glabra* L. rhizomes and roots. *RSC Adv.* **2015**, *5*, 26991–26997. [CrossRef]
40. Damiano, S.; Forino, M.; De, A.; Vitali, L.A.; Lupidi, G.; Taglialatela-Scafati, O. Antioxidant and antibiofilm activities of secondary metabolites from *Ziziphus jujuba* leaves used for infusion preparation. *Food Chem.* **2017**, *230*, 24–29. [CrossRef] [PubMed]
41. Likhitwitayawuid, K.; Sritularak, B. A new dimeric stilbene with tyrosinase inhibitory activity from *Artocarpus gomezianus*. *J. Nat. Prod.* **2001**, *64*, 1457–1459. [CrossRef] [PubMed]
42. Masuda, T.; Yamashita, D.; Takeda, Y.; Yonemori, S. Screening for tyrosinase inhibitors among extracts of seashore plants and identification of potent inhibitors from *Garcinia subelliptica*. *Biosci. Biotechnol. Biochem.* **2005**, *69*, 197–201. [CrossRef] [PubMed]

Permissions

All chapters in this book were first published in MDPI; hereby published with permission under the Creative Commons Attribution License or equivalent. Every chapter published in this book has been scrutinized by our experts. Their significance has been extensively debated. The topics covered herein carry significant findings which will fuel the growth of the discipline. They may even be implemented as practical applications or may be referred to as a beginning point for another development.

The contributors of this book come from diverse backgrounds, making this book a truly international effort. This book will bring forth new frontiers with its revolutionizing research information and detailed analysis of the nascent developments around the world.

We would like to thank all the contributing authors for lending their expertise to make the book truly unique. They have played a crucial role in the development of this book. Without their invaluable contributions this book wouldn't have been possible. They have made vital efforts to compile up to date information on the varied aspects of this subject to make this book a valuable addition to the collection of many professionals and students.

This book was conceptualized with the vision of imparting up-to-date information and advanced data in this field. To ensure the same, a matchless editorial board was set up. Every individual on the board went through rigorous rounds of assessment to prove their worth. After which they invested a large part of their time researching and compiling the most relevant data for our readers.

The editorial board has been involved in producing this book since its inception. They have spent rigorous hours researching and exploring the diverse topics which have resulted in the successful publishing of this book. They have passed on their knowledge of decades through this book. To expedite this challenging task, the publisher supported the team at every step. A small team of assistant editors was also appointed to further simplify the editing procedure and attain best results for the readers.

Apart from the editorial board, the designing team has also invested a significant amount of their time in understanding the subject and creating the most relevant covers. They scrutinized every image to scout for the most suitable representation of the subject and create an appropriate cover for the book.

The publishing team has been an ardent support to the editorial, designing and production team. Their endless efforts to recruit the best for this project, has resulted in the accomplishment of this book. They are a veteran in the field of academics and their pool of knowledge is as vast as their experience in printing. Their expertise and guidance has proved useful at every step. Their uncompromising quality standards have made this book an exceptional effort. Their encouragement from time to time has been an inspiration for everyone.

The publisher and the editorial board hope that this book will prove to be a valuable piece of knowledge for researchers, students, practitioners and scholars across the globe.

List of Contributors

Daria Śmigiel-Kamińska, Piotr Stepnowski and Jolanta Kumirska
Faculty of Chemistry, University of Gdańsk, ul. Wita Stwosza 63, 80-308 Gdansk, Poland

Jolanta Wąs-Gubała
Institute of Forensic Research, Microtrace Analysis Section, Westerplatte 9, 31-033 Krakow, Poland

René González-Albarrán, Josefina de Gyves and Eduardo Rodríguez de San Miguel
Departamento de Química Analítica, Facultad de Química, UNAM, Ciudad Universitaria, 04510 Cd. Mx., Mexico

Ana Lobo-Prieto and Diego L. García-González
Instituto de la Grasa (CSIC), Campus Universidad Pablo de Olavide-Edificio 46, Ctra. de Utrera, Km. 1, 41013 Sevilla, Spain

Noelia Tena and Ramón Aparicio-Ruiz
Department of Analytical Chemistry, Universidad de Sevilla, C/Prof. García González 2, 41012 Sevilla, Spain

Ewa Sikorska
Institute of Quality Science, The Poznan University of Economics and Business, al. Niepodleglosci 10, 61-875 Poznan, Poland

Yi-Fei Pei
Institute of Medicinal Plants, Yunnan Academy of Agricultural Sciences, Kunming 650200, China
College of Traditional Chinese Medicine, Yunnan University of Traditional Chinese Medicine, Kunming 650500, China

Qing-Zhi Zhang
College of Traditional Chinese Medicine, Yunnan University of Traditional Chinese Medicine, Kunming 650500, China

Zhi-Tian Zuo and Yuan-Zhong Wang
Institute of Medicinal Plants, Yunnan Academy of Agricultural Sciences, Kunming 650200, China

Sergio Ghidini, Maria Olga Varrà and Emanuela Zanardi
Department of Food and Drug, University of Parma, Strada del Taglio 10, 43126 Parma, Italy

Ying Xue
School of Pharmacy, Chengdu Medical College, Chengdu 610500, China
Chengdu Institute of Biology, Chinese Academy of Sciences, Chengdu 610041, China
Sichuan Provincial Center for Disease Control and Prevention, Chengdu 610041, China

Lin-Sen Qing
Chengdu Institute of Biology, Chinese Academy of Sciences, Chengdu 610041, China

Li Yong, Xian-Shun Xu and Bin Hu
Sichuan Provincial Center for Disease Control and Prevention, Chengdu 610041, China

Ming-Qing Tang
School of Pharmacy, Chengdu Medical College, Chengdu 610500, China
Sichuan Provincial Center for Disease Control and Prevention, Chengdu 610041, China

Jing Xie
School of Pharmacy, Chengdu Medical College, Chengdu 610500, China

F. Anguebes-Franseschi, A. V. Córdova Quiroz and M. A. Ramírez-Elias
Faculty of Chemistry, Autonomous University of Carmen, Street 56 No. 4 Esq. Av. Concordia, Col. Benito Juárez, Z. C. 24180 Ciudad del Carmen, Campeche, Mexico

M. Abatal and A. Flores
Faculty of Engineering, Autonomous University of Carmen, Campus III, Avenida Central s/n, Esq. Con Fracc. Mundo Maya, C. P. 24115 Ciudad del Carmen, Campeche, Mexico

Lucio Pat
South Frontier College, Av. Rancho Polígono 2-A, Ciudad Industrial, 24500 Lerma, Campeche, Mexico

L. San Pedro, O. May Tzuc and A. Bassam
Faculty of Engineering, Autonomous University of Yucatan, Av. Industrias no Contaminantes Periférico Norte, Cordemex, Z.C. 97310 Mérida, Yucatan, Mexico

Tao Shen
Yunnan Herbal Laboratory, Institute of Herb Biotic Resources, School of Life and Sciences, Yunnan University, Kunming 650091, China
The International Joint Research Center for Sustainable Utilization of Cordyceps Bioresouces in China and Southeast Asia, Yunnan University, Kunming 650091, China
College of Chemistry, Biological and Environment, Yuxi Normal University, Yu'xi 653100, China

Hong Yu
Yunnan Herbal Laboratory, Institute of Herb Biotic Resources, School of Life and Sciences, Yunnan University, Kunming 650091, China
The International Joint Research Center for Sustainable Utilization of Cordyceps Bioresouces in China and Southeast Asia, Yunnan University, Kunming 650091, China

Yuan-Zhong Wang
College of Traditional Chinese Medicine, Yunnan University of Chinese Medicine, Kunming 650500, China

Luya Li, Yuting Chen, Xue Feng, Jintuo Yin, Yupeng Sun and Lantong Zhang
School of Pharmacy, Hebei Medical University, Shijiazhuang 050017, China

Shenghao Li
School of Pharmacy, Hebei University of Chinese Medicine, Shijiazhuang 050000, China

Xingping Zhang and Hualin Yang
College of Life Science, Yangtze University, Jingzhou 434025, China
Research and Development Sharing Platform of Hubei Province for Freshwater Product Quality and Safety, Yangtze University, Jingzhou 434025, China

Jiujun Wang, Qinghua Wu, Li Li and Yun Wang
College of Life Science, Yangtze University, Jingzhou 434025, China

Yehong Li
School of Pharmacy, China Pharmaceutical University, Nanjing 210009, China

Yiming Zhang, Zejun Zhang and Yupiao Hu
Faculty of Life Science and Technology, Kunming University of Science and Technology, Kunming 650500, China

Xiuming Cui and Yin Xiong
Faculty of Life Science and Technology, Kunming University of Science and Technology, Kunming 650500, China
Yunnan Key Laboratory of Panax notoginseng, Kunming University of Science and Technology, Kunming 650500, China
Institute of Biology Leiden, Leiden University, 2333BE Leiden, The Netherlands

Andrei Mocan, Cadmiel Moldovan and Gianina Crişan
Department of Pharmaceutical Botany, "Iuliu Haţieganu" University of Medicine and Pharmacy, 400012 Cluj-Napoca, Romania

Alina Diuzheva and Vasil Andruch
Department of Analytical Chemistry, Pavol Jozef Šafárik University, SK-04180 Košice, Slovakia

Sabin Bădărău
Department of Environmental Sciences, Babeş-Bolyai University, 400084 Cluj-Napoca, Romania

Simone Carradori, Cristina Campestre, Angela Tartaglia, Marta De Simone and Marcello Locatelli
Department of Pharmacy, University "G. D'Annunzio" of Chieti-Pescara, 66100 Chieti, Italy

Dan Vodnar
Department of Food Science, University of Agricultural Sciences and Veterinary Medicine, 400372 Cluj-Napoca, Romania

Matteo Tiecco and Raimondo Germani
Department of Chemistry, Biology and Biotechnology, University of Perugia, 06132 Perugia, Italy

Index

Printed in the USA
CPSIA information can be obtained
at www.ICGtesting.com
JSHW051409091023
49903JS00006B/345